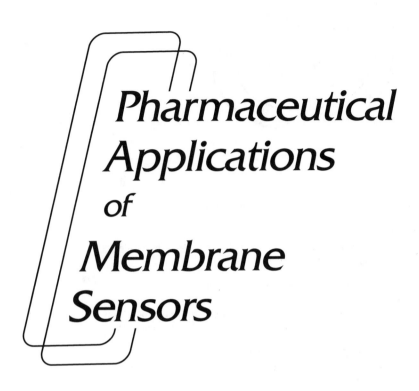

Pharmaceutical
Applications
of
Membrane
Sensors

Vasile V. Coşofreţ
Institute of Chemical and Pharmaceutical Research
Bucharest, Romania

Richard P. Buck
Department of Chemistry
University of North Carolina
Chapel Hill, North Carolina

CRC Press
Boca Raton   Ann Arbor   London   Tokyo

Library of Congress Cataloging-in-Publication Data

Coşofreţ, Vasile V.
   Pharmaceutical applications of membrane sensors/by Vasile
V. Coşofreţ and Richard P. Buck.
     p. cm.
   Includes bibliographical references and index.
   ISBN 0-8493-4406-9
   1. Biosensors. 2. Electrochemical sensors. 3. Electrodes. Ion
-selective. 4. Pharmaceutical technology. I. Buck, Richard P.
II. Title.
   [DNLM: 1. Biosensors. 2. Cell Membrane—drug effects.
3. Electrodes. 4. Monitoring, Physiologic—instrumentation.
5. Pharmacology—methods. QV 26 C834h]
R857. B54C67 1992
615′. 1901—dc20
DNLM/DLC
for Library of Congress   91-36578
                                  CIP

This book represents information obtained from authentic and highly regarded sources. Reprinted material is quoted with permission, and sources are indicated. A wide variety of references are listed. Every reasonable effort has been made to give reliable data and information, but the author and the publisher cannot assume responsibility for the validity of all materials or for the consequences of their use.

Direct all inquiries to CRC Press, Inc., 2000 Corporate Blvd., N.W., Boca Raton, Florida 33431.

©1992 by CRC Press, Inc.

International Standard Book Number 0-8493-4406-9

Library of Congress Card Number 91-36578

Printed in the United States of America   1 2 3 4 5 6 7 8 9 0

Printed on acid-free paper

ABOUT THE AUTHORS

Vasile V. Coşofreţ is senior researcher at the Institute of Chemical and Pharmaceutical Research Bucharest. He received his B.S. degree in analytical chemistry from University of Bucharest in 1970 and his Ph.D. degree from Polytechnic Institute of Bucharest in 1975.

He acted as laboratory and R & D department manager with KHD Eng.GmbH-Koln (Germany) from 1981 to 1983. Dr. Coşofreţ worked as an associate researcher in the laboratory of Professor R. P. Buck at the University of North Carolina, Chapel Hill for six months from 1983 to 1984. He was also an invited researcher and lecturer in the Chemistry Department of Shanghai Teacher's University, Shanghai, China, for three months in 1988. Presently he is visiting research professor in the Department of Chemistry at the University of North Carolina, Chapel Hill.

Dr. Coşofreţ is author or co-author of over 60 technical articles in analytical chemistry, especially electroanalytical chemistry, with an emphasis on the role of membrane sensors in organic and pharmaceutical analysis as well as of various reviews and book chapters on membrane electrodes. He is co-author, with Professor G. E. Baiulescu, of *Applications of Ion-Selective Membrane Electrodes in Organic Analysis* and author of *Membrane Electrodes in Drug-Substances Analysis*. He is also a member of the Board of Advisory Editors of the journal *Selective Electrode Reviews*.

His area of interest is in pharmaceutical analysis and drug-release monitoring by various analytical techniques, drug-membrane sensors, and the development of microchemical sensors for acute cardiovascular applications.

Richard P. Buck is professor of chemistry and biomedical engineering at William Rand Kenan, Jr. Laboratories of Chemistry at the University of North Carolina, Chapel Hill. He received his B.S. and M.S. at the California Institute of Technology in 1950 and 1951 and his Ph.D. degree at the Massachusetts Institute of Technology in 1954.

From 1989 to 1991 he was visiting professor at Bundeswehr University of Munich, Germany, von Humboldt Preis of the West German von Humboldt Stiftung from 1989 to 1990, and visiting professor at the Imperial College in London in 1987. His fellowships include the Pogue

Fellowship and the Fogarty Fellowship from the National Institutes of Health, both in 1987.

Dr. Buck served as chairman of the Charles N. Reilley Award Committee, Society of Electroanalytical Chemistry from 1987 to 1989. From 1982 to 1987 he was a member of the David C. Grahame Award Committee of the Electrochemical Society, serving as chairman from 1984 to 1986. He was also a member of the Advisory Board of the National Institute of Health's Semiconductor Chemical Transducer Resource in the Electronics Design Center at Case Western Reserve University from 1983 to 1986 and was a former member of the Advisory Board of the Center for Chemical Electronics (now Sensor Technology) at the University of Pennsylvania.

Dr. Buck has published over 250 research articles, reviews, and book chapters. His main topics of research include theory and tests of ion-selective electrode responses, contributions to various analytical methods, design and practice of ion-selective electrodes, amperometric and biosensors, transport of charge in pure ionic conductors, and impedance, voltage-step, and steady-state theory, and experiment for thin-layer and membrane systems.

CONTENTS

PART II

ANALYSIS OF PHARMACEUTICALS BY MEMBRANE SENSORS

Contents

Pharmaceutical Applications *of* Membrane Sensors

INTRODUCTION

In the past few years a large amount of research has been done in the field of selective membrane sensors and associated highly sensitive analytical techniques.[1-14] The developments and various applications of new electrochemical sensors continues to be a rapidly growing area of analytical chemistry. Many researchers are currently working on constructing new drug-sensitive membrane sensors or on applying the well-known commercially available electrodes to monitor certain drugs in pure form, complex pharmaceutical formulations, and biological materials.

Several recent reviews,[15-35] conference proceedings,[36-44] and monographs[45-53] give comprehensive accounts of this work. Many specialized Chinese papers on drug sensors show, once more, that increasingly authors from the People's Republic of China are becoming involved in this area of research.

Japanese scientists have constructed a computer data base on electrochemical sensors in which numerical data and other relevant, important information on electrochemical sensors are compiled.[54]

For analytical control of pharmaceuticals, most of the pharmacopeias describe accurate methods but, in some cases, these are lengthy and difficult. Membrane sensor techniques offer several advantages in terms of simplicity, rapidity, and accuracy over many known official methods. As to the simplicity and rapidity, the entire determination in some cases takes less than 15 minutes and the procedures can be directly applied to drug determinations in pharmaceutical preparations without prior separation; in many cases, the excipients are inactive in sensor response and less cleanup is needed. The rapidity with which the assay can be carried out using such devices makes it practical to perform the procedure on a single pharmaceutical preparation (e.g., tablet, capsule, etc.) so that pharmaceutical preparation variation can be followed if desirable.

In most pharmaceutical applications of membrane electrodes, four main types of sensors are used.

1. *Primary electrodes containing crystalline membranes* prepared from either a single compound (e.g., Ag_2S) or a homogeneous mixture of sparingly soluble compounds (e.g., AgX/Ag_2S, where $X =$

halogen): Most have been commercially available for many years, and their characteristics and performances are still very good (see Appendix 1).

2. *Primary electrodes containing non-crystalline membranes*—glass membrane electrodes (e.g., H^+, Na^+) and electrodes with membranes containing a mobile carrier: In the latter case the electroactive material is dissolved either in a hydrophobic polymer (e.g., PVC) or a hydrophobic liquid solvent (e.g., nitrobenzene). Only a few electrodes in this category are commercially available (e.g., BF_4^-, Ca^{2+}, and K^+) and most of them are laboratory-made. When a hydrophobic cation (e.g., Aliquat 336S) is used, sensitive sensors for various anions (organic acids, amino acids) can be obtained; whereas, when a hydrophobic anion (e.g., tetraphenylborate, dinonylnaphthalene sulfonate) is used, sensitive membrane sensors for cations (organic bases, alkaloids) can be obtained.

3. *Gas-sensing electrodes and probes* (e.g., NH_3, CO_2) that are commercially available: Most of these are based on a sensitized pH electrode.

4. *Bio-selective electrodes (Potentiometric Biosensors)*, based on enzyme–substrate reactions: These are laboratory-made and are very selective for the respective substrate; most of them were created when stable and reliable potentiometric sensors for NH_3, CO_2, and H_2S became commercially available on a routine basis. Such sensors combine the technology of ion-selective electrodes with that of microporous synthetic membranes. Microbial or plant-tissue electrodes can also be included in this category. In these cases the cells are held on the surface of the electrochemical sensor by a dialysis membrane. Their general principle of operation is similar to that of conventional enzyme sensors that utilize isolated enzymes as the biocatalytic component.

Some of the electrochemical potentiometric membrane sensors presented in this book, even those that are laboratory-made, are feasible for drug monitoring and we strongly recommend them for this purpose. This recommendation is correlated with their characteristics as follows:

1. The sensor is, in most cases, specific for the drug of interest.
2. The linearity of the calibration curve is relatively large, generally covering a 10^{-2} to 10^{-5} M range. Detection limits of 10^{-6} or 10^{-7} M are also reported for some drug-sensitive sensors. With previous preconcentration of the sample, the detection limits of potentiometric techniques using membrane sensors may equal or surpass those of some expensive and sophisticated techniques such as radioimmunoassay, gas–liquid chromatography, high-performance liquid chromatography, chemical ionization mass spectrometry, etc.

3. With some exceptions, the electrochemical sensors have fast response times, usually within 30 s, depending on the analyte concentration. Among the exceptions are enzyme electrodes as well as microbial and plant-tissue electrodes. There is, as yet, no general theoretical formulation for the steady-state and time-dependent behavior of these electrodes in terms of geometric and kinetic parameters.

4. Many sensors are amenable to miniaturization (e.g., for ease of intravascular insertion) and can be constructed of material that is physiologically compatible, non-toxic, and easily sterilized. A new type of ion-selective device, called the ion-selective field effect transistor (ISFET), promises to be adequate for biomedical analysis *in vivo*.

5. The time and cost of one determination with selective membrane sensors are substantially reduced.

This book has been made possible only as a result of the remarkable contributions by numerous researchers from all over the world. It contains many new drug sensors that have found real applications in pharmaceutical analysis. Some sensors can be applied for monitoring a drug during its release from a given drug delivery system. The marriage of drug membrane sensors with drug delivery technologies is already underway. The next prospect is that a laboratory-made drug-sensitive membrane sensor will become commercially available.

This book is divided into three main parts. The first concerns design and principles of various membrane drug sensors and contains the basic theoretical considerations as well as the basic characteristics of such sensors; here are discussed parameters such as electrode function, limits of detection, selectivity, response time, etc. The analyst is provided with the necessary information to assess whether a method is suitable for use in a particular analysis. Some details are given on standardization of membrane electrodes as well as on various analytical techniques involving them (e.g., direct potentiometry, standard addition and subtraction methods, potentiometric titrations, Gran plots, etc.). The second part of the book refers to the analysis of pharmaceuticals by membrane sensors, and the third part refers to drug release monitoring by membrane sensors.

In the second part of the book, which is the largest one, many analytical procedures are described for the assay of more than 350 compounds with biological activity. Both commercially available membrane electrodes as well as laboratory-made sensors are successfully applied for drug analysis, mainly by potentiometric techniques. In most cases, the membrane sensor method is comparatively discussed with the official method, such as that included in the *United States Pharma-*

copeia (USP), the *British Pharmacopeia*, or other official monographs. The authors did not intend to give more details on the physical or pharmacological properties of the drugs discussed. They considered it necessary and sufficient for the reader to include only therapeutical activity of the named drug compound. The main sources for these properties were the *Merck Index*, USP, British Codex, or other pharmacopeias. No attempt has been made to represent analytical recipes for electrode construction because these may be easily found in the cited reference literature.

References

1. M. E. Meyerhoff and Y. M. Fraticelli, *Anal. Chem.*, **54**, 27R (1982).
2. J. Koryta, *Anal. Chim. Acta*, **139**, 1 (1982).
3. J. Koryta, *Anal. Chim. Acta*, **159**, 1 (1984).
4. J. Koryta, *Anal. Chim. Acta*, **183**, 1 (1986).
5. J. Koryta, *Anal. Chim. Acta*, **206**, 1 (1988).
6. M. A. Arnold and M. E. Meyerhoff, *Anal. Chem.* **56**, 20R (1984).
7. M. A. Arnold and R. L. Solsky, *Anal. Chem.*, **58** 84R (1986).
8. R. L. Solsky, *Anal. Chem.*, **60**, 106R (1988).
9. G. J. Moody and J. D. R. Thomas, *Ion-Selective Electrode Rev.*, **6**, 209 (1984).
10. G. J. Moody and J. D. R. Thomas, *Ion-Selective Electrode Rev.*, **7**, 209 (1984).
11. G. J. Moody and J. D. R. Thomas, *Ion-Selective Electrode Rev.*, **8**, 209 (1986).
12. G. J. Moody and J. D. R. Thomas, *Ion-Selective Electrode Rev.*, **9**, 197 (1987).
13. G. J. Moody and J. D. R. Thomas, *Selective Electrode Rev.*, **10**, 231 (1988).
14. G. J. Moody and J. D. R. Thomas, *Selective Electrode Rev.*, **11**, 265 (1989).
15. V. V. Coşofreţ, *Ion-Selective Electrode Rev.*, **2**, 159 (1980).
16. M. S. Ionescu and V. V. Coşofreţ, *Rev. Chim.* (*Bucharest*), **31**, 1005 (1980).
17. M. S. Ionescu and V. V. Coşofreţ, *Rev. Chim.* (*Bucharest*), **31**, 1088 (1980).
18. R. K. Gilpin, L. A. Pachla, and R. S. Ranweiler, *Anal. Chem.*, **53**, 142R (1981).
19. R. K. Gilpin, L. A. Pachla, and R. S. Ranweiler, *Anal. Chem.*, **55**, 70R (1983).
20. R. K. Gilpin and L. A. Pachla, *Anal. Chem.*, **57** 29R (1985).
21. R. K. Gilpin and L. A. Pachla, *Anal. Chem.*, **57**, 174R (1987).
22. R. K. Gilpin and L. A. Pachla, *Anal. Chem.*, **61**, 191R (1989).

23. Z.-Z. Chen, *Huaxue Chuan Gan Qi*, **5**, 11 (1985).

24. E. Pungor, Z. Feher, G. Nagy, E. Lindner, and K. Tóth, *Anal. Proc. (London)*, **19**, 79 (1982).

25. T. C. Pilkington and B. L. Lawson, *Clin. Chem.*, **28**, 1946 (1982).

26. R. L. Solsky, *CRC Crit. Rev. Anal. Chem.*, **14**, 1 (1982).

27. V. V. Coşofreţ and R. P. Buck, *Ion-Selective Electrode Rev.*, **6**, 59 (1984).

28. Z.-R. Zhang and V. V. Coşofreţ, *Selective Electrode Rev.*, **12**, 35 (1990).

29. R. Vytras, *Ion-Selective Electrode Rev.*, **7**, 77 (1985).

30. S.-Z. Yao, *Yaoxue Xuebao*, **20**, 552 (1985).

31. S.-Z. Yao and L. Nie, *Anal. Proc. (London)*, **24**, 338 (1987).

32. C.-X. Xu, *Yaowu Fenxi Zazhi*, **7**, 185 (1987).

33. R.-Q. Yu, *Ion-Selective Electrode Rev.*, **8**, 153 (1986).

34. D. Feng, *Ion-Selective Electrode Rev.*, **9**, 95 (1987).

35. Z.-R. Zhang, *Anal. Chem. Symp. Ser.*, **22**, 695 (1985).

36. D. W. Lubbers, H. Acker, R. P. Buck, G. Eisenman, M. Kessler, and W. Simon, Eds., *Progress in Enzyme and Ion-Selective Electrodes*, Springer-Verlag, Berlin, 1980.

37. G. J. Moody, Ed., *Int. Symp. Electroanalysis in Clinical, Environmental and Pharmaceutical Chemistry*, Royal Society of Chemistry, UWIST, Cardiff, Wales, 1981.

38. D. W. Lubbers, H. Acker, and R. P. Buck, Eds., *Progress in Enzyme and Ion-Selective Electrodes*, Springer-Verlag, Berlin, 1981.

39. Abstracts Int. Symp. Ion-Selective Electrodes, Shanghai, China, June, 1985.

40. Abstracts 4th State Conf. Ion-Selective Electrodes, Hunan, China, Sept. 1988.

41. Proc. 1987 Electroanalytical Conf., Guanzhou, China, Oct. 1987.

42. Proc. 2nd Beijing Conference and Exhibition on Instrumental Analysis, Oct. 1987.

43. Abstracts 2nd State Meeting on Drug Electrodes, Xiamen, China, May 1988.

44. R. P. Buck and V. V. Coşofreţ, Design of sensitive drug-sensors: principles and practice, in *Fundamentals and Applications of Chemical Sensors*, D. Schuetzle and R. Hammerle, Eds., American Chemical Society, Washington, DC., 1986, chap. 22.

45. G. E. Baiulescu and V. V. Coşofreţ, *Applications of Ion-Selective Membrane Electrodes in Organic Analysis*, John Wiley & Sons, Chichester, 1977, and MIR, Moscow, 1981.

46. J. Koryta, Ed., *Medical and Biological Applications of Electrochemical Devices*, John Wiley & Sons, New York, 1980.

47. V. V. Coşofreţ, *Membrane Electrodes in Drug-Substances Analysis*, Pergamon Press, Oxford, 1982.

48. T.-S. Ma and S. S. M. Hassan, *Organic Analysis Using Ion-Selective Electrodes*, Vols. 1 and 2, Academic Press, London, 1982.

49. J. Koryta, *Ions, Electrodes and Membranes*, John Wiley & Sons, Chichester, 1982.

50. J. Koryta and K. Stulik, *Ion-Selective Electrodes*, 2nd ed., Cambridge University Press, Cambridge, 1983.

51. J. G. Schindler and M. M. Schindler, *Bioelectrochemical Membrane Electrodes*, Walter de Gruyter, Berlin, 1983.

52. Z.-Z. Chen and Z. Qin, *Applications of Ion-Selective Electrodes in Pharmaceutical Analysis*, Renmin Weisheng Publishing House, Beijing, 1985 (in Chinese).

53. M. A. Arnold and G. A. Rechnitz, *Biosensors: Fundamentals and Applications*, Oxford Science Publications, Oxford, 1987.

54. S. Okozaki, K. Nozaki, and H. Hara, *Database on Electrochemical Sensors*, San-Ei Publishing Co., Kyoto, 1987.

Part I

DESIGN AND
PRINCIPLES OF
MEMBRANE DRUG
SENSORS

Chapter 1

THEORETICAL CONSIDERATIONS FOR PRIMARY ELECTRODES AND ION ASSOCIATION DRUG SENSORS

A phase that separates two other phases to prevent overall mass movement between them, but allows passage with various degrees of restriction of one or several species of the external phases, may be defined as a membrane.[1] When membranes are used as recognition or recognition–amplification elements in conjunction with electron or ion conductor electrodes, the combination becomes a "membrane electrode." The electrode normally uses inner and outer reference electrodes to form an electrochemical cell. The behavior of the membrane electrode will be determined mainly by the properties of the membrane, which can be solid or liquid containing ionized or ionizable groups. The membrane, with asymmetric bathing, may generate a voltage, or the membrane may simply be a barrier that controls access of material to another sensor. Reference inner and outer electrodes contribute an additional constant "offset" voltage. A completely gaseous membrane will not be discussed here, although they occur in cells designed to measure electronic work functions. Membrane electrodes respond directly or indirectly to gases, and they will be described in Chapter 2.

Ion-selective membrane electrodes may be roughly classified in the following way, according to the physical state of the substances (the electroactive materials) that form the electrode membrane.

(i) Ion-selective electrodes with solid membranes: The membrane may be homogeneous as in a monocrystal, a sparingly soluble, ionic crystalline substance, or a glass that is considered to be a solid because of the immobility of the ionic components. Alternatively, the membrane may be heterogeneous, by the incorporation of the electroactive substance within an inert matrix.

(ii) Ion-selective electrodes with liquid membranes: Here the electrode membrane is represented by an organic liquid immiscible with water. The organic liquid contains a charged electroactive substance that serves as "sites" for exchange of ions between membrane and solution. The membrane is responsive and may be selective for the exchangeable ions.

This classification is also useful from the theoretical point of view. In this respect, ion-selective electrodes should be classified according to the homogeneity or heterogeneity of the membrane, because these terms refer to the composition and not to operation.

The electrodes in categories (i) and (ii) are *primary* in the sense that they have one sensing reaction or one main function. The following are included with (i).

1. *Crystalline electrodes* that may be homogeneous or heterogeneous: They contain mobile ions of one sign and fixed sites of opposite sign. The fluoride electrode made from LaF_3 is an example.

   a. *Homogeneous membrane electrodes* are ion-selective electrodes in which the membrane is a crystalline material prepared from either a single compound or a homogeneous mixture of compounds (i.e., Ag_2S, AgI/Ag_2S).

   b. *Heterogeneous membrane electrodes* are formed when an active substance or a mixture of active substances is mixed with an inert matrix (such as silicone rubber or PVC) or is placed on hydrophobized graphite, to form the sensing membrane, which is heterogeneous in nature.

   c. *Metal contact or all-solid-state electrodes* are formed from membrane materials with both ionic and electronic conductivities. They need not be interposed between two electrolyte solutions. The inner reference electrode can be replaced with an electronic conductor (e.g., AgBr on Ag, or a cation radical salt on Pt). This configuration contrasts with normal membrane usage in which electrolyte solutions (inner filling solution and outer "test" solution) contact membranes.

Within category (ii) we find the following.

2. *Non-crystalline electrodes:* In these electrodes, a support matrix, containing an ion exchanger (either cationic or anionic), a plasticizer solvent, and possibly an uncharged, selectivity-enhancing species, form the ion-selective membrane, which is usually interposed be-

tween two aqueous solutions. The support used can be either macroporous (e.g., Millipore filter, glass frit, etc.) or microporous (e.g., Vycor, glass, or inert polymeric material such as PVC), yielding with the ion-exchanger and the solvent a "solidified" homogeneous mixture. These electrodes exhibit a response due to the presence of the ion-exchange material (typically selected from category 2b) in the membrane.

a. *Rigid, self-supporting, matrix electrodes* (e.g., synthetic cross-linked polymer or glass electrodes) are ion-selective electrodes in which the sensing membrane is a thin polymer with fixed sites or a piece of glass. The chemical composition of the polymer (e.g., polystyrene sulfonate, Nafion, amino-poly(vinylchloride)) or of the glass determines the selectivity of the membrane. In this group are hydrogen-ion-selective electrodes and monovalent-cation-selective electrodes.

b. *Electrodes with mobile charged sites:*

   (1) *Positively charged*, hydrophobic cations (e.g., those of quaternary ammonium salts or salts of substitutionally inert transition-metal complexes such as the derivatives of 1,10-phenanthroline), which, when dissolved in a suitable organic solvent and held on an inert support (e.g., Millipore filter or PVC), provide membranes that are sensitive to changes in the activities of anions. The hydrophobic cations are "trapped," mobile sites that are mainly confined to the membrane phase.

   (2) *Negatively charged* hydrophobic anions (e.g., of type $(RO)_2PO_2^-$, tetra-*p*-chlorophenylborate, dinonylnaphthalene sulfonate), which, when dissolved in a suitable organic solvent and held in an inert support (e.g., Millipore filter or PVC), provide membranes that are sensitive to changes in the activities of cations.

   (3) *Uncharged "carrier"* electrodes based on solutions of molecular complexing agents of cations (e.g., ion-dipole formers—antibiotics, macrocyclic compounds, or other sequestering agents) and anions (e.g., adduct formers—organotin compounds, activated carbonyl compounds, and some porphyrins), which can be used in fixed sites of mobile trapped-site ion exchanger membrane preparations to give sensitivity and selectivity to certain cations and anions.

   (4) *Hydrophobic ion pair* electrodes of plasticized polymers (e.g., PVC) containing a dissolved hydrophobic ion pair and "ion association complex" (e.g., a cation drug tetraphenylborate or tetraalkylammonium surfactant anion) respond to component ion activities in bathing electrolytes. Responses can be Nernstian to bathing electrolytes of the cation drug chloride or sodium tetraphenylborate.

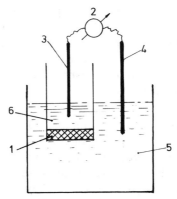

Figure 1.1 Schematic representation of membrane electrode cell assembly: (1) membrane; (2) potentiometer; (3) internal reference electrode; (4) external reference electrode; (5) sample solution; (6) internal filling solution. (Reproduced from Coşofreţ, V. V., *Membrane Electrodes in Drug-Substances Analysis*, Pergamon Press, Oxford, 1982, 8. With permission.)

A compilation of equations for membrane transport has been published.[2] Discussion of membrane types and equations for membrane potentials in the context of ion-selective electrodes was given by Buck.[3-5]

A schematic representation of the cell assembly is shown in Figure 1.1. The membrane, selective to a particular ion, is the basic component of the electrochemical cell and separates two electrolyte solutions having different ionic activities. The potential difference, established between the two sides of the membrane, is measured by the potentiometer by means of the internal and external reference electrodes introduced into the internal filling and external solutions, respectively. Usually the membrane is held in a compact unit containing the internal filling solution and the internal reference electrode to constitute the ion-selective membrane electrode. In some cases the internal filling solution is dispensed with and electrical contact is made by connecting a wire directly to the inner face of the membrane. The use of ion-selective membrane electrodes depends on the determination of membrane potentials that represent the electrical potentials arising across membranes when they separate two electrolyte solutions. The individual potential components cannot be determined directly, but their changes can be deduced from the EMF values for complete electrochemical cells illustrated in Figure 1.1.

1.1   Solid Membranes

Relations based on investigations by Nicolsky[6] have been derived from experimental data for the EMF of cells with liquid or solid ion-exchange membranes.[7-10] These relations are as follows for glass and pure solid ion-exchange membranes[11]:

$$E = E_0 + \frac{nRT}{F} \ln \frac{a_A^{1/n} + \left(k_{A,B}^{pot} a_B\right)^{1/n}}{(a_A')^{1/n} + \left(k_{A,B}^{pot} a_B'\right)^{1/n}} \tag{1.1}$$

where R = gas constant, T = absolution temperature, F = Faraday constant, a_A, a_B = ion activities in the sample solution (monovalent ions), a_A', a_B' = ion activities in the internal filling solution (monovalent ions), n = constant depending on the ions A and B and on the membrane; $k_{A,B}^{pot}$ = selectivity coefficient (preference of sensor for ion B in relation to ion A).

For an ion-selective electrode having a given inner reference electrode system, (a_A', a_B' = constant) we have

$$E = E_0 + \frac{nRF}{F} \ln\left[a_A^{1/n} + \left(k_{A,B}^{pot} a_B\right)^{1/n}\right] \tag{1.2}$$

The quantity n is a descriptor for the concentration dependence of the *monovalent* ionic activity coefficients in a solid or glassy phase. This quantity appears mainly in the literature for *monovalent ion exchanger glass* and is not necessary, experimentally, for other ion-selective electrode responses. In fact, the ionic charge is the dominant factor for most electrodes, and we write the response

$$E = \pm \frac{RT}{F} \ln \frac{\displaystyle\sum_{i=1}^{N} k_{A,i}^{pot} a_i^{1/z_i}}{\displaystyle\sum_{i=1}^{N} k_{A,i}^{pot} (a_i')^{1/z_i}} \tag{1.3}$$

in general, and + is taken for cations or − for anions. The charges z_i are absolute values.

$$E = E_0 \pm \frac{RT}{F} \ln\left[\sum_{i=1}^{N} k_{A,i}^{pot} a_i^{1/z_i}\right] \tag{1.4}$$

for a given inner reference electrode system, and z_i is the charge on each potential-determining ion. In this notation, the first ion ($i = 1$) is ion A, and it is the ion for which the electrode is normally most sensitive. Any other interfering ion for which $i > 1$ is called B when comparing its selectivity with that of A. Of course, the selectivity of A with respect to A is unity. The selectivity coefficient $k_{A,B}^{pot}$, which characterizes the preference of the sensor for the ion B as compared to the ion A, is given by

$$k_{A,B}^{pot} = \frac{u_B}{u_A} K_{A,B} \tag{1.5}$$

where $K_{A,B}$ is the ion exchange equilibrium constant for the exchange reaction

$$\frac{1}{z_B} B_{solution} + \frac{1}{z_A} A_{membrane} = \frac{1}{z_B} B_{membrane} + \frac{1}{z_A} A_{solution} \quad (1.6)$$

and u_A and u_B are the mobilities of the ions in the membrane; z_A and z_B are absolute values of ion charge. $K_{A,B}$ is roughly related to the individual ion partition coefficients by

$$K_{A,B} = \frac{k_B^{1/z_B}}{k_A^{1/z_A}} - = \frac{K_{so}(\text{MA or AX})^{1/z_A}}{K_{so}(\text{MB or BX})^{1/z_B}} \quad (1.7)$$

If the sensor responds to a divalent cation (a_A) and a monovalent cation (a_B), Equation 1.4 becomes

$$E = E_0 + \frac{RT}{F} \ln\left[a_A^{1/2} + k_{A,B}^{pot} a_B \right] \quad (1.8)$$

For a completely reversible cell assembly, $k_{A,B}^{pot}$ can, in principle, be determined approximately by measurements carried out in a solution of the ion A and, separately, in a solution of the ion B. A number of solid membrane electrode configurations are shown in Figure 1.2.

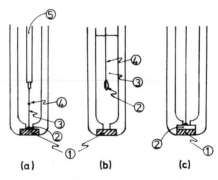

Figure 1.2 Examples of selective electrodes based on inorganic salts: (a) All-solid-state electrode—(1) solid, nonporous silver salt membrane; (2) silver-loaded epoxy cement; (3) silver wire; (4) solder to copper wire; (5) shielded cable. (b) Ion-selective electrode with internal solution and internal reference electrode—(1) silver salt membrane (solid and nonporous); (2) AgX coating; (3) internal reference solution; (4) Ag internal reference electrode. (c) Electrode of the second kind with porous metal salt containing silver metal—(1) metal salt; (2) metal billet. (Reproduced from Buck, R. P., Electrochemistry of ion-selective electrodes in *Comprehensive Treatise of Electrochemistry*, Vol. 8, White, R. E., Bockris, J., Conway, B., and Yeager, E., Eds., Plenum Press, New York, 1984, 182. With permission.)

For the theoretical interpretation of the behavior of precipitate-based ion-selective electrodes, the electrodes based on silver halide may be considered as a model.[11] By using either a heterogeneous or a homogeneous ion-selective electrode at zero current in a solution containing the ion to which the electrode is reversible, the equilibrium between the solution and the solid phase is attained when the difference of the electrochemical potentials of the solvated ion and the ion bonded to the solid phase is equal to zero. The electrochemical potential of the appropriate ith ion in the solution is

$$\tilde{\mu}_s = \mu_s + z_i F \phi_s \qquad (1.9)$$

whereas in the membrane it is

$$\tilde{\mu}_m = \mu_m + z_i F \phi_m \qquad (1.10)$$

where $\tilde{\mu}$ is the electrochemical potential (an energy), μ is the ordinary chemical potential, ϕ is the local electrostatic potential (Galvani), z_i is the valency of ion i with sign, F is the Faraday constant, and s and m denote the solution and membrane phases, respectively. At equilibrium,

$$\tilde{\mu}_s = \tilde{\mu}_m \qquad (1.11)$$

and

$$\phi_m - \phi_s = \frac{1}{z_i F} \left[\frac{\mu_s - \mu_m}{RT} \right]$$

$$= \frac{RT}{z_i F} \left[\frac{\mu_s^0 - \mu_m^0}{RT} + \ln \frac{(a_i)_s}{(a_i)_m} \right] \qquad (1.12)$$

$$E = E_0 + \frac{RT}{z_i F} \ln \frac{(a_i)_s}{(a_i)_m} = \frac{RT}{z_i F} \ln \frac{k_i (a_i)_s}{(a_i)_m} \qquad (1.13)$$

where E is the electrode potential, E_0 is the standard electrode potential, μ^0 is the standard chemical potential of ion i in solution s and in membrane m, and $(a_i)_s$ and $(a_i)_m$ are the activities of the ith ion in the solution and membrane phases, respectively. Note that E_0 is a measure of the standard energy difference between the exchanging ion in the membrane relative to the solution energy. In fact, E_0 is defined by the energy difference or by the *single-ion partition coefficient* k_i that often appears in the descriptions of liquid and synthetic solid ion-exchanger membranes.

$$k_i = \exp\left[-\frac{\mu^0_{i(s)} - \mu^0_{i(m)}}{RT}\right] \qquad (1.14)$$

In most practical cases, the membrane is prepared from, and contains, a salt of the same permeable ion that is being detected and measured. This means that as quickly as the electrode is dipped into solutions of the permeable ion salts, the equilibrium, Equation 1.11, will occur. Only motion of a small amount of space charge is needed to create the interfacial potential difference in Equation 1.12 in about 1 μs. A little more time may be needed to actually mix the bathing solution sufficiently to make the surface concentrations become the same as the bulk value. Usually 1 s is enough. In this deduction of the electrode potential,[11] the ion diffusion across the membrane is not considered because it normally has no effect on the membrane potential at zero current.

If the concentration of the appropriate ion is relatively low in the solution examined, then the electrode potential approaches a limiting value that can be expressed for the univalent cations in the following way:

$$E = E_0 + \frac{RT}{F}\ln\left\{\frac{(a_+)_s + \left[(a_+)^2_s + 4\gamma_+ K_{so}/\gamma_-\right]^{1/2}}{2}\right\} \qquad (1.15)$$

where K_{so} is the solubility product of the precipitate used as the electrode and a_i is the formal activity added to the solution. Of course, the actual concentration exceeds the apparent (formal) value by the amount dissolving from the electrode surface. For univalent anions, the sign of the ln term is negative and subscripts are changed from + to −.

In the solution containing not only the ion to which the electrode is reversible, but another ion (X), which also forms a precipitate with one of the components of the membrane matrix, the following precipitate exchange reaction is established:

$$AgI + X = AgX + I \qquad (1.16)$$

$$K_{I,X} = \frac{(a_I)_s(a_X)_m}{(a_X)_s(a_I)_m} = \frac{K_{so}(AgI)}{K_{so}(AgX)} = \frac{k_X}{k_I} \qquad (1.17)$$

The membrane species take their constant bulk values in a short time if the reactant concentrations in solution are not too low. Normally, X^- is

Cl^- or Br^-. Then $K_{I,X}$ is smaller than unity and the electrode is attacked to a small extent by the interference. On the other hand, an AgCl electrode has the selectivity coefficient $K_{so}(AgCl)/K_{so}(AgX)$. If $X = Br^-$ or I^-, the selectivity coefficient is greater than unity and the electrode could be attacked, destroyed, and converted entirely to AgX. To avoid this event, the interference concentration must be controlled to low values such that $(X^-)K_{IX} < (Cl^-)$ in this case.

On the basis of this equilibrium in Equation 1.16 the following equation can be derived for the potential of the membrane electrode if the intramembrane diffusion phenomena are neglected.[11] Except for a few "superionic conductors," permeable ions in solid electrodes usually move much less rapidly than the same ions in electrolyte solutions. This means that interior (intramembrane) processes in solid electrodes can be neglected because they do not contribute an important part of the measured potential. Using the exchange equilibrium Equation 1.16 for ions of the same charge z_i, any interference ion k will cause a change in the measured membrane and cell potentials according to

$$E = E_0 + \frac{RT}{z_i F} \ln\left[(a_i)_s + \sum_k K_{i,k}(a_k)_s \right] \qquad (1.18)$$

that is:

$$E = E_0 + \frac{RT}{z_i F} \ln\left[(c_i)_s f_{+i} + \sum_k K_{i,k}(c_k)_s f_k \right] \qquad (1.19)$$

where a_i, c_i and a_k, c_k are the activities and concentrations of the ith and kth ions, respectively; if N is the number of ions taking part in the ion exchange reaction, then summing over $N = 1$ to $N = k$ is required to cover all possible responses. The summation subscript is simply shown as k. The mean activity coefficients are f_i and f_k that apply to the ions i and k at ionic strengths determined by the solution electrolyte composition including other inert salts.

1.2 Liquid Membranes

1.2.1 Electrically Charged Ligands (Liquid Ion Exchangers)

Eisenman and co-workers[7-9] have derived relations for the EMF of a cell assembly consisting of a liquid ion-exchange membrane electrode and an external reference electrode. In the ideal case that all ions and charged ligands in the membrane are completely ionized,

$$E = \frac{RT}{z_i F} \ln \frac{\sum\limits_{i=1}^{N} k_{A,i}^{pot} a_i}{\sum\limits_{i=1}^{N} k_{A,i}^{pot} a'_i} = E_0 + \frac{RT}{z_i F} \ln\left(\sum\limits_{i}^{N} k_{A,i}^{pot} a_i \right) \qquad (1.20)$$

where z_i is the valence of the ith counter-ion species, a_i and a'_i are its activities in the solutions on each side of the membrane, with $k_{A,i}^{pot}$ expressed by

$$k_{A,i}^{pot} = \frac{u_i f_{Am}}{u_A f_{im}} K_{A,i} \qquad (1.21)$$

The mobilities and activity coefficients are, as before, u_i and f_i. The first ion ($i = 1$) is called A, as before. Its selectivity with respect to itself is unity. Usually all other selectivity coefficients are less than unity, because the electrode is designed to respond best to ion A. Responding "best" means that the response curve is most positive or most negative for ion A compared to all other ions. Most positive occurs when positive ions are being compared and most negative when negative ions are compared. The difference between Equations 1.21 and 1.3 is inclusion of the internal membrane activities coefficients in Equation 1.21 for the ions that are present in the membrane. This difference arises because the transport problem (the internal diffusion potential) depends on concentrations rather than on activities. When the membrane contains ion pairs, as it mostly generally does because of the low dielectric constant of the plasticizer, there is a more general solution,

$$E = \frac{RT}{z_i F} \ln \frac{\sum\limits_{i=1}^{N} k_{A,i}^{pot} a_i}{\sum\limits_{i=1}^{N} k_{A,i}^{pot} a'_i} - \int_1 - \int_2 \qquad (1.22)$$

The integrals (over the membrane thickness) generate additional terms and contain especially formation constants and mobilities of ion pairs. When all ions have the same charge, the results simplify, in the cases of strong complexation, to an equation like Equation 1.20, but now

$$k_{A,i}^{pot} = \frac{u_{is} k_i K_{is}}{u_{As} k_A K_{As}} = \frac{u_{is} K_{is}}{u_{As} K_{As}} K_{A,i} \qquad (1.23)$$

where the mobilities and formation constants (K_{is}) apply to the ion pairs is and As. The activity coefficients are omitted.

$$\int_1 = \int_0^d t \, d \ln \frac{\sum\limits_i (u_{is}/K_i)}{\sum\limits_i u_i c_i} \qquad (1.24)$$

$$\int_2 = \int_0^d \frac{(u_s J_s^*/RT) \, dx}{\left(u_s + \sum\limits_i (u_{is} c_i/K_i) \right) \sum\limits_i u_i c_i + u_s c_{is} \sum\limits_i (u_{is} c_i/K_i)} \qquad (1.25)$$

in which the subscripts s refer to the dissociated site species and the subscripts is refer to the undissociated ion pairs. Thus u_s is the mobility of the dissociated site species, u_{is} is the mobility of the undissociated ion pair, K_i is the dissociation constant of this pair, and c_{is},

c_i, and c_s are the concentrations of undissociated pairs, dissociated counter ions, and dissociated sites within the membrane. J_s^* in $\int_2$ is the total flux of sites regardless of whether in a dissociated or undissociated state), whereas the parameter t in $\int_1$ is given by the expression

$$t = \frac{u_s c_s}{\left(u_s c_s / \sum_i u_{is} c_{is} + 1\right) \sum_i u_i c_i + u_s c_s} \tag{1.26}$$

The parameter t varies between 0 for complete dissociation and $u_s c_s / (u_s c_s + u_i c_i)$ for strong association.

The general form of Equation 1.22 is valid only under the following conditions:

1. The ligands are situated exclusively within the membrane.
2. There are no co-ions in the membrane (Donnan exclusion applies).
3. Equilibrium exists between the ion i and the ligand at all points in the membrane.
4. Ion aggregates of higher order do not occur.
5. The activities can be equated to the concentrations inside the membrane.
6. The system is at open circuit (zero current), and the interior of the membrane has reached steady state.

The form of Equation 1.22 with the second integral omitted is actually a more normal form, although not so general. In addition to requirements 1–6, the following holds:

7. at steady state, the total flux of all species including s is zero, e.g., the sum of all ion-pair and free-ligand fluxes is zero.

The simple form of Equation 1.20 requires the same conditions, but requires the following in addition:

8. the ligand concentration is so small, or formation of ion pairs so weak, that complex formation just does not occur;
9. the internal filling solution containing the principal detectable and measurable ion.

When the ion pairing is nearly complete in the membrane, the general problem is difficult, and Equation 1.23 seems to be a limiting case. There are very few, if any reliable tests of Equation 1.23 because ion-pairing constants are so difficult to determine. The requirements listed are presumed to apply. If the membrane plasticizer has a high dielectric constant (over 10 and up to about 30, where miscibility with water occurs), ion pairing decreases and Equation 1.20 is appropriate. Then the selectivity depends on the ion-exchange constant. With the low dielectric constants, ion pairing is favored and the ion-pairing dissociation constants (or the reciprocal formation constants) determine the selectivity.

Some popular electrode configurations for liquid membranes are shown in Figure 1.3.

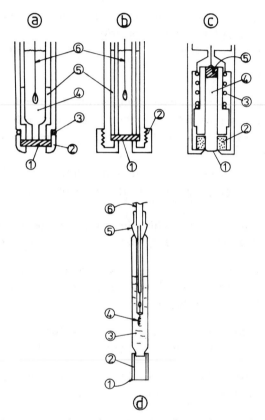

Figure 1.3 Various constructions of sensing tips of ion-selective electrodes based on organic ion-exchangers or neutral carriers. (a) Liquid membrane (similar to the Corning type)—(1) hydrophobic porous membrane; (2) membrane-retaining cap; (3) O-ring seal; (4) internal reference solution; (5) solution of active material; (6) internal reference electrode. (b) Liquid membrane (similar to the Corning type)—(1) cellulose acetate membrane; (2) screw-on cap that holds membrane; (5) and (6) are in (a). (c) Liquid membrane (similar to the Orion type)—(1) hydrophobic porous membrane; (2) plastic foam saturated with active material; (3) spring; (4) internal reference solution; (5) internal reference electrode. (d) PVC membrane according to Davies, Moody, and Thomas, *Analyst*, 97, 87, 1972, and Griffiths, Moody and Thomas, *Analyst*, 97, 420, 1972. (Reproduced from Buck, R. P., Electrochemistry of ion-selective electrodes, in *Comprehensive Treatise of Electrochemistry*, Vol. 8, White, R. E., Buckris, J., Conway, B., and Yeager, E., Eds., Plenum Press, New York, 1984, 1983. With permission.)

1.2.2   Electrically Neutral Ligands (Neutral Carriers)

In these systems, a plasticized membrane contains a dissolved complex-forming neutral charged carrier. The system is intentionally hydrophobic so that the plasticizer and carrier do not leach out and dissolve in the aqueous bathing solutions. The principal carriers are macrocyclic compounds that form selective complexes with ions in solution and extract those ions preferentially into the membrane to satisfy the electroneutrality condition inside the membrane. The membranes contain fixed or mobile sites. Eisenman and co-workers[12-15] derived an equation for the EMF of lipid bilayers loaded with the carrier valinomycin. Electroneutrality that is required for descriptions of thick plasticized ion-selective electrodes was not needed or considered. The general solution to the membrane potential equations, analogous to Equation 1.22, covering the possibility of free as well as complexed ions in the membrane, has never been given. However, in the practical limit that ions extracted into the membrane are complexed, the few remaining free ions can be ignored. For a plasticized membrane electrode bathed with variable activities a_1 of the principal ion and a_2 of a typical interfering ion, and for a given internal reference electrode system, the response function is

$$E = \frac{RT}{z_i F} \ln \frac{\sum_{i=1}^{N} k_{A,i}^{\text{pot}} a_i}{\sum_{i=1}^{N} k_{A,i}^{\text{pot}} a'_i} = E_0 + \frac{RT}{z_i F} \ln \left(\sum_{i}^{N} k_{A,i}^{\text{pot}} a_i \right) \quad (1.27)$$

Equation 1.23 gives expressions for k^{pot} in terms of formation constants corresponding to the reaction between free ions in the membrane and the lipophilic complexing agent:

$$K^+ + \text{valinomycin (val)} \rightarrow \text{Kval}^+ \quad (1.28)$$

for example.

Equation 1.27 is based on a number of observations concerning the nature of the negative sites in thick, electroneutral membranes used as potassium sensors. An original suggestion was the existence of negative sites as impurities in poly(vinyl chloride) (PVC) membranes.[16] Experimental observation of Donnan failure[17] and functional improvement by adding negative surface sites[18] and bulk sites[19-21] suggested by plausibility of impurity (fixed and/or mobile) sites. Extraction of anions from bathing solutions posed a theoretical problem, unless it could be shown that extracted anions became immobile in the membranes.[22-23] Experimental proof of the existence of ion-carrier complexes and fixed sites, and inference of mobile sites appeared.[24, 25] Also, evidence for extraction and immobilization of small anions was found[26-28] and a recent summary of the carrier mechanism has been published.[29]

Equations 1.26 and 1.27 are valid when the following hold:

1. Mixtures of monovalent ions are used in the bathing solutions.
2. The complexing agent in the membrane is present at total concentrations (free agent and complex) in excess of the site concentrations.
3. Free simple extracting and exchanging counterions can be ignored because formation constants of complexes are large.
4. Activities can be replaced by concentrations; activity coefficients are ignored.
5. Ion pairing between complexes and sites is ignored; formation of aggregates is ignored.
6. The system is at zero current (potentiometric).

Interference occurs by the overall reaction

$$B^+(aqueous) + AS^+(membrane) \rightarrow BS^+(membrane) + A^+(aqueous)$$

$$(1.29)$$

This equilibrium is expressed by the ion exchange and complex formation parameters in Equation 1.23. Because the mobilities of the complex species are primarily determined by their large size, the "isosteric" assumption is often made: the mobility ratio in Equation 1.23 is unity and can be eliminated. The construction of the electrodes is illustrated in Figure 1.3. Likewise, in the next section, ion association drug sensors are constructed according to the examples in this figure.

1.2.3   Ion Association Drug Sensors

Design principles for sensors of ionic drugs follow from application of the concepts used to describe the theoretical voltage–activity response relations for the various primary ion-sensitive sensors. The so-called trapped sites of the mobile-site, liquid ion-exchanger electrodes belong to a category of compounds known as ion association extractants. Examples are long-chain diesters of phosphoric acid and tricaprylylmethylammonium (Aliquat) ions. The latter cation was studied extensively by Freiser and co-workers[30-32] in the design of anion sensors. The former were the original class of hydrophobic, liquid ion-exchanger species used in the Ca^{2+} sensors of Ross.[33]

In 1970 Higuchi, et al.[34] and Liteanu[35] introduced liquid-membrane electrodes responsive both to organic and inorganic ions, including both cations and anions of the species constituting the membrane. Because they responded to bathing activities of inorganic counterions (as expected) and also to any site species dissolved in the bathing electrolyte (not expected), these electrodes seemed to be neither conventional mobile-site, liquid ion-exchanger- nor neutral-carrier-based. At the 1973 IUPAC International Symposium on Ion-Selective Electrodes, J. R. Cockrell, Jr., presented an unpublished example of a liquid membrane responsive to both *organic cations and organic anions*, not unlike responses of silver halide membranes, i.e., that membrane electrode responded to its component cations or anions depending on which was in excess in the bathing electrolyte. If a membrane contained a hydrophobic salt M^+X^- in a solvent-plasticized membrane, then a response was found for bathing salts M^+Cl^- or Na^+Y^- in solution. A membrane containing a detergent ion pair was responsive to either cationic or anionic detergent species in solution, depending on which was in excess. At the time, this effect seemed anomalous, because the idea of trapped sites was violated. It is now known that hydrophobic liquid membrane component ions are not really trapped, but only thoroughly partitioned into the membrane by a favorable salt extraction equilibrium. Nevertheless, there must always be some salt remaining in the bathing solution, and this residual concentration determines the detection limit of the electrode in the same way that the solubility product of a silver halide membrane determines its ultimate response sensitivity.

1.2.3.1 *Conditions for Two Opposite-Charge Ion Responses*

The local interfacial equilibrium principles based on equality of electrochemical potential of each ion that reversibly equilibrates across an immiscible interface were available for analysis of drug-sensing electrodes prior to 1970. However, the theory and consequences for membranes based on ion association or ion pair hydrophobic salts partitioned into membranes were worked out later than the theory for other primary electrodes.[36, 37] To develop an interfacial potential difference, two ions M^+ and Y^- that partition are generally involved, although in many simplified cases it appears that only ions of one sign are partitioning. This condition can be considered necessary, but not sufficient, to produce a sensor with good response for a single ionic species. The reason is that salt partitioning, e.g., extraction of equal concentrations of M^+ and Y^-, produces a potential difference (pd) that is independent of bathing salt concentration in either phase.

To develop a pd dependent solely on a single ion activity, say M^+, three ions are required: M^+, X^-, and Y^-, of which Y^- is very oil soluble, X^- is mainly water soluble, and M^+ is soluble in both phases. The salt MY is typically an organic ion association species that may be isolated and dissolved in an organic solvent or prepared *in situ* by extraction. The anion is typically picrate, tetraphenylborate, or an even more oil-soluble species. The salt MX, where $X^- = Cl^-$, is the sample whose M^+

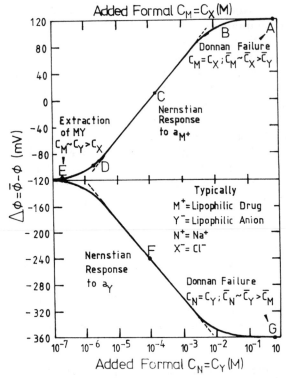

Figure 1.4 Calculated response curves for drug membrane sensors: (upper curve) response to M^+; (lower curve) response to Y^-. See text for values of the constants k_M, k_Y, and k_X. (Reproduced from Buck, R. P. and Coşofreţ, V. V., Design of sensitive drug sensors: principles and practice, in *Fundamentals and Applications of Chemical Sensors*, ACS Symp. Ser. No. 309, Schuetzle, D. and Hammerle, R., Eds., American Chemical Society, Washington, DC, 1986, 365. With permission.)

activity is to be measured at variable values in the aqueous bathing electrolyte. When MX is varied, the interfacial pd is overall S-shaped (millivolts vs. log[MX]), but contains a linear section, the so-called Nernstian region.

It is merely an extension of these ideas to demonstrate the conditions that the same membrane, containing MY, should also be responsive, in a Nernstian fashion, to Y^- activities in solution. These conditions again call for consideration of three ions: M^+, Y^-, and N^+. The salt NY is the aqueous sample whose Y^- activity is to be measured. N^+ is typically a hydrophilic cation such as Na^+. When aqueous NY activity is varied, the interfacial pd is again S-shaped (millivolts vs. log[NY]). These responses are illustrated from a theoretical calculation in Figure 1.4. The assumed extraction parameters are given in a later section. The similarities with silver halide membrane electrode responses are summarized here.

Saturated Salt AgBr (aq)	Partitioned Salt MY (very oil soluble)
Anion Responsive	Anion Responsive
(aq)Na$^+$Br$^-$ at $Cs > K_{so}^{1/2}$	(aq)Na$^+$Y$^-$ at $Cs > C_{MY}$(aq)
Cation Responsive	Cation Responsive
(aq)Ag$^+$NO$_3^-$ at $Cs > K_{so}^{1/2}$	(aq)M$^+$Cl$^-$ at $Cs > C_{MY}$(aq)

It is the hyperbolic relations $(\text{Ag}^+)(\text{Br}^-) = K_{so}$ and $(\text{M}^+)(\text{Y}^-)_s = (\text{M}^+)(\text{Y}^-)_m/K^2$ that provide the basic analogy between the two kinds of system. In the latter, K^2 is the ionic salt partition coefficient relating membrane (overbar) and bathing solution activities at an equilibrium interface. The latter form can also be derived for insoluble salt membranes. However, the salt phase ion activities are constant and so are hidden in the value of the solubility product K_{so}.

1.2.3.2 Local (Interface) Equilibrium Theory

Each ion M$^+$, N$^+$, X$^-$, and Y$^-$ will generally have different energies in water and in the organic phase within the solvent-plasticized membrane. The plasticizer is usually a low dielectric constant compound such as dibutylphthalate or dioctyladipate. The partition free energy for each ion is written

$$\Delta G_i = \mu_{im}^0 - \mu_{is}^0 = -RT \ln k_i \qquad (1.30)$$

for the reaction

$$\text{species } i \text{ (solution } s) = \text{species } i \text{ (organic membrane } m) \qquad (1.31)$$

is a measure of the intrinsic ionic oil-solubility of an ion, where k_i is the single-ion partition coefficient and m and s denote organic membrane and aqueous bathing phase quantities. For oil-soluble (hydrophobic) ions, ΔG is more negative and k_i is larger than for water-soluble (hydrophilic) ions.

In a typical two-phase water–organic system containing equilibrated salts MX and MY, some MX and MY will be present in each phase. This two-salt–three-ion system provides the necessary conditions to obtain the upper curve for the M$^+$ sensor in Figure 1.4. The concentrations in each phase depend on the salt partition coefficients $k_M k_X$ and $k_M k_Y$, with mass and charge balance equations to provide the additional required relationships. The link between energies and equilibrium concentrations (activities) is the electrochemical potential for each ion at equilibrium across the interface (see Equations 1.9, 1.10, and 1.11).

By eliminating the potential ϕ from the electrochemical potential equations for pairs of oppositely charged species, one has

$$\frac{(a_{M^+})(a_{X^-})_m}{(a_{M^+})(a_{X^-})_s} = K^2 = k_M k_X \tag{1.32}$$

$$\frac{(a_{M^+})(a_{Y^-})_m}{(a_{M^+})(a_{Y^-})_s} = K^2 = k_M k_Y \tag{1.33}$$

Because MY is intentionally more oil soluble than MX, MY maintains virtually constant activity in the organic, membrane phase when MX_s is varied. Only a very small amount of X^- (along with additional M^+) appears in the membrane because M_m^+ is already large and fixed by the MY extraction. This is a manifestation of Donnan exclusion. However, at very high bathing concentrations of MX_s, an increasing amount of MX will be extracted and find its way into the organic phase. In principle, at high enough MX_s levels, the amount of extracted MX could overwhelm and exceed the MY already present. This is the condition of Donnan failure because the co-ion X^- enters the membrane at concentrations comparable to or greater than Y_s^-.

There is a similar, converse argument that applied to the fate of M_s^+ in the aqueous bathing phase as MX_s concentration is continually decreased. Because MY maintains a very low equilibrium concentration in water, as MX_s is decreased *below* the equilibrium value of MY_s, M_s^+ approaches and remains at the constant concentration of MY_s. The value of M_s^+ cannot fall below this value set by the leaching out of MY into water. A constant interfacial pd is approached that is independent of further decreases in MX_s. In Figure 1.2 the single-ion partition coefficients were chosen to be $k_X = 10^{-2}$, $k_M = 10^2$, and $k_Y = 10^6$. Consequently, the equilibrium aqueous MY concentration is 10^{-6} M.

In the lower part of Figure 1.2, the corresponding two-phase–two-salt extraction system is MY and NY. The values of partition coefficients selected for illustrated were, as before, $k_M = 10^2$, $k_Y = 10^6$, and $k_N = 10^{-2}$. Consequently, mirror-image potential characteristics result, and Donnan failure occurs when NY_s exceeds about 10^{-2} to 10^{-1} M NY_s.

1.2.3.3 Potential Theory

The interfacial potential difference (pd) for each interface is found from the ionic partition equilibrium. The value must be the same whether calculated from electrochemical potentials of M^+, N^+, X^-, or Y^- using

Equations 1.9, 1.10, and 1.11:

$$\Delta\phi = \phi_m - \phi_s = \frac{RT}{z_i F} \ln \frac{k_i a_{is}}{a_{im}} \qquad (1.34)$$

However, for convenience in calculation, the interfacial pd is determined from those species whose activities are known or easily calculated. For the upper curve in Figure 1.4, in the linear range MX is predominantly in water whereas MY is predominantly in the organic phase, because $k_X \ll k_Y$. Consequently, M^+ activities are known and used in Equation 1.34, to give

$$\Delta\phi = \frac{RT}{z_i F} \ln \frac{k_M a_{Ms}}{a_{Mm}} = \Delta\phi^0 + \frac{RT}{F} \ln a_{Ms} \qquad (1.35)$$

This is a Nernstian response over a wide activity range. However, at very low MY bathing activities, the pd becomes insensitive to further decreasing M^+ activities and the response levels off at a value

$$\Delta\phi = \frac{RT}{2F} \ln \frac{k_M}{k_Y} \quad \text{for low MX} \qquad (1.36)$$

whereas at very high MX activities (generally only seen when $X^- = I^-$, NO_3^-, or ClO_4^-) the pd again levels off because Donnan exclusion is violated by encroaching X^- entering the organic phase and

$$\Delta\phi = \frac{RT}{2F} \ln \frac{k_M}{k_X} \quad \text{for high MX} \qquad (1.37)$$

These limiting, single-salt pds are derived from the more general expression

$$\Delta\phi = \frac{RT}{2F} \ln \frac{k_M a_{Ms}/a_{Mm} + k_N a_{Ns}/a_{Nm}}{k_X a_{Xs}/a_{Xm} + k_Y a_{Ys}/a_{Ym}} \qquad (1.38)$$

The values of activities to be used are those calculated from equilibrium theory. For one salt, say MX,

$$\Delta\phi = \frac{RT}{2F} \ln \frac{k_+ a_{+s}/a_{+m}}{k_- a_{-s}/a_{-m}} \qquad (1.39)$$

Equation 1.39 is suitable even in the cases of unequal concentrations of M^+ and X^- in each phase. However, if only MX (or only MY) is present at equilibrium in both phases, a simplification is possible because there are equal concentrations of $+$ and $-$ ions in each phase.

$$\Delta\phi = \frac{RT}{2F} \ln \frac{k_+ a_{+s}/f_{+m}}{k_- a_{-s}/f_{-m}} = \frac{RT}{2F} \ln \frac{k_+ f_{+s}/f_{+m}}{k_- f_{-s}/f_{-m}} \qquad (1.40)$$

These values are, in some sense derived from corrosion theory, "mean" or "mixed" potentials because they are determined by exchange of two charged species. When activity coefficients are ignored, Equation 1.40 reduces to Equation 1.37. By the same analysis, a single salt MY partitioned gives a constant potential in Equation 1.36. These two limiting values are shown in Figure 1.4, points A and E. Likewise, Donnan failure upon addition of excess aq. NY gives a constant negative limit of

$$\Delta\phi = \frac{RT}{2F} \ln \frac{k_N}{k_Y} \qquad (1.41)$$

Details of the curvature regions in Figure 1.4 have been given in Melroy and Buck.[36]

1.2.3.4   Detection Limit and Selectivity

The lower detection limit for ions M^+ or Y^- is often given as the intersection of the Nernstian region with the limiting potential of Equation 1.36. This value depends on the membrane loading MY_m and is given by

$$\frac{a_{Ys}(\text{lower limit})}{a_{Ym}} = \frac{a_{Ms}(\text{lower limit})}{a_{Ym}} = (k_M k_Y)^{-1/2} \qquad (1.42)$$

which is 10^{-6} M M^+ in Figure 1.4. The total range of available potentials for M^+ measurements is determined by the two limiting potentials in Equations 1.36 and 1.37. This is the "availability window" for M^+

$$\Delta\phi = \frac{RT}{2F} \ln \frac{k_Y}{k_X} \qquad (1.43)$$

There is also a window for Y^- measurements.

The definition of selectivity of the electrode for two ions of the same sign requires consideration of the responses to MY and IX when the membrane is initially loaded with MY. M^+ is the principal ion and I^+ is the interfering ion. Defining the fraction fr as

$$\text{fr} = \frac{k_M a_M + f_I}{k_I a_I + f_M} \tag{1.44}$$

then ion-exchange equilibrium requires formation of both MY and IY in the membrane with activities

$$a_{Mm}^+ = \frac{f_{Mm}(Y_m)\text{fr}}{1 + \text{fr}} \tag{1.45}$$

$$a_{Im}^+ = \frac{f_{Im}(Y_m)}{1 + \text{fr}} \tag{1.46}$$

Consequently, the interfacial pd is related to both a_{M^+} and to a_{I^+} according to

$$\Delta\phi = \frac{RT}{F} \ln\left[\frac{k_M}{f_{Mm}(Y_m)} \left(a_M + \frac{k_I f_M}{k_M f_I} a_I \right) \right] \tag{1.47}$$

The interfacial pd selectivity coefficient, the factor multiplying a_I is determined by the ion exchange ratio $K_{iex} = k_I/k_M$, by the activity coefficient ratio, and by the mobility ratio, after the internal diffusion potential contribution is added. This result is identical to Equation 1.21. Clearly, interferences should correlate with the ion-exchange constant, which can be determined from salt extraction coefficients $k_I k_X/k_M k_X$ for a series of positive drugs, comparing salts with common anions. This result is well documented in early ion-selective electrode response studies.[38, 39] A commentary on this topic is given by Koryta[40] and Koryta and Stulik.[41] For drug electrodes there is a correlation of improved selectivity of N-based drugs with the extent of nitrogen substitution:

$$RNH_3^+ < R_2NH_2^+ < R_3NH^+ < R_4N^+$$

in which quaternary drugs of the same carbon number are most sensitively detected.

Omitted from this elementary theory are effects of plasticizer and ion pairing. No theory of ion pairing specifically for membrane-based drug sensors—other than the mobile, liquid ion-exchanger response theory in this chapter—has been published. Presumably, there are no special effects arising from hydrophobic pairs that are not already accounted for in the earlier theory. Effects of plasticizers in membranes have received attention from two points of view: first, the design of plasticizers for emphasizing divalent ion selectivity over monovalent selectivity, or the reverse; second, the design of hydrophobicity character to minimize leaching loss of plasticizers and to prolong lifetime of electrodes. Plasticizers are principally chosen for compatibility with the plastic membrane, low water solubility, high boiling temperature, and chemical inertness. Low water solubility implies a low dielectric constant—well below the value about 40, when water miscibility sets in. Long-chain carboxylic esters are low-dielectric plasticizers that favor extraction of monovalent ions and encourage ion pairing. Nitroaromatic compounds are at the higher end of the permitted dielectric constant range, and these encourage extraction of divalent species.[42]

Intrinsic compatibility of plasticizers and membranes, from the point of view of prolonged lifetime and freedom from loss of plasticizer to surroundings, was treated experimentally and theoretically by Oesch and Simon.[43, 44]

References

1. N. Lakshminarayanaiah, *Membrane Electrodes*, Academic Press, New York, 1976.

2. N. Lakshminarayanaiah, *Equations of Membrane Biophysics*, Academic Press, New York, 1984.

3. R. P. Buck, Theory and principles of ion selective electrodes, in *Ion Selective Electrodes in Analytical Chemistry*, Vol. 1, H. Freiser, Ed., Plenum Press, New York, 1978, pp. 1–141.

4. R. P. Buck, Electrochemistry of ion-selective electrodes, in *Proc. NATO Advanced Study Institute On Chemically Sensitive Electronic Devices*, P. Bergveld, J. Zemel, and S. Middelhoek, Eds., Elsevier, Amsterdam, 1981, pp. 197–260.

5. R. P. Buck, Electrochemistry of ion-selective electrodes, in *Comprehensive Treatise of Electrochemistry*, Vol. 8, R. E. White, J. Bockris, B. Conway, and E. Yeager, Eds., Plenum Press, New York, 1984, pp. 137–248.

6. B. P. Nicolsky, *Zh. Fiz. Khim.*, **10**, 495 (1937).

7. G. Eisenman, *Anal. Chem.*, **40**, 310 (1968).

8. J. Sandblom, *J. Phys. Chem.*, **73**, 249 (1969).

9. J. Sandblom, G. Eisenman, and J. L. Walker, *J. Phys. Chem.*, **72**, 3862 (1967).

10. W. Simon, H. R. Wuhrmann, M. Vasak, L. A. R. Pioda, R. Dohner, and Z. Stefanac, *Angew. Chem., Int. Ed. (English)*, **9**, 445 (1970).

11. E. Pungor and K. Tóth, *Analyst*, **95**, 625 (1970).

12. G. Eisenman, in *Ion-Selective Electrodes*, R. A. Durst, Ed., National Bureau of Standards Special Publ. No. 314, National Bureau of Standards, Washington, DC, 1969, chap. 1.

13. S. Ciani, G. Eisenman, and G. Szabo, *J. Membrane Biol.*, **1**, 1 (1969).

14. G. Eisenman, S. Ciani, and G. Szabo, *J. Membrane Biol.*, **1**, 294 (1969).

15. G. Szabo, G. Eisenman, and S. Ciani, *J. Membrane Biol.*, **1**, 346 (1969).

16. O. Kedem, M. Perry, and R. Bloch, Paper 44, IUPAC Int. Symp. Selective Ion-Sensitive Electrodes, Cardiff, Wales, 1973.

17. J. H. Boles and R. P. Buck, *Anal. Chem.*, **45**, 2057 (1973).

18. S. B. Lewis and R. P. Buck, *Anal. Lett.*, **9**, 439 (1976).

19. W. E. Morf, G. Kahr, and W. Simon, *Anal. Lett.*, **7**, 9 (1974).

20. W. E. Morf, D. Ammann, and W. Simon, *Chimia (Switzerland)*, **28**, 54 (1974).

21. W. E. Morf and W. Simon, in *Ion-Selective Electrodes in Analytical Chemistry*, H. Freiser, Ed., Plenum Press, New York, 1978.

22. H.-R. Wuhrmann, W. E. Morf, and W. Simon, *Helv. Chim. Acta.*, **56**, 1011 (1973).

23. W. E. Morf, P. Wuhrmann, and W. Simon, *Anal. Chem.*, **48**, 1031 (1976).

24. A. P. Thoma, A. Viviani-Nauer, S. Arvanitis, W. E. Morf, and W. Simon, *Anal. Chem.*, **49**, 1567 (1977).

25. E. Lindner, E. Graf, Zs. Niegreisz, K. Tóth, E. Pungor, and R. P. Buck, *Anal. Chem.*, **60**, 295 (1988).

26. G. Horvai, E. Graf, K. Tóth, E. Pungor, and R. P. Buck, *Anal. Chem.*, **58**, 2735 (1986).

27. R. P. Buck, K. Tóth, E. Graf, G. Horvai, and E. Pungor, *J. Electroanal. Chem.*, **223**, 51 (1987).

28. M. L. Iglehart, R. P. Buck, and E. Pungor, *Anal. Chem.*, **60**, 290 (1988).

29. M. L. Iglehart and R. P. Buck, *Talanta*, **36**, 89 (1989).

30. C. J. Coetzee and H. Freiser, *Anal. Chem.*, **40**, 2071 (1968).

31. C. J. Coetzee and H. Freiser, *Anal. Chem.*, **41**, 1128 (1969).

32. H. J. James, G. P. Carmack, and H. Freiser, *Anal. Chem.*, **44**, 853 (1972).

33. J. W. Ross, Jr., *Science*, **156**, 3780 (1967).

34. T. Higuchi, C. R. Illian, and J. L. Tossounian, *Anal. Chem.*, **42**, 1674 (1970).

35. C. Liteanu and E. Hopirtean, *Talanta*, **17**, 1067 (1970).

36. O. R. Melroy and R. P. Buck, *J. Electroanal. Chem.*, **143**, 23 (1983).

37. R. P. Buck and V. V. Coşofreţ, Design of sensitive drug sensors: principles and practice, in *Fundamentals and Applications of Chemical Sensors*, ACS Symp. Ser. No. 309, D. Schuetzle and R. Hammerle, Eds., American Chemical Society, Washington, DC, 1986, pp. 363–372.

38. G. Baum, *J. Phys. Chem.*, **76**, 1872 (1972).

39. R. P. Scholer and W. Simon, *Helv. Chim. Acta*, **24**, 372 (1970).

40. J. Koryta, *Ion-Selective Electrodes*, Cambridge University Press, Cambridge, 1975, chap. 6.

41. J. Koryta and K. Stulik, *Ion-Selective Electrodes*, Cambridge University Press, Cambridge, 1983, chap. 7.

42. W. E. Morf, *The Principles of Ion-Selective Electrodes and of Membrane Transport, Elsevier Studies in Analytical Chemistry*, Vol. 2, Elsevier, Amsterdam, 1981, pp. 289, 294–295, 320–322.

43. U. Oesch and W. Simon, *Helv. Chim. Acata*, **62**, 754 (1979).

44. U. Oesch and W. Simon, *Anal. Chem.*, **52**, 692 (1980).

Chapter 2

THEORETICAL CONSIDERATIONS FOR MULTILAYER, POTENTIOMETRIC GAS SENSORS AND BIOSENSORS

2.1 Proposed Definitions: Chemical Sensors and Biosensors

Chemical sensors are those that use chemical processes in the recognition and transduction steps. Biosensors are a subsection of chemical sensors that use biological recognition processes. Both are intrinsically "chemical." These statements are under consideration for official sanction by the International Union of Pure and Applied Chemistry (IUPAC). For the primary sensors of Chapter 1, the recognition process is built into the ion-exchange mechanism. The principles are purely classical electrostatic, e.g., based on ion–ion and ion–dipole interactions and interaction energies.[1] At the simplest level, for a model with a homogeneous charge sphere in a continuous dielectric medium, the selectivity information is contained in the single-ion partition coefficient, e.g., in the free energy of ion transfer between phases of differing dielectric constant as expressed by the Born equation. For each ion of different size and different charge, the free energy of transfer is a different number that cannot be directly measured. Only products or ratios of partition coefficients (or sums or differences of transfer free energies) can be measured by real methods that are thermodynamically defined. For liquid membranes, the result of electrostatic theory is an interpretation of anion selectivity order as expressed by the Hofmeister lyotropic (lipophilicity) series and the cation series.

For solid electrodes there is a classical theory by Mott and Littleton[2] for the calculation of ion energies in ionic crystals. These can be combined to interpret solubility products, for example. The theory is hardly practical to use, in comparison with the Born equation. Yet modern improvements continue to be made,[3] because of the need to calculate the energies to create ionic defects in solid-state materials and to calculate the resulting conductivity parameters. The aim of theory is to predict materials with permselectivity for a particular ion or group of ions with the same charge. Theory has been a guide, but most ion-selective electrodes were discovered by intuition and by use of materials with experimentally known permselectivities, e.g., charge transport properties.

Use of biorecognition has been a challenging part of sensor design for many years, and the classical paper by Clark and Lyons[4] using enzymes disclosed an amperometric sensor. An oxygen electrode is modified to sense dissolved glucose by the catalytic enzyme oxidation reaction with glucose oxidase, during which the ambient oxygen concentration is decreased. The indirect glucose measurement became possible by combination of the oxygen decrement measurement with the enzyme-catalyzed reaction on the electrode surface. Similarly, Guilbault and Montalvo[5] applied urease to the recognition of urea. The hydrolysis reaction was measured potentiometrically with the cationic-sensing glass electrode, in this case, by following the generated ammonium ion concentration.[5]

2.2   Potentiometric Gas-Sensing Membrane Electrodes

Gas-sensing membrane sensor "electrodes" are not single electrodes, but are combination electrodes with a base indicating electrode and an external reference electrode combined into one unit. They form a complete electrochemical cell. The indicator portion monitors changes in the cell composition arising from gases that penetrate into the cell.[6] Gas-sensing electrodes are sensors that use a gas-permeable membrane to separate the sample gas (or solution) from a thin film of an intermediate solution, which is held between the gas membrane and the ion-sensing membrane part of the sensor.

In the early examples, the intermediate solution interacts with the gaseous species in such a way as to produce a change in a measured constituent (e.g., H^+ or NH_4^+) of the intermediate solution. This change is then sensed by the ion-selective electrode and is related to the partial pressure of the gaseous species in the sample. (*Note:* An exception to this classification is the hydrogen gas electrode, which responds both to the partial pressure of hydrogen and to the pH. The Clark oxygen electrode does not conveniently fit under this classification. In contrast to most other gas sensors, it is an amperometric not a potentiometric device.)

Gas-monitoring electrodes, which sense gas indirectly, involve the use of a gas-permeable, but not ion-permeable, membrane to separate the thin-film internal solution from the sample to be analyzed. The gas molecules, such as hydrogen sulfide, hydrogen cyanide, sulfur dioxide, carbon dioxide, or ammonia, diffuse through the membrane to effect a change in the activity level of an ion contained in the thin-film internal solution. In these instances, the pS^{2-}, pCN^- or pH change is detected by the indicator ion-selective electrode, which might be a primary Ag_2S, AgI, glass, or an iridium oxide electrode.

The behavior of a gas-sensing electrode, particularly its response, sensitivity or slope, and limit of detection, depend in a complex way on the variables of geometry, membrane properties, and the internal electrolyte used. In order to determine the relative effect and importance of these variables, Ross et al.[7] considered an electrode that is in equilibrium with a sample solution having a dissolved gas concentration C_1 for a species permeable to the gas-separator membrane. The concentration of the diffusing species will also be C_1 in the internal electrolyte (see Figure 2.1). The concentration of the species in the membrane will be $\overline{C}$, where

$$\overline{C} = k \times C_1 \tag{2.1}$$

and k is the partition coefficient of the species between the aqueous sample, internal electrolyte phase, and the membrane phase.[7]

At time $t = 0$, the concentration in the membrane solution is suddenly changed to C_2. Now it is assumed that the partition equilibrium at the membrane–gas interface is very rapid. In that case the concentration C in the membrane at the interface will immediately change to a new value C_2. In general, this assumption is valid.[7] A concentration gradient will exist in the membrane. As a result, the concentration C_1 of the species in the internal electrolyte will change from its initial value in the direction of the new equilibrium value C_2. For an electrode of area A, according to Fick's first law,

$$\text{flux} = \frac{AD\,\Delta\overline{C}}{m} \tag{2.2}$$

where m is the membrane thickness, D the diffusion coefficient in the membrane phase, and $\Delta\overline{C}$ at the inside surface and $\overline{C}_2$ at the outside surface.

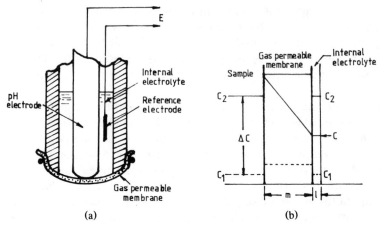

Figure 2.1 (a) Severinghaus carbon dioxide electrode. (Reproduced from Ross, J. W., Riseman, J. H., and Kreuger, J. A., *Pure Appl. Chem.*, 36, 473, 1973. With permission.) (b) Steady-state model for gas-sensing membrane electrode response. (Reproduced from Ross, J. W., Riseman, J. H., and Krueger, J. A., *Pure Appl. Chem.*, 36, 473, 1973. With permission.)

The species diffusing through the membrane can exist in a number of forms in the internal electrolyte. It can be either a neutral species or as various ionized or complexed species. As a result of the flux, a change in the total number of moles $(dAlC_T)$ of the species will occur in the internal electrolyte given by

$$Al\frac{dC_T}{dt} = \text{flux} \qquad (2.3)$$

$$C_T = C + C_B \qquad (2.4)$$

where C is the concentration of the neutral species, C_B is the sum of the concentrations of all the other forms and C_T is the total concentration of the diffusing species in the internal electrolyte of thickness l with the same cross-sectional area A. Combining Equations 2.2 through 2.4 gives

$$\frac{(1 + dC_B/dC)\,dC}{C_2 - C} = -\frac{Dk}{lm}\,dt \qquad (2.5)$$

Considering ϵ as the fractional approach to equilibrium, i.e.,

$$\epsilon = \frac{C_2 - C}{C_2} \qquad (2.6)$$

Substitution into Equation 2.5 yields

$$\left(1 + \frac{dC_B}{dC}\right) d \ln \epsilon = -\frac{Dk}{lm} dt \qquad (2.7)$$

Integration of Equation 2.7 requires a knowledge of dC_B/dt that, in principle, can be obtained from the internal electrolyte composition and all of the equilibrium constants describing the species contributing to C_B.[7] In practice this is extremely difficult to do, hence the simple cases will be considered when either (i) $dC_B/dC \ll 1$ or (ii) the range $C_2 - C$ is sufficiently small so that dC_B/dC can be considered a constant.

Under these restrictions, the integration of Equation 2.7 gives

$$t = \frac{lm}{Dk}\left(1 + \frac{dC_B}{DC}\right)\ln\frac{\Delta C}{\epsilon C_2} \qquad (2.8)$$

The model predicts the effect of geometry (lm), membrane characteristics (Dk), electrolyte composition (dC_B/dC), and experimental conditions $(\Delta C/\epsilon C_2)$ on the response time of the electrode.

Generally, the relationship between the emf and the gas concentrations is given by

$$E = E_0(\text{gas}) \pm \frac{k}{n} \log C_{\text{gas}} \qquad (2.9)$$

where $E_0(\text{gas})$ contains all the constant terms as for a Nernst-type equation. The sign " $+$ " is valid for acidic gases and the sign " $-$ " for basic gases, when using a pH-indicator electrode.

2.3 Potentiometric Enzyme-Based Electrodes

One of the most interesting and potentially far-reaching combinations of a chemical reaction combined with electrometric measurements is in the use of enzymatic reactions to produce an electroactive species,

$$\text{substrate} \xrightarrow{\text{enzyme}} \text{electroactive species} \qquad (2.10)$$

Enzyme substrate electrodes are sensors in which an ion-selective electrode is covered with a coating containing an enzyme that causes the catalytic reaction of an organic or inorganic substance (substrate) to produce a species to which the electrode responds. These devices are used most often to measure a substrate concentration, but they can be used to measure an inhibitor substance. Alternatively, the sensor could be covered with a layer of substrate that reacts with the enzyme,

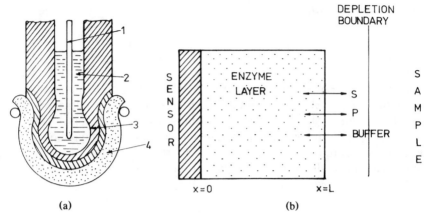

Figure 2.2 (a) Layout of an immobilized enzyme supported on an ion-selective electrode: (1) internal reference electrode; (2) reference solution; (3) ion-selective membrane; (4) enzyme matrix layer. (Reproduced from Coşofreţ, V. V., *Membrane Electrodes in Drug-Substances Analysis*, Pergamon Press, Oxford, New York, 1982, 17. With permission.) (b) Schematic, one-dimensional diffusion-reaction coordinate system for simple enzyme electrodes. (Reproduced from Janata, J., *Principles of Chemical Sensors*, Plenum Press, New York, 1989, 17. With permission.)

co-factor, or inhibitor to be assayed. Operationally, the enzyme-based-sensor combinations for substrate analysis involve interposing the enzyme system between the test solution and the species-selective electrode. This can be done several ways, for example:[8–10]

1. by immobilizing the enzyme in a preformed polymeric or other matrix, which is placed over the sensing element of the electrode;
2. by coating the electrode surface by placing the enzyme-containing solution mixture and polymer or gelation precursors on the electrode (polymerization or gelation occurs in place; the principle of assembly in that instance is illustrated in Figure 2.2);
3. by direct covalent bonding of the enzyme to the electrode surface, usually through surface oxygens (this has been recommended).

The earliest and simplest approach is to confine a thin layer of enzyme-containing solution next to the electrode surface by means of a polymer membrane (typically dialysis film) that is permeable to electrolyte and small substrate species, but impermeable to the enzymes and other large molecules.[11] The theoretical problem that is immediately recognized is that the pH of the enzyme layer must change during the sensing process so that the indicator electrode may respond. Quite often the enzyme layer operates optimally at only one pH value. This requirement suggests that two active chemical layers are needed: the buffered enzyme layer and the unbuffered pH sensor region.

When an enzyme electrode is dipped into a solution containing the corresponding substrate, the substrate and other permeable ions diffuse through the external barrier layer (if it is used) into the polymer layer where the enzyme initiates the desired reaction to produce the ion to be measured. As a result of the increase in activity of the product species, ions to be sensed diffuse farther to the indicator electrode surface.

The enzymatic approach has the obvious advantages of general specificity of substrate–enzyme reactions and reasonable sensitivity for many applications. The principal limitations are the following:

1. availability of suitable sensor electrodes;
2. interference with the sensor response by species other than the reaction product to be measured;
3. relatively short operational lifetime of sensitized sensor electrodes;
4. relatively narrow concentration response range, e.g., from 10^{-4} to 10^{-1} M;
5. generally, a lack of so-called Nernstian response, e.g., a decade change in substrate activity is not necessarily accompanied by a decade change in H^+.

The potentiometric response of pH measurement-based, enzyme-catalyzed substrate sensor electrodes requires a model shown in Figure 2.2b. The preferred model makes several reasonable assumptions:

1. At the sample–enzyme layer interface, there is partition equilibrium between the substrate in solution and in the enzyme layer. When sample-containing solutions are stirred, there is good evidence that mass transport limitation external to the film (slow transport or concentration polarization) does not limit response.
2. Buffer components initially present in the sample are assumed to be partitioned in an equilibrium way and that protonation–deprotonation reactions of buffer and enzyme are rapid.
3. The enzyme is uniformly accessible to the inward-diffusing substrate.
4. The reaction between substrate and enzyme follow Michaelis–Menten kinetics and defines the maximum catalytic velocity v_m and the enzyme–substrate complex dissociation constant K_M. The kinetics are assumed to be pH dependent.
5. Simultaneous partial differential equations for one-dimensional diffusion, coupled with homogeneous chemical reaction of substrate, buffer components, and protons have to be solved.

Contributions to the theory have been made by a number of workers.[12-14] Most recently, Janata has summarized theories and results concerning optimization of enzyme-containing membranes.[15] For a classical system,

$$E \text{ (enzyme)} + S \text{ (substrate)} \underset{k_{-1}}{\overset{k_1}{\rightleftharpoons}}$$

$$ES \text{ (complex)} \xrightarrow{k_2} E + P \text{ (products including protons)} \quad (2.11)$$

whose rate in homogeneous solution is assumed to be governed by

$$\frac{dC_S}{dt} = \frac{k_2(E_T)C_S}{k_M + C_S} = \frac{v_m C_S}{(C_S + K_M)f(\text{pH})} \quad (2.12)$$

where C_S is the substrate concentration (in moles per cubic centimeter), $k_2(ES)$ is the enzyme activity (in micromoles per second per cubic centimeter) or velocity v, and K_M is the Michaelis constant $((k_2 + k_{-1})/k_1$ in moles per cubic centimeter). The maximum velocity is defined $k_2(E_T)$, where $E_T = E + ES$ and would be observed if all E were combined as ES. The velocity at any time is related to v_m by

$$v = \frac{v_m C_S}{C_S + K_M} \quad \text{or} \quad \frac{v_m C_S}{(C_S + K_M)f(\text{pH})} \quad (2.13)$$

and the latter applies in the usual case that the rate constant k_2 is a function of pH. An example of a realistic $f(\text{pH})$ was explicitly used for a solution of this system by Janata and co-workers.[16]

The diffusion equation (Fick's second law) with a homogeneous reaction is

$$\frac{\partial C_S}{\partial t} = D_S \frac{\partial^2 C_S}{\partial x^2} - \text{homogeneous loss rate} \quad (2.14)$$

Incorporation of this homogeneous chemical rate into the diffusional mass transport equation is simplified by using dimensionless variables:

$$t = \frac{\mathbf{t}L^2}{D_S} \qquad C_S = \mathbf{C_S}K_M \qquad x = \mathbf{x}L$$

$$\frac{\partial \mathbf{C_S}}{\partial \mathbf{t}} = \frac{\partial^2 \mathbf{C_S}}{\partial \mathbf{x}^2} - \Phi^2 \frac{\mathbf{C_S}}{1 + \mathbf{C_S}} \quad (2.15)$$

where the relevant lumped parameter is the Thiele modulus function Φ given by

$$\Phi = L\left[\frac{v_m}{K_M D_S f(\text{pH})}\right]^{1/2} \tag{2.16}$$

This equation can be, and has been, solved in the transient and in the steady state for blocking at the sensor interface (concentration gradients are zero) and partition equilibrium at the solution—membrane interface.

Conclusions from the derivation and calculations include the following:

1. The optimum thickness of an enzyme layer depends on the values of the kinetic parameters.
2. Given typical diffusion coefficients, the lower detection limit is about 10^{-4} M substrate because the products (protons) diffuse away about as fast as the substrate can generate them!
3. Response range depends somewhat on enzyme loading because v_m and K_M are in the Thiele modulus.
4. The upper limit of response (where the response levels off and becomes insensitive to further increase in substrate concentration), can be improved by added enzyme, but only to the limits of solubility. Eventually any level of enzyme will become kinetically "saturated." A typical limit is 0.1 M substrate.
5. Response curves are S-shaped and the linear portion need not be linear in the log substrate activity.

2.4 Compound or Multiple Membrane (Multiple Layer), Potentiometric Ion-Selective Electrodes

In contrast with the layered structure of gas-sensing ion-selective electrodes, the enzyme sensors originally used a single immobilized coating on a glass pH electrode (for catalytic enzymatic reactions generating or using portions) or monovalent cation-sensing electrodes (for catalytic enzymatic reactions generating ammonium ions.) The latter were used for urea electrodes on which the immobilized urease converted urea to ammonium ions[17]. As pointed out previously, it was soon found that the pH change of the enzyme layer could be avoided by massive buffering at the ideal pH for enzyme catalysis, and the resulting product CO_2 or NH_3 could be detected optimally with a gas-sensing electrode. Two-layer electrodes were made with the enzyme layer on the outside surface of a gas sensor.[18] Response times are substantially lengthened, but there is no new fundamental or theoretical principle involved when using two or more membrane systems.

Use of purified enzymes proved to be unnecessary and bacteria that generate the required enzyme could be substituted in the enzyme layer.[19] Similarly, tissue sections could be substituted for the enzyme layer. Problems of growing bacteria in media that emphasize the enzymes required for the electrode function were addressed. Bacteria may well contain 3500 enzymes, of which only a few are desirable. To construct an electrode with high selectivity and sensitivity for L-histidine, the principle of "forced feeding" is used during the culturing to be sure that the desired enzymes are present in large amounts. For example, to emphasize the growth of histidine ammonia-lyase, the *Pseudomonas* sp. ATCC 11299b bacterium was cultured in a buffered broth consisting mainly of L-histidine and some L-tryptophane.[20, 21] Recent reviews cover these electrodes and other biosensors: sensors that use biochemical recognition processes.[22-25]

2.5   Amperometry and Potentiometry Compared

In contrast with potentiometric sensors, where the working electrode *potential* conveys the information on species activities, "amperometric sensors" is an equally generic term for sensors where the working electrode current (at a constant applied potential) is measured and conveys the concentration of the species reacting at the electrolyte–electrode interface. The relation between these methods comes from the fact that any electrochemical cell is characterized by a function of three variables (at constant temperature and pressure): voltage, current, and species activities. Other parameters such as stirring rate and species mobilities can also be involved. Potentiometry is a technique at which current is held constant, usually at the value zero. Potential differences and cell voltages are then a logarithmic function only of activities. Conversely, for amperometry, applied cell voltage (or better the working electrode potential difference) is fixed and the measured current is a linear function of concentration in many cases. To illustrate this relationship, Figure 2.3 shows typical, schematic current–voltage (I–V) curves for a solid electrode, such as platinum, in a stirred solution of a soluble redox pair such as ferrocyanide or ferricyanide ions. When the ferricyanide concentration is fixed and the ferrocyanide concentration is varied, for example, the potentiometric response moves along the voltage axis as the logarithm of the ratio (ox)/(red). On the other hand, for positive applied voltages at the working electrode, the oxidation current increases linearly with the concentration of red, because the ferrocyanide ions are oxidized as quickly as they diffuse to the working electrode

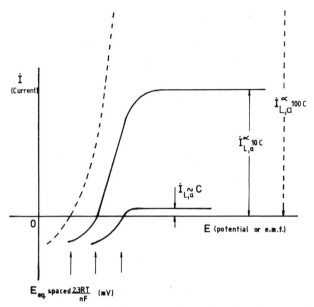

Figure 2.3 Current–voltage $(I–V)$ curves for redox couple, e.g., ferrocyanide–ferricyanide solution mixtures. Limiting anodic currents, $I_{L,a}$, illustrated schematically for increasing ferrocyanide in units of C, $10C$, and $100C$; ferricyanide is at constant activity.

surface. This result is an example of diffusion-limited current found generally in amperometric methods using a single concentration-polarized electrode.

The historic references for amperometry are given by Lingane.[26] The method was discovered by Heyrovsky and was a natural application of the polarographic method. Here, the working electrode was dropping mercury whose surface is constantly renewed and does not suffer the high level of contamination found for solid electrodes. The method is not restricted to one working electrode, but two may be used as a differential pair (the so-called "dead stop" method or amperometry with two polariz-able electrodes). One electrode at a time controls the total current and the theory is an extension of the $I–V$ curve example given previously, but there are now two $I–V$ curves, one for each electrode. Some skill is required to show the interrelationships that control the net current through the pair. Another method uses only one working electrode, but the current is *not* zero. This process of "potentiometry at finite current" is also based on the analysis of $I–V$ curves, and it may have some applications in the drug sensor field.

Outside the sensor field, the amperometric process lends itself to titrimetry. If one were titrating Cl^- with Ag^+, the titration could be followed either by potentiometry or by amperometry. In the latter case, the applied voltage–current relationship for various points during the

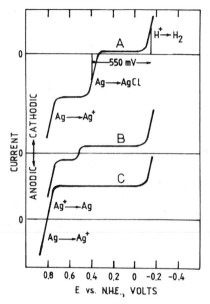

Figure 2.4 Current–voltage curves of a silver electrode at various stages of a chloride–silver ion titration. The zero of current is indicated, but the current scale is qualitative and dependent on the electrode size as well as the concentrations. The voltage scale is approximately quantitative. (Reproduced from Lingane, J. J., *Electroanalytical Chemistry*, 2nd ed., Interscience, New York, 1958, 283. With permission.)

titration is illustrated in Figure 2.4. Curve A is early in the titration, where Cl^- is in excess. Curve B is near the end point, and curve C is after the end point, when Ag^+ is in excess. Between voltages of $+0.4$ and $+0.8$ V vs. the normal hydrogen electrode (NHE) the current is controlled by diffusion of Cl^- to the Ag surface with formation of AgCl. The electrode reaction process is reversible and is written

$$AgCl + e = Ag(m) + Cl^-$$

At the end point this current reaches zero and changes sign because Ag^+ is reduced to Ag metal at this same voltage in the absence of Cl^-. If the potential were controlled between 0 and $+0.4$ V, the current remains zero until the end point and then increases linearly with excess titrant.

The interpretation of an amperometric redox titration proceeds along the same lines, based on the I–V curves in Figure 2.5. In this case the chemical reaction is

$$Ce^{4+} + Fe^{2+} = Ce^{3+} + Fe^{3+}$$

and the amperometric electrode can respond to any one principle species,

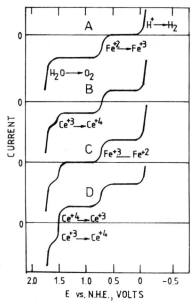

Figure 2.5 Current–voltage curves of a platinum electrode at various stages during the titration of ferrous ion with ceric ion in the presence of a sufficient concentration of sulfuric acid (3 M or greater) so that the ceric–cerous couple behaves reversibly. The current scale is qualitative, whereas the voltage scale is quantitative. (Reproduced from Lingane, J. J., *Electroanalytical Chemistry*, 2nd ed., Interscience, New York, 1958, 288. With permission.)

depending on the applied voltage and which species is the sample and which the tritrant. Normally Fe^{2+} is the sample and Ce^{4+} is the tritrant. The half-cell reactions are

$$Fe^{3+} + e = Fe^{2+}$$

and/or

$$Ce^{4+} + e = Ce^{3+}$$

Fe^{2+}/Fe^{3+} couple passes from oxidation to reduction at about $+0.7$ V. The Ce^{3+}/Ce^{4+} couple requires a larger voltage for oxidation and so passes from oxidation to reduction at about $+1.5$ V. Thus, for titration of Fe^{2+}, the I–V curve A shows the initial condition and the voltage range between $+0.7$ and $+1.5$ V would be a region for monitoring the concentration of unreacted Fe^{2+}. Curve B is half-titration, where currents for both remaining Fe^{2+} and the Fe^{3+} formed (between 0. and $+0.7$ V) are shown. Curve C is at the end point and shows the presence of Ce^{3+}, being oxidized above $+1.5$ V and the product Fe^{3+} being reduced. Curve D is past the end point and demonstrates the presence of excess Ce^{4+}.

The currents are given by a simple expression:

$$I = \pm nFAD[(C)_{\text{bulk}} - (C)_{\text{surface}}]/d \qquad (2.17)$$

where n is the number of electrons transferred per mole; A is the electrode area; D is the species diffusion coefficient, and d is the thickness of the Nernst boundary layer at the stirred electrode. The limiting current I_L occurs when the surface concentration is reduced to zero. Thus, the limiting current measures the bulk concentration of the reacting species:

$$I_L = \pm nFAD[(C)_{\text{bulk}}]/d \qquad (2.18)$$

The present IUPAC convention of signs of currents tells us that a cathodic current (reduction of an oxidized species) is $-$ and an anodic current (oxidation of a reduced species) is $+$. Thus $+$ (oxidization) currents occur as the applied working electrode potential difference pd is moved in the increasing positive direction. Negative (reduction) currents increase negatively as the pd is moved to more negative values. On a current–voltage plot (I–V curve), anodic currents usually occur on the right of the zero-voltage axis, in the upper right quadrant. Reduction currents occur on the lower left quadrant. The I–V curve is S-shaped and moves from the negative limit, $I_{L,c}$ (a negative number in the lower left), through zero to the positive $I_{L,a}$ (a positive number in the upper right quadrant). For a reversible electron- or ion-transfer process at the underlying metal electrode surface, the shape of the I–V curve (often called a wave) is found by substituting the surface concentrations of oxidized form (O) and reduced form (R), from Equation 2.17, into the Nernst equation

$$E = E_{1/2} + \frac{RT}{nF} \ln \frac{(I - I_{L,c})}{(I_{L,a} - I)} \qquad (2.19)$$

The $E_{1/2}$ is the so-called half-wave potential, which is the potential value at the current midway between the two limiting values. If there were only a reduced species present, and therefore only an anodic current, $E_{1/2}$ would correspond with the current halfway up the wave from zero to the limiting current value. The $E_{1/2}$ is related to transport parameters according to the mechanism for establishing the transport-controlled limiting current. For a simple membrane barrier with steady-state diffusion,

$$E_{1/2} = E_0 + \frac{RT}{nF} \ln \frac{D_R}{D_O} \qquad (2.20)$$

However, for static electrodes in stirred solution, the Nernst layer thickness is not a reproducible quantity. The thickness d is inversely proportional to the stirring rate to a power between -0.4 and -0.7. Only for certain cases, such as a rotating disk of infinite extent, can it be proved that the exponent is exactly -0.5. Certain other geometries also have exact transport equation solutions, including a flat plate with laminar flow and the hollow cylinder under laminar flow. The finite rotating disk is a good approximation to the exact theory when one is careful to keep the rotation rate less than a value that causes nonlaminar (turbulent) flow at the outer edges of the disk. In this rotating disk electrode configuration, the two-soluble-species redox process half-wave potential expression, Equation 2.20, is modified by raising the diffusion coefficient ratio to the power $2/3$.

2.6 Amperometric Sensors — The Oxygen Electrode

Reproducibility of the amperometric process requires close control of the thickness of the diffusion layer (see Equation 2.17). In ordinary stirred solutions, transport conditions were not reproducible and amperometric electrodes were used to follow the course of titrations, rather than to make direct quantitative measurements of species concentrations. It was known that dissolved oxygen could be measured amperometrically using mercury electrodes and that there were two two-electron reduction steps:

$$O_2 + 2H_2O + 2e = H_2O_2 + 2OH^- \qquad (2.21)$$

$$H_2O_2 + 2e = 2OH^- \qquad (2.22)$$

On Pt, Au, and C single four-electron reductions (the sum of these two reactions) could be observed although an overpotential was required to drive the reaction into the limiting current condition. Some progress was made by producing sensors with recessed metal electrodes that required an electroactive species to diffuse from solution bulk down a static solution channel to the recessed electrode.[27] Diffusion down the channel defined the distance parameter d and improved the calibration and individual measurement reproducibilities. However, in 1956 Clark[28] coated noble-metal electrodes with a gas-permeable membrane and included a thin film of KCl-containing electrolyte and a large (nearly non-polarizable) Ag/AgCl reference electrode in the region behind the membrane and the metal electrode, shown in Figure 2.6. Membranes of Teflon, silicon rubber, and polyethylene can be used because they are

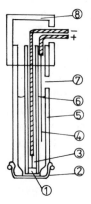

Figure 2.6 Schematic diagram of a Clark oxygen electrode: (1) platinum cathode; (2) polyethylene membrane; (3) indium solder; (4) silver wire; (5) Lucite tube; (6) electrolyte; (7) filling port; (8) nylon cap. (Reproduced from Fatt, I., *Polarographic Oxygen Sensor*, CRC Press, Cleveland, OH, 1976, 26. With permission.)

oxygen permeable, but not ion permeable. There is no solvent in the membrane and the oxygen transports as an unsolvated gas. The oxygen in the bathing solution sample partitions rapidly into the membrane. When the metal is polarized negatively (-0.6 to -0.9 V vs. Ag/AgCl), dissolved oxygen is removed from the thin layer of solution and the membrane develops an oxygen gradient. The thickness of the membrane defines d and the diffusion of oxygen is determined by the relatively small diffusion coefficient in the membrane compared with the value in the underlying solution or in the bathing solution. The current–concentration relation is determined from Equations 2.17 and 2.18, using the concentration of oxygen at the membrane–bulk-solution interface and the concentration at the membrane–metal interface. The *I–V* curve is different from Equation 2.19 because the reduced form does not back diffuse from the electrode surface to the bulk solution. Furthermore, the electrode reaction is irreversible so the Nernst equation does not apply. However, the *I–V* curve is well represented by

$$E = E_0 + \frac{RT}{\alpha nF} \ln \frac{nFAD}{d} + \frac{RT}{\alpha nF} \ln(I_{L,c} - I) \qquad (2.23)$$

$$= E_{1/2} + \frac{RT}{\alpha nF} \ln(I_{L,c} - I) \qquad (2.24)$$

The experimental parameter α, running from 0 to 1, measures the irreversibility. The half-wave potential now depends on the oxygen

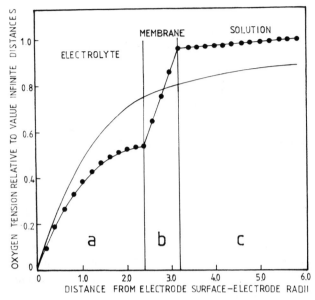

Figure 2.7 Calculated concentration of oxygen (a) in the internal bathing electrolyte, (b) in the membrane and (c) in the external solution. The metal electrode surface is the coordinate 0. (Reproduced from Fatt, I., *Polarographic Oxygen Sensor*, CRC Press, Cleveland, OH, 1976, 16. With permission.)

concentration:

$$E_{1/2} = E_o + \frac{RT}{\alpha nF} \ln \frac{C}{2} \qquad (2.25)$$

This result means that the I–V curves show a concentration dependence of the half-wave potential values.

If the transport through the membrane is *not* limiting, then oxygen depletion can occur in the exterior bathing solution unless it is rapidly stirred to bring dissolved oxygen to the membrane surface by forced convection. Often this is not possible. Figure 2.7 shows a calculated profile of the diffusing dissolved oxygen for a typical flat-form Clark electrode.

Involvement of diffusion and concentration polarization in the outer solution is indicated by stirring dependence of the measured current. To avoid all concentration polarization, the electrode area must by reduced to a size that the oxygen diffusion field is virtually hemispherical. Diffusion to a sphere or a hemisphere can be independent of time and show no depletion effects. Even bare ring electrodes, which can be microfabricated to have bandwidth dimensions between 100 and 500 Å, show no stirring dependence of limiting current.[30]

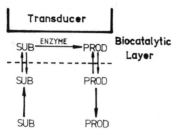

Figure 2.8 Schematic diagram of a biocatalytic-based biosensor. (Reproduced from Arnold, M. A. and Meyerhoff, M. E., *Crit. Rev. Anal. Chem.*, 20, 149, 1988. With permission.)

2.7 Amperometric Biosensors

The idea of building specificity into potentiometric sensors via enzymes was introduced in Section 2.3. Amperometric sensors also can use enzyme layers to convert an electrochemically inert species (the substrate or determinand, such as glucose) into an electroactive species. Instead of measuring the electrode potential difference, an applied voltage is used to oxidize or reduce a product or catalytic intermediate and the current is monitored. Frequently, some electroactive species is either generated or destroyed by the catalytic enzymatic reaction. When the electroactive species can be identified and its measurable concentration related to the electrically inert substrate concentration, then an amperometric sensor becomes feasible.

The general case of a biosensor is simply described in Figure 2.8. Any transducer that recognizes and amplifies the product species could serve as the basis. Of course, the simplest possible system would use a redox enzyme, such as glucose oxidase (GOD) on an electrode surface, in contact with a substrate, glucose. Ideally, the reduced form of GOD would donate electrons back to the electrode and complete the catalytic cycle. Unfortunately, this process does not occur reversibly on metallic electrodes, and other strategies are required to complete the reoxidation. One method is to use a chemically modified surface layer on a metallic electrode. Materials such as N-methylphenazinium (NMP^+) tetracyanoquinodimethane anion ($TCNQ^-$) are known to catalyze the electron transfers from enzymes to other molecules. The surface coating improved the electron transfer rate to a great extent, but is not a total cure-all. In nature, the reoxidation is achieved by dissolved oxygen, so electrode designers have focused on measuring oxygen decrement or electroactive product, hydrogen peroxide.

The surface modification layer is applied first. This layer is frequently referred to as the promoter layer, to distinguish it from the "mediator" species, such as ferrocene and ferrocinium that replace oxygen and peroxide in this example.

The enzyme is applied to the electrode in four ways:

1. The enzyme is applied as a solution and held in contact with the electrode by dialysis membrane. The pore size is selected so that substrates and small reactants can pass the membrane.
2. The enzyme is partially immobilized within a thick gel layer on the electrode surface.
3. The enzyme is adsorbed or covalently attached to the electrode surface.
4. The enzyme is compounded with the electrode material (initially in a powder form) and pressed into an electrode shape. Much of the enzyme remains inside the electrode.

The simplest reaction scheme for an enzyme-catalyzed sensor that produces an electroactive product is

$$S + E \underset{k_{-1}}{\overset{k_1}{\rightleftharpoons}} ES \overset{k_2}{\longrightarrow} P + E \qquad (2.26)$$

and

$$P \pm ne \rightleftharpoons Q_1 \qquad (2.27)$$

The reaction product P is electroactive and can be oxidized or reduced to Q. Equation 2.26 is the same as Equation 2.11 for a simple enzyme-catalyzed system. Consequently, Equations 2.13 and 2.14 apply to S (as before), but also to a product P after care is exercized to obtain the correct stoichiometry. Signs of the homogeneous reaction terms change from minus for a consumed species, to plus for a generated species. Of course, the boundary conditions change according to the concentrations of species initially present in the membrane and the equilibria at the solution–membrane and membrane–electrode interfaces.

There is no fundamental added complication when the scheme is amplified to include other reacting participants on the left-hand side of Equation 2.26. Typically for the practical system S = glucose, E = glucose oxidase (frequently called GOD), another reactant O_2 is a participant. As long as the ambient dissolved O_2 remains uniformly distributed (but not necessarily constant) the previous equations apply with the second-order forward rate constant replaced by its product with the concentration of oxygen:

$$O_2 + S + E \underset{k_{-1}}{\overset{k_1(O_2)}{\rightleftharpoons}} ES \overset{k_2}{\longrightarrow} P_1 + H_2O_2 + E \qquad (2.28)$$

and

$$H_2O_2 \pm ne \rightleftharpoons Q_1 \qquad (2.29)$$

or

$$O_2 \pm ne \rightleftharpoons Q_2 \qquad (2.30)$$

Reaction 2.28 should be understood as a catalytic cycle:

$$S + E \rightleftharpoons ES \rightleftharpoons E'P \rightarrow P_1 \qquad (2.31)$$

$$O_2 + E' \rightleftharpoons E + H_2O_2 \qquad (2.32)$$

in which enzyme is actually reduced and then reoxidized by dissolved oxygen. The decrement of oxygen can be monitored, or the generated peroxide can be monitored. The loss-of-oxygen method is the oldest of all the methods and is achieved by building the reactant layer directly on a Clark oxygen-sensor membrane. When the membrane is exposed to glucose and oxygen, the reaction occurs with a consequent steady-state decrement of oxygen. This value at the Clark electrode surface is related to the ambient oxygen concentration, and the bulk membrane and bathing solution glucose concentrations. The Clark sensor draws sufficiently small current to sense the oxygen level without seriously perturbing its value. Steady-state solutions for the species and responses were given in detail[31] and are also discussed by Carr and Bowers.[32]

The problem with peroxide generation and monitoring is that peroxide eventually destroyed the GOD and shortens the useful electrode lifetime. When oxygen is monitored, catalase can be added to remove the peroxide. Nevertheless there is an advantage in monitoring the membrane peroxide by oxidation, because the membrane oxygen remains sensibly constant. Steady-state solutions by Mell and Maloy[33] and transient solutions by Schulmeister and Scheller[34] show the membrane species concentration profiles and the current vs. concentration of substrate responses. Some generalizations, in addition to those itemized for potentiometric enzyme electrodes, are as follows:

1. The current–substrate calibration response curves are not linear. The linear response range can be controlled by either slow mass transport of substrate or by enzyme kinetics. At low substrate concentrations the transport-controlled slope is observed. At high substrate concentrations the slope is always diminished as concentration increases.

2. Limiting maximum currents are approached as the substrate concentration is increased.

3. A high upper limit of current and long range of linear response is promoted by low mobility of substrate, a high K_M, and a high activity of enzyme in the electrode.

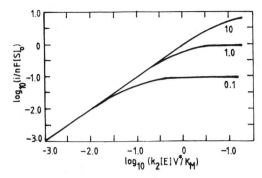

Figure 2.9 Sensitivity of amperometric enzyme electrodes to enzyme loading and membrane permeability of substrate. Numbers on the curves are dimensionless Michaelis constants. Results apply when the bulk substrate concentrations are less than K_M (Reproduced from Carr, P. W. and Bowers, L. D., *Immobilized Enzymes in Analytical and Clinical Chemistry*, John Wiley & Sons, New York, 1980, 230. With permission.)

4. Addition of enzyme increases the high concentration range response slope and prolongs the linear range until there is no further effect and mass transport is controlling over the whole substrate concentration range.
5. Currents at low substrate concentrations increase as the membrane thickness decreases if mass transport is rate limiting.

Figure 2.9 shows a theoretical set of calibration responses.

2.8 Recent Improvements and Extensions of Amperometric Biosensor Principles

Most recent work on enzyme-based sensors has overcome two limiting defects of the early glucose sensors. This work has improved the glucose analyses and has extended the application of amperometric enzyme electrodes to entire new groups of substrates. There have been two major directions:

1. to improve the rate of electron transfer of the enzyme itself at surface modified electrodes using improved promoters;
2. to replace the oxygen–peroxide mediators by other redox couples.

The first topic, pioneered by Albery et al.,[35] was developed experimentally and theoretically. A number of promoters were investigated, such as tetrathiafulvalene (TTF$^+$) and quinolinium TCNQ$^-$ salts, although quite a long list can be compiled for electronic catalysis.[36] The theory is steady state and gives an expression for the current (flux) in terms of rate constants for the sequence.

$$S + E \underset{k_{-1}}{\overset{k_1}{\rightleftharpoons}} ES \underset{k_{-2}}{\overset{k_2}{\rightleftharpoons}} E'P \underset{k_{-3}}{\overset{k_3}{\rightleftharpoons}} E' + P \qquad (2.33)$$

Determination of the rate-controlling step has been the aim of the work, and for the glucose electrode, as designed, the mass transport through the membrane was response controlling.

The desire to replace dissolved oxygen with some other anaerobic oxidant that could function to regenerate the reduced enzyme E' in Reactions 2.28 and 2.29 has lead to a number of new amperometric sensors and improvements (or alternative configurations) of older enzyme electrodes. Because the enzyme GOD contains the oxidizing center flavin adenine dinucleotide (FAD), the act of oxidizing glucose causes formation of the reduced enzyme containing the group $FADH_2$. Various fast oxidants can penetrate the enzyme and reoxidize E' to E, including the soluble species ferricyanide, ferricinium, and ruthenium hexammine and the polymer films species oxidized polypyrrole, whose oxidation potentials are more positive than 0.05 V at pH 7 for the $FADH_2/FAD$ couple. The GOD molecule is sufficiently robust that ferrocene can be covalently attached so that the modified enzyme can transmit and accept electrons from an electrode.

Instead of the sequence

$$glucose + FAD \rightleftharpoons gluconolactone + FADH_2 \qquad (2.34)$$

$$FADH_2 + O_2 \rightleftharpoons FAD + H_2O_2 \qquad (2.35)$$

Equation 2.35 is replaced by ferrocene–ferrocinium $(Fe(Cp)_2^+/Fe(Cp)_2)$ (Case et al.[37]):

$$FADH_2 + 2\,Fe(Cp)_2^+ \rightleftharpoons FAD + 2\,Fe(Cp)_2 + 2\,H^+ \qquad (2.36)$$

There are over 250 enzymes that use the mediating co-enzyme $NAD^+/NADH$. Sensors for reduced, neutral species, such as ethanol (which is difficult to directly oxidize in a clean way), can be catalytically oxidized to acetaldehyde with the combination alcohol dehydrogenase (ADH) and NAD^+. The product is anodically oxidized back to the starting material at the modified electrode surface[38]:

$$SH_2 + NAD^+ \overset{dehyrogenase}{\rightleftharpoons} S + NADH + H^+ \qquad (2.37)$$

$$NADH \rightleftharpoons NAD^+ + H^+ + 2e \qquad (2.38)$$

In the more complicated schemes, several enzymes can be combined in one layer, or into multiple layers in series. If co-enzymes are involved, it is usual to oxidize or reduce these species when their concentrations are directly related to the substrate concentration under investigation.[40]

References

1. F. Helfferich, *Ion Exchange*, McGraw-Hill, New York, 1962, p. 156.
2. N. F. Mott and M. J. Littleton, *Trans. Faraday Soc.*, **34**, 485 (1938).
3. P. W. M. Jacobs, *J. Chem. Soc., Faraday Trans. 2*, **85**, 413 (1989).
4. L. C. Clark and C. Lyons, *Ann. N.Y. Acad. Sci.*, **102**, 29 (1962).
5. G. G. Guilbault and J. G. Montalvo, *Anal. Lett.*, **2**, 283 (1969).
6. D. Midgley and K. Torrance, *Potentiometric Water Analysis*, John Wiley & Sons, Chichester, 1978.
7. J. W. Ross, J. H. Riseman, and J. A. Krueger, *Pure Appl. Chem.*, **36**, 473 (1973).
8. G. G. Guilbault, in *Comprehensive Analytical Chemistry*, Vol. 8, G. Svehla, Ed., Elsevier, Amsterdam, 1977.
9. G. G. Guilbault, Use of enzyme electrodes in biomedical investigations, in *Medical and Biological Applications of Electrochemical Devices*, J. Koryta, Ed., John Wiley & Sons, New York, 1980.
10. G. G. Guilbault, *Analytical Uses of Immobilized Enzymes*, M. Dekker, New York, 1984.
11. P. J. Elving, *Bioelectrochem. Bioenerg.*, **2**, 251 (1975).
12. P. W. Carr, *Anal. Chem.* **49**, 799 (1977).
13. J. E. Brady and P. W. Carr, *Anal. Chem.* **52**, 977 (1980).
14. W. E. Morf, *Mikrochim. Acta 2*, 317 (1980).
15. J. Janta, *Principles of Chemical Sensors*, Plenum Press, New York, 1989, pp. 15–21.
16. S. D. Caras, J. Janata, D. Saupe, and K. Schmidt, *Anal. Chem.*, **57**, 1917 (1985).
17. G. G. Guilbault and J. Montalvo, *J. Am. Chem. Soc.*, **92**, 2533 (1970).
18. G. G. Guilbault and W. Stokbro, *Anal. Chim. Acta*, **76**, 237 (1975).
19. P. D'Orazio, M. E. Meyerhoff, and G. A. Rechnitz, *Anal. Chem.*, **50**, 1531 (1975).
20. R. R. Walters, B. E. Moriarty, and R. P. Buck, *Anal. Chem.*, **52**, 1680 (1980).
21. R. R. Walters, P. A. Johnson, and R. P. Buck, *Anal. Chem.*, **52**, 1684 (1980).
22. M. Arnold, *Ion-Selective Electrode Rev.*, **8**, 85 (1986).
23. M. A. Arnold and M. E. Meyerhoff, *Crit. Rev. Anal. Chem.*, **20**, 149 (1988).
24. R. K. Kobos, in *Ion-Selective Electrodes in Analytical Chemistry*, Vol. 2, H. Freiser, Ed., Plenum Press, New York, 1980, p. 1.

25. J. Koryta and K. Stulik, *Ion-Selective Electrodes*, 2nd Ed. Cambridge University Press, London, 1983.

26. J. J. Lingane, *Electroanalytical Chemistry*, 2nd ed., Interscience, New York, 1958, chap. 12.

27. P. W. Davies and F. Brink, Jr., *Rev. Sci. Instrum.*, **13**, 524 (1942).

28. L. C. Clark, Jr., *Trans. Am. Soc. Artif. Intern. Organs*, **2**, 41 (1956).

29. I. Fatt, *Polarographic Oxygen Sensor*, CRC Press, Cleveland, OH, 1976.

30. Y. Saito, *J. Appl. Physiol.* **23**, 979 (1967).

31. D. A. Gough, J. K. Leypoldt, and J. C. Armour, *Diabetes Care*, **5**, 190 (1982).

32. P. W. Carr and L. D. Bowers, *Immobilized Enzymes in Analytical and Clinical Chemistry*, John Wiley & Sons, New York, 1980.

33. L. D. Mell and J. T. Maloy, *Anal. Chem.*, **47**, 299 (1975); **48**, 1597 (1976).

34. T. Schulmeister and F. Scheller, *Anal. Chim. Acta*, **170**, 279 (1985); **171**, 111 (1985); **198**, 223 (1987).

35. W. J. Albery, P. N. Bartlett, A. E. G. Cass, D. H. Craston, and B. G. D. Haggett, *J. Chem. Soc., Faraday Trans. 1*, **82** 1033 (1986).

36. L. Gorton, *J. Chem. Soc., Faraday Trans. 1*, **82** 1245 (1986).

37. A. E. G. Cass, G. Davis, G. D. Francis, H. O. A. Hill, W. J. Aston, I. J. Higgins, E. V. Plotkin, L. D. I. Scott, and A. P. F. Turner, *Anal. Chem.*, **56**, 667 (1984).

38. W. J. Albery and P. N. Bartlett, *J. Chem. Soc., Chem. Commun.*, 234 (1984).

39. A. P. F. Turner, G. S. Wilson, and I. Karube, Eds., *Biosensors*, Oxford University Press, Oxford, 1987.

40. Proc. 5th Int. Conf. Solid-State Sensors and Actuators, Sensors and Actuators, B1 Nos. 1–6 (1990).

Chapter 3

BASIC CHARACTERISTICS OF POTENTIOMETRIC AND AMPEROMETRIC MEMBRANE ELECTRODES

3.1   Electrode Function

3.1.1   Potentiometric Sensors

The most useful representation of the performance of a potentiometric membrane electrode is the functional relation between the dependent variable, the electrode potential (E) when current $I = 0$, and the independent variable: the ion activity of that ion for which the electrode is most selective (Figure 3.1). In addition, the characteristics of electrode response include a number of frequently measured parameters: selectivity, response time, operative potentiometric response range, temperature dependence of response, and operating lifetime, among the more important.

The slopes S of the curves presented in Figure 3.1 show that electrode sensitivity is constant for the linear portions, but at low activities the sensitivity is reduced asymptotically to zero. Slope and sensitivity are equivalent terms appearing in the literature. All are expressed in millivolts per decade of the activity and represent the slope of the response curve plotted from the difference in measured potential that is obtained when the electrode is transferred from, for example, a 10^{-2} M solution to a 10^{-3} M solution (at constant ionic strength so that activity coefficients are constant) containing the ion for which the electrode is selective. At standard temperature, the theoretical slopes of electrode function are numerically equal to 59.16 and 29.58 mV decade^{-1} for monovalent and divalent cations, respectively. The anion slopes are the same magnitude with opposite sign. These slopes are the coefficients

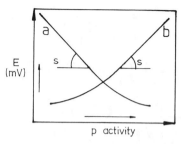

Figure 3.1 Typical potentiometric membrane electrode functions: (curve a) cation-selective electrode; (curve b) anion-selective electrode. (Reproduced from Coşofreţ, V. V., *Membrane Electrodes in Drug-Substance Analysis*, Pergamon Press, Oxford, 1982, 21. With permission.)

multiplying the ln terms in the response functions of Chapters 1 and 2, as corrected to conventional logarithms (to base 10) by the multiplier 2.303.

Slope and sensitivity are also measured as millivolts per decade of the *concentration*. When the ionic strength is not maintained constant, these quantities have different values in every concentration range. For practical work, when electrode responses are being compared, these latter definitions of slope and sensitivity are often used. Care must be taken when reading the literature to determine which definitions are being used.

3.1.2   Amperometric Electrodes

The most useful representation of the performance of an amperometric membrane electrode is the functional relation between the dependent variable, the current flowing through the electrode (I) when the potential is fixed, and the independent variable: the concentration of that ion for which the electrode is most selective (Figure 3.2). In addition, the characteristics of electrode response include a number of frequently measured parameters: selectivity, response time, operative amperometric response range, temperature dependence of response, and operating lifetime, among the more important.

The slopes S of the curves presented in Figure 3.2 show that electrode sensitivity is constant for the linear portions, but at low activities the sensitivity is reduced asymptotically to zero. Sensitivity is expressed in amperes (concentration)$^{-1}$ or in amperes (cm^2 $\times$ concentration)$^{-1}$ using currents or current densities, respectively. These quantities represent or measure the slope of the response curve plotted from the difference in measured current that is obtained when the electrode is transferred from, for example, a 10^{-3} M solution to a 10^{-4} M solution (at constant ionic strength so that transport parameters—single-ion diffusion coefficients or mobilities—are constant) containing the species

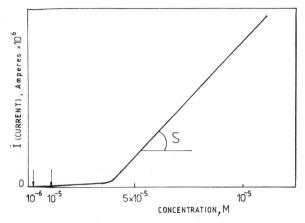

Figure 3.2 Typical amperometric membrane electrode functions: current (I) vs. concentration (a linear scale, in contrast with the logarithmic scale in Figure 3.1). Currents are positive and increase with concentration of reduced species; currents are negative and become more negative with concentration of oxidized species.

for which the electrode is selective. At standard temperature, the theoretical slopes of electrode function are dependent mainly on electrode area, stirring rate, electrons transferred and diffusion coefficient, or mobility and temperature. The exact numerical value cannot be stated, but must be empirically measured. The theoretical slope is given by Equation 2.18 as the coefficient of the independent variable. Temperature changes affect the limiting current and the slopes in Figure 3.2 two ways: mainly from the temperature coefficient of the diffusion coefficient, and, to a much lesser extent, from expansion or contraction of the solution volume with consequent small change in species concentrations.

3.2   Limits of Detection

3.2.1   Potentiometric Limits

Normally, ion-selective membrane electrodes respond to several ion activities when present alone in simple solutions of each ion's salt. There will be a range of concentrations over which the relationship between the potential of the electrode and the concentration of the ion follows the Nernst–Horowitz–Nicolsky–Eisenman equation. In dilute solutions, however, the response deviates from the ideal and as mentioned previously, the slope of the plot of potential vs. log concentration becomes smaller with decrease in concentration of the ion, until finally the potential remains constant below a certain concentration, i.e., the slope becomes zero, which is regarded as the limit of detection.

In the IUPAC definition, the limit of detection is taken as the concentration at the point of intersection of the extrapolated high-concentration, linear segment with the low-concentration, zero-slope segment of a plot of cell EMF against the logarithm of the concentration.[1] In an earlier recommendation,[2] the limit of detection was the concentration at which the calibration deviated by $S \log 2/z$ mV from the extrapolated Nernstian response.

The slope S is the Nernst value $(2.303 RTz/F)$ for a univalent electrode and z is the ionic charge. These detection limits are the same for well-behaved electrodes that obey model equations from Chapters 1 and 2. Although the definitions may be useful as a rule-of-thumb characteristic for electrode performance, they are not rigorously related to analytical performance, in contrast to the statistically based definitions of limit of detection that have been applied to other techniques of chemical analysis. Thus, the limit of detection should not be arbitrarily determined from the calibration graph without considering the random errors associated with the measurement, but should be decided on a statistical basis that allows a solution containing a given concentration of determinand to be discriminated, with a specified degree of confidence, from a blank solution.[3]

For solid-state ion-selective electrodes, the limit of detection is considered to depend ultimately on the solubility product K_{so} of the membrane sensor material and has been noted many authors.[4-16] This is a thermodynamic limit that is often obscured by other events. Important extraneous factors include the adsorption of primary ions on container wall,[17] adsorption of the principal component ions on the sensor electrode,[18, 19] interferences from supporting electrolytes,[5, 9, 10, 12, 13] and spontaneous surface reactions by dissolved oxygen.[14, 16] These have been suggested quite early and found to be perpetual problems. However, other suggested effects, such as slow precipitation–dissolution kinetics[20, 21] and defects in the solid state,[22] seem not very significant.

The concentration or activity corresponding to the thermodynamic limit was already stated in Equation 1.15 in terms of the solubility product of the membrane salt. On the other hand, for a liquid membrane or a polymer-based liquid ion-exchanger electrode, the limit of detection was stated to be governed by the equilibrium leakage concentration of the liquid exchanger into the bathing solution. There are a number of studies that confirm this hypothesis.[3, 23-27] Improved sensitivity should result from increased hydrophobicity of a liquid ion exchanger. Růžička et al.[28] and Moody and Thomas[29] proved this point by synthesizing improved dialkylphosphates related to the original exchanger of Ross.[30] Kamo et al.[26] generalized Equation 1.15 to give

$$E = E_0 - \frac{RT}{F} \ln \frac{\left[C + \left(C^2 + A_x\right)\right]^{1/2}}{2} \qquad (3.1)$$

where E_0 is a constant depending only on the concentration of the internal reference electrode solution; A_x is a parameter governing the limit of detection and which depends on the species of anion, x. Their equation was derived for a positively charged exchanger that exchanges anions (negative charge counterions). However, the equation is correct for negative site exchangers of cations (positive charge counterions), by changing the sign in front of the ln term. When $C^2 \gg A_x$, Equation 3.1 can be simplified to give the familiar Nernst-type of equation. Under the condition that $C^2 \ll A_x$, the potential tends asymptotically to a constant value that depends only on A_x. Thus, the parameter A_x is related to the limit of detection. If the ion-exchanger dissociates nearly completely in the membrane phase, the parameter A_x is given by

$$A_x = \frac{4C_{\text{iex}}^2}{k_{\text{iex}}k_x} = \frac{4C_{\text{iex}}^2}{b_x} \tag{3.2}$$

where C_{iex} is the concentration of liquid ion exchanger in the membrane. Kamo et al. used $b_x = k_{\text{iex}}k_x$ for the thermodynamic partition coefficient of the exchanger salt with ion x,

$$K_{\text{iex}, x} = \frac{a_{\text{iex}}(s)a_x(s)}{a_{\text{iex}}(m)a_x(m)} = \frac{1}{k_{\text{iex}}k_x} \tag{3.3}$$

The single-ion partition coefficients were defined in the same way by Kamo et al. as in Equation 1.14. The selectivity coefficient for two species x and y were given as

$$k_{x,y}^{\text{pot}} = \frac{u_y b_y}{u_x b_x} \tag{3.4}$$

and in the case that the ion exchanger is bulky, so all ion salts move with about the same mobility in a membrane, then

$$k_{x,y}^{\text{pot}} = \frac{A_x}{A_y} \tag{3.5}$$

and the dependency of selectivity coefficient on the A's is clear.

A particularly clear comparison of the different low-level limiting potential values for different cations with the same anion (NO_3^-) was compiled by Nielson and Hansen.[31] These responses follow the prediction that the most oil-soluble cation, in a series of liquid anion exchangers, will provide the lowest detection limit for a given anion. These are illustrated in Figure 3.3.

The statistical approach for expressions of detection limit has been adopted by Midgley[3] and by Liteanu et al.[32, 33] Detection limit can be

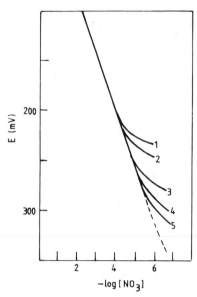

Figure 3.3 Calibration curves of NO_3^- ion-selective electrode containing (1) trioctyl-methylammonium, (2) tetraheptylammonium, (3) tetraoctylammonium, (4) tetradecylam-monium, and (5) tetradodecylammonium ions as mobile cation sites for anion exchange in dibutylphthalate plasticizer. All electrode responses have been adjusted to the same E_0 value. (Reproduced from Nielsen, H. J. and Hansen, E. H., *Anal. Chim. Acta*, **84**, 1, 1976. With permission from Elsevier Science Publishers, Physical Sciences & Engineering Division.)

expressed as functions of the factors determining the deviation of the electrode response from the Nernstian value, e.g., reagent blanks, solubility products, partition constants, and interferences. The equations enable one to predict (a) the degree of precision with which the EMF has to be measured if an electrode is to attain a desired limit of detection in specified conditions or (b) whether changing the conditions might bring the desired limit of detection within reach of a given precision of a solubility product or partition equilibrium constant, e.g., by working at a low temperature in order to reduce a solubility product. Practical examples with ion-selective electrodes justified the proposed statistical treatment of the limit of detection and demonstrated that the errors for electrodes operating in the Nernstein range were normally distributed.[3]

3.2.2 Amperometric Limits

Amperometric membrane electrodes may respond to several ion concentrations when present alone in simple solutions of each ion's salt. There will be a range of interfacial potential differences (and overall applied cell voltages) over which the relationship between the current and concentration of each ion follows the various equations, Equations 2.17 through 2.25. These were illustrated in Figures 2.4 and 2.5 for simple systems with two electroactive ions in each case. It is control of the applied voltage that most often allows the selectivity of the classical amperometric experiment. In many cases, a range of applied voltages can be found so that the limiting current is proportional to the one species that is of interest for measurement. On the other hand, there are cases when the entire cell chemistry must be modified, by adjustment of pH or addition of complexing agents that can mask the presence of an interfering ion while allowing the desired ion concentration to be followed amperometrically.

In dilute solutions, however, the response deviates from the ideal and as mentioned previously for potentiometric sensor responses. The slope of the current or current density vs. concentration plot becomes smaller with decrease in concentration of the ion, until finally the current stays constant below a certain concentration, which is regarded as the limit of detection.

3.3 Interferences

An interfering substance is any species (other than the ion being measured) whose presence in the sample solution affects the measured EMF or current of a cell. Interfering substances fall into two classes: "electrode" interferences and "method" interferences. Examples of the first class are those substances that give a response similar to that of the ion being measured and whose presence generally results in an apparent increase in the activity (or concentration) of the ion to be determined (e.g., Na^+ for the K^+ electrode, or certain hydrophobic buffer cations for the Ca^{2+} electrode) and those species that interact with the membrane so as to change its chemical composition (i.e., organic solvents for the liquid or polyvinylchloride [PVC] membrane electrodes) or electrolytes present at a high concentration, giving rise to appreciable liquid junction potential changes or incipient Donnan exclusion failure of a membrane.

The second class of interfering substance covers substances that interact with the ion being measured so as to decrease its activity or apparent concentration. However, the electrode continues to report the true activity (e.g., CN^- present in the measurement of Ag^+).

3.4 Selectivity

Membrane electrodes have the advantage that they can, in principle, be produced for many ionic and neutral charge species. The factors that can be controlled include membrane properties, bathing electrolyte properties, and external experimental control:

1. membrane plastic material, with characteristic dielectric constant, intrinsic charge site density (if any) and functional groups, cross-linked property and characteristic solubility parameters for different plasticizers and compounds;

2. membrane plasticizer, especially functional groups and dielectric constant;

3. ion exchanger material, neutral carrier material;

4. enzyme, co-enzyme, homogeneous mediators, heterogeneous mediators or promoters;

5. overcoating materials with predetermined transport selectivity properties;

6. chemical composition control of the bathing solution, by prior separations or addition of masking reagents;

7. control of temperature, applied voltage, etc.

Specific problems arise when the analyte and interference are very closely related chemically. A classical example in potentiometry is the difficulty of distinguishing different ions, particularly those of the same sign. Hence, considerable effort has been devoted to assessing the selective performance of electrode in the presence of interfering species. For potentiometry this has been done in terms of the selectivity coefficient, $k_{A,B}^{pot}$, defined variously.

3.4.1 The Modified Nicolsky or Nicolsky – Eisenman Equation

E = constant

$$+(2.303RT/z_A F)\log\left[a_A + k_{A,B}^{pot}a_B^{z_A/z_B} + k_{A,C}^{pot}a_C^{z_A/z_C}\dots\right] \quad (3.6)$$

E	is the experimentally observed EMF of a cell (in volts) when the only variables are activities in the test solution;
R	is the gas constant and is equal to 8.31441 J K^{-1} mol^{-1};
T	is the absolute temperature (in degrees Kelvin);
F	is the Faraday constant and is equal to 9.64867×10^4 C mol^{-1};
a_A	is the activity of ion A;
a_B and a_C	are the activities of the interfering ions, B and C, respectively;
$k_{A,B}^{pot}$	is the potentiometric selectivity coefficient for ion B with respect to the principal ion A;

z_A is the charge number—an integer with magnitude and sign corresponding to the absolute charge of the principal ion A;

z_B and z_C are charge numbers corresponding to the charge of interfering ions, B and C, respectively. The sign of these charge numbers is the same as the principal ion.

3.4.2 Buck Modification of the Nicolsky – Eisenman Equation

$$E = \text{constant}$$

$$\pm (2.303 RT/F) \log \left[a_A^{1/|z_A|} + k_{A,B}^{pot'} a_B^{1/|z_B|} + k_{A,C}^{pot'} a_C^{1/|z_C|} \cdots \right]$$

$$(3.7)$$

Here the charge numbers are absolute values. When anion sensors are studied, the sign of the log term is negative, otherwise the sign is positive for cation sensors. The constant term includes the formal or zero potential difference of the indicator electrode, $E_{ISE}^{0'}$, the reference electrode potential difference E_{ref}, and the junction potential difference E_J (all in volts or millivolts).

These equations can be derived using non-equilibrium thermodynamics only when charge numbers are exactly the same, e.g., all univalent cations or all divalent cations, for example. Equations 3.6 and 3.7 then become the same, as well. However, for ions of the same sign, but different magnitudes, the second form is more nearly correct, but the selectivity coefficients depend on the site concentration of the ion-selective electrode (ISE). Dependence of selectivity coefficients on concentration is clear from dimensional analysis of both equations. The equations are not appropriate for Donnan failure systems responsive to ions of different sign.

The difference between the two forms of response equation arises from a consistency argument: Buck's variation gives parallel response plots for pure A and pure B when plotted as millivolts vs. $\log a_A^{1/|z_A|}$ or millivolts vs. $\log a_B^{1/|z_B|}$. The vertical spacing is $(2.303 RT/F) \log (K_{A,B}^{pot'})$, for all combinations of ion charge. The Nicolsky–Eisenman equation can give ideal parallel responses for pure A and pure B provided the plots are in millivolts vs. $\log a_A$ and millivolts vs. $\log a_B^{z_A/z_B}$, respectively. The problem now is that the spacing is $(2.303 RT/z_A F) \log (K_{A,B}^{pot})$. As formulated,

$$\text{Buck's } K_{A,B}^{pot'} = \left(\text{Nicolsky–Eisenman } K_{A,B}^{pot} \right)^{1/|z_A|} \qquad (3.8)$$

In principle the Buck formulation limits the range of K's when z_A is greater than 1.

3.4.3   Methods for Determining Selectivity Coefficients

(i) *Fixed interference method* The EMF of a cell comprising an ion-selective electrode and a reference electrode is measured with solutions of constant activity of interference, a_B, and varying activity of the primary ion. The EMF values obtained are plotted vs. the activity of the primary ion. The intersection of the extrapolation of the linear portions of this plot will indicate the values of a_A that are to be used to calculate $k_{A,B}^{pot}$ from the Nicolsky–Eisenman equation

$$k_{A,B}^{pot} = (a_A)/(a_B)^{z_A/z_B} \tag{3.9}$$

Buck's value

$$k_{A,B}^{pot} = (a_A)^{1/|z_A|}/(a_B)^{1/|z_B|} \tag{3.10}$$

is smaller than the Nicolsky–Eisenman value when $|z_A| > |z_B|$.

(ii) *Separate solution method I* The EMF of a cell comprising an ion-selective electrode and a reference electrode is measured with each of two separate solutions, one containing the ion A at the activity a_A (but no B), the other containing the ion B at the same activity $a_B = a_A$ (but no A). If the measured values are E_A and E_B, respectively, the value of $k_{A,B}^{pot}$ may be calculated from the equation

$$\log k_{A,B}^{pot} = (E_B - E_A)z_A F/2.303RT + (1 - z_A/z_B)\log a_A \tag{3.11}$$

(iii) *Separate solution method II* The EMF of a cell comprising an ion-selective electrode and a reference is measured with each of two separate solutions, one containing the ion A at the activity a_A (but no B), the other containing the ion B (but no A) at a different activity such that $a_B^{1/|z_B|} = a_A^{1/|z_A|}$ (or equally $a_B = a_A^{z_B/z_A}$, $a_A = a_B^{z_A/z_B}$, etc.). If the measured values are E_A and E_B, respectively, the value of $k_{A,B}^{pot}$ may be calculated from the equation

$$\log k_{A,B}^{pot} = (E_B - E_A)z_A F/2.303RT \tag{3.12}$$

(iv) *Separate solution method III* The concentrations of a cell comprising an ion-selective electrode and a reference electrode are adjusted with each of two separate solutions, one containing the ion A at the activity a_A (but no B), the other containing the ion B (but no A) to achieve the same measured cell voltage. From any pair of activities a_A and a_B giving the same cell voltage, the value of $k_{A,B}^{pot}$ may be calculated from the equation

$$k_{A,B}^{pot} = (a_A)/(a_B)^{a_A/z_B} \tag{3.13}$$

These methods are recommended only when the electrode exhibits a Nernstian response to both principal and interfering ions. They are less desirable than the fixed interference method because they do not represent as well the actual conditions under which the electrodes are used. These methods are based on the assumption that plots of E_1 vs. $\log(a_A^{1/|z_A|})$ and E_2 vs. $\log(a_B^{1/|z_B|})$ will be parallel, and the vertical spacing is $(2.303RT/F)\log(K_{A,B}^{pot})$.

3.4.4   Correlations and Estimation of Selectivity Coefficients

In Chapter 1 a number of expressions were derived for the formal selectivity coefficient of an interfering ionic species relative to the base ion for which the electrode is most sensitive. The results show that selectivity is determined by and can be expressed by a series of factors:

1. the free energy of transfer of an ion from bathing solution (usually aqueous) to a hydrophobic membrane phase—the quantity in exponential form is the single-ion partition coefficient; the ratio of the partition coefficients occur in the expression, and this quantity is the ion exchange constant;
2. ratios of the complex formation constants for all species in either phase that add stability to a charge species in each phase;
3. The mobility of the interfering species in the membrane relative to the base ion in the membrane—This ratio applies only in the case that the two ions are permeable and can exist together in the membrane;
4. the activity coefficient ratio of the interfering and base species in the membrane.

Usually, activity coefficient and mobility coefficient ratios are near to unity in comparison with the partition coefficient ratios (the ion exchange constant). This result means that magnitudes and orders of selectivities can be roughly compared by determining ion exchange constants for pairs of ions. For solid electrodes Equation 1.17 predicted selectivity coefficients for precipitate-based electrodes should correlate with ratios of solubility products for two salts with a common counterion. For liquid membranes, selectivity coefficients should correlate with ratios of salt extraction (salt partition) coefficients using a common ion.

Proof that selectivity coefficients of solid-state electrodes may be assessed from solubility product ratios was given by Morf et al.[34] Their correlation is presented in Figure 3.4, which highlights common interference of silver-based crystal membranes and precipitate-impregnated membrane electrodes. Here the interfering ion reacts at the membrane surface to form a new insoluble compound. A knowledge of solubility products of the membrane salt and of the interfering salt permits

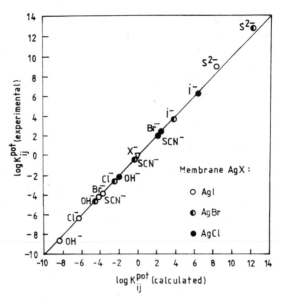

Figure 3.4 Comparison of the experimental and calculated anion selective coefficients of different silver halide membrane electrodes. (Reprinted with permission from Morf, W. E., Kahr, G., and Simon, W., *Anal. Chem.*, **46**, 1538, 1974, Copyright 1974 American Chemical Society.)

prediction of the selectivity coefficient. For example, the selectivity coefficient $k_{Br,SCN}^{pot}$ for the thiocyanate interference of the silver bromide electrodes is given by

$$k_{Br,SCN}^{pot} = \frac{K_{so}(AgBr)}{K_{so}(AgSCN)} = \frac{3.3 \times 10^{-13}}{10^{-12}} = 3.3 \times 10^{-1} \quad (3.14)$$

Similarly, the selectivity coefficients for a series of cations were compared with respect to cesium ion, by measuring the partition coefficients of the different cation salts and a common anion. Then the ratios of the salt partition coefficients divided by the partition coefficient of the cesium salt gives the selectivity coefficient of each cation relative to cesium. The effects of the common anion are presumed to be the same for all cations and therefore cancelling when the ratio is formed. On a logarithmic scale, the numbers on the ordinate become free energies of transfer. The results are illustrated in Figure 3.5.

Another example is a sequence of selectivity coefficients for neutral carrier-based monovalent ion sensors. The selectivities are referenced to potassium ion. The correlation is with respect to ratios of the arbitrary ion–valinomycin complex formation constant and the potassium valinomycin complex formation constants using a common anion in water. These results are shown in Figure 3.6.

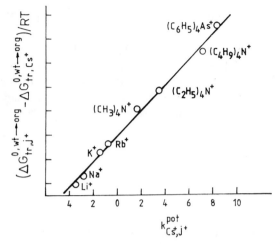

Figure 3.5 Dependence of the difference between standard Gibbs transfer energies on the logarithm of the selectivity coefficient for ISEs based on 2-nitro-*p*-cymene. The arbitrary ion is called J^+, and Cs^+ is the reference ion. (Reproduced from Scholer, J. and Simon, W., *Helv. Chim. Acta*, **55**, 1801, 1970. With permission.)

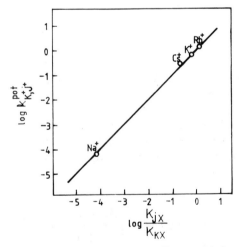

Figure 3.6 Correlation between the selectivity coefficient and the ratio of the stability constants in water for a liquid-membrane ion-selective electrode based on valinomycin dissolved in nitrobenzene. (Reproduced from Morf, W. E., *The Principles of Ion-Selective Electrodes and of Membrane Transport*, Elsevier Scientific, Amsterdam, 1981, 290. With permission.)

3.5   Ionic Strength and Activity Coefficients

The various equations for calculating selectivity coefficients depend upon a knowledge of activities, and it follows that consistency and exact reporting of selectivity data ought also to include information on the calculation of the activity coefficient, f, which is used to calculate activity from concentration, because $a = fC$. The Debye–Hückel limiting law is

$$\log f = Az^2(\text{I})^{1/2} \tag{3.15}$$

where A is a constant that for water is 0.511 at 25°C, z is the ionic charge, and I is the ionic strength. The equation gives adequate results when the ionic strengths are less than about 10^{-3} M. Above this I, where most experiments are performed, a variety of more or less empirical extensions of the limiting law have been proposed to fit the data (with varying success) at moderate, practical values of I. Ionic strength of a solution is defined by

$$\text{I} = \tfrac{1}{2}\sum_i C_i z_i^2 \tag{3.16}$$

C_i is the concentration in moles per liter of an ion i; z_i is the charge of the ion i. For simple $1:1$ salts I $= C$; for $1:2$ and $2:1$ salts, I $= 3C$; for $2:2$ salts I $= 4C$.

Some calculated values are reported in Figure 3.7.

The variation of the Debye–Hückel equation used extensively in ion-selective electrode work to calculate single-ion activity coefficients is

$$\log f = -\frac{Az^2(\text{I})^{1/2}}{\left[1 + (\text{I})^{1/2}\right]} \tag{3.17}$$

The principal use is calculating activity coefficients for ions present in calibration solution. Each standard solution, *of known concentration*, can be converted to a solution of approximately known activity by using the calculated correction factor.

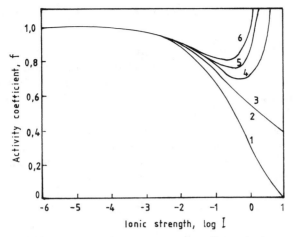

Figure 3.7 Computer simulation of various activity coefficient–ionic strength equations for univalent ions: (1) $\log f = -Az^2\sqrt{I}$; (2) $\log f = -Az^2\sqrt{I}/(1 + \sqrt{I})$; (3) $\log f = -Az^2\sqrt{I}/(1 + 0.329R\sqrt{I})$ (R taken as 3 Å); (4) $\log f = -Az^2[(\sqrt{I}/(1 + 1.5\sqrt{I})]$; (5) $\log f = -z^2[[A/\sqrt{I}/(1 + \sqrt{I})] - 0.2I]$; (6) $\log f = -z^2[[AI/(1 + 1.5I)] - 0.2I]$. (Reproduced from Moody, G. J. and Thomas, J. D. R., *Talanta*, **19**, 623, 1972. With permission.)

3.6 Range and Span

A potentiometric or an amperometric species-selective electrode in conjunction with a reference electrode is said to have a range of response (in activity or concentration units) between the lower and upper detection limits, determined from a plot of the cell potential difference vs. the logarithm of responsive ionic activity or from a plot of current vs. the concentration. The span is defined (by IUPAC) as the absolute measured voltage change between $E(t = 0)$ and E(steady state), or the absolute measured current change for a step change in primary ion activity. These quantities depend on the initial and final concentrations of the primary ion and interferences. Experimentally, sensitivity and slope are the same quantity; each may be different from the ideal slope behavior expressed in the model response equations. The experimental conditions used should be stated, especially the stirring rate, the composition of the solution in which the quantities are measured, the composition of the solution to which the electrode was exposed prior to this measurement, the history and preconditioning of the electrode, and the temperature.

3.7   Response Time

When ion-selective membrane electrodes are employed (either in a small volume of some solution or in a flowing system) it is assumed that on a step change in the concentration (activity) of the measured sample, a fast response follows. Once the response is completed, a time-independent reading can be safely postulated. In all measurements, it is rarely found that these expectations are met. The speed of response of membrane electrodes has many variations and must be considered on a system-by-system basis. It is one of the more important electrode–solution system characteristics.

3.7.1   Conventional Response Times

Previously defined response times t_{95} (to 95% of span) and t^* (to 1 mV from the steady value) require prior knowledge of steady-state E values that may not be available. These descriptive quantities underestimate practical response times of ion-selective electrodes in clinical applications where the total span may be less than 10 mV. A specified value of $\Delta E/\Delta t$ can be related to t_{95} and t^* through mathematical models, provided the long-time potential-determining processes have been identified.

3.7.2   Practical Response Times

The practical response time is the length of time that elapses between the instant at which an ion-selective electrode and a reference electrode are brought into contact with a sample solution (or at which the concentration of the ion of interest in a solution in contact with an ion-selective electrode and a reference electrode is changed) and the first instant at which the EMF/time slope ($\Delta E/\Delta t$) becomes equal to 0.6 mV min^{-1}, or 0.01 mV s^{-1}. This is an ideal and desirable, span-independent definition. Electrodes for clinical applications with small EMF spans may choose a more strict time constant of 0.06 mV min^{-1}.

3.7.3   Effects of Concentration on Response Time

Concentration of the ion to which the electrode responds affects the response time when the mass transport to the electrode surface is controlling. Steps up in concentration are always faster than steps down. However, the ratio of final concentration divided by initial concentration is the determining factor. All steps up with the same ratio (e.g., 10^{-3} M to 10^{-2} M or 10^{-4} M to 10^{-3} M) give the same response time. Likewise, all steps down with the ratio give a common response time that is different from the step-up value. Principal papers confirm these conclusions.[38–43] However, the apparently conflicting literature reports are usually found to arise from electrodes with still slower processes, e.g., transport in surface layers on the electrode, rather than transport in the solution. There are a few convenient rules that follow from the mathematical analysis of diffusion to a wall[43]:

1. From a common starting concentration, steps up to increasingly larger final values give continuously shorter apparent time constants. Steps down to continuously lower final concentrations give always longer apparent time constants.
2. Steps up in concentration to final values with the same final/initial ratio give the same time constant. Likewise, steps down with the same final/initial ratio give a constant value.
3. In flowing systems, the time response will be less than in static systems. Generally, in solutions containing high electrolyte concentrations in excess of about 0.1 M, the steady-state measured EMF is the same as the unstirred solution. The motion of the solution near the electrode is not rapid enough to perturb the space charge double layer because the latter can relax rapidly. However, in very dilute solutions, stirring dependence on the rest potential can be expected.
4. For ideally homogeneous, compacted, leak-free electrodes, the size, shape, and thickness of the device are not expected to influence response time when the electrode is exposed to a salt solution with the single, permselective salt, e.g., a calibration solution.

Morf et al.[43] worked out the time response for any membrane that does not absorb salts, e.g., is permselective. The key process is diffusion of fresh material from bulk solution to the electrode surface, through the static layer, the so-called Nernst layer. The boundary condition at the wall is zero slope of the potential-determining ion concentration. When fresh solution is added to a recently emptied beaker with an electrode and a previous bathing solution, there is some remaining solution adsorbed on the electrode surface. This static layer is typically 10 μm thick. The fresh solution must diffuse through the old solution layer before the concentration in the whole beaker becomes uniform and the new potential is reached. This is a well-known diffusional boundary value problem with zero slope of potential-determining ion at the electrode surface. The solution for the interfacial potential difference is found from the Nernst–Donnan equation using the time-dependent surface concentration:

$$E_t = E_\infty + S \log\left[1 - \left(1 - a_i^0/a_i\right)e^{-t/\tau}\right] \qquad (3.18)$$

with

$$\tau = \delta^2/2D \qquad (3.19)$$

There is some experimental evidence, from neutral-carrier-based membrane electrodes, for an equation of somewhat different form[44]:

$$E_t = E_\infty + S \log\left[1 - \left(1 - a_i^0/a_i\right)\left(1 + (t/\tau)^{1/2}\right)^{-1}\right] \quad (3.20)$$

with

$$\tau = \delta^2 D K^2/D' \quad (3.21)$$

where the D and D' are mean diffusion coefficients of the potential-determining species in bathing solution and in the membrane of thickness δ, respectively. The K is a salt partition coefficient. The idea was advanced that the external change of electrolyte concentration caused a similar change in the neutral carrier concentration inside the membrane and that both phases required diffusion of potential-determining species to reach the steady response. Because the diffusion was slow in the membrane, relative to diffusion in the bathing solution, it seemed reasonable that the external solution reached steady state while the interior species were still diffusing and obeying a $t^{-1/2}$ type of law. More recent examination of these problems suggests that Equation 3.18 applies in most cases, even for liquid ion-exchanger and neutral-carrier membranes, provided the permselectivity is maintained.[45] The typical diffusion control ($t^{-1/2}$ behavior) requires some further process, such as onset of Donnan exclusion failure, where whole salts permeate a membrane.

3.7.4   Other Factors Affecting Response Times

The presence of membrane-interfering ions and of non-ionic surfactants cause prolonged response times. Membrane-interfering ions include those that adsorb on an electrode surface and those that exchange ions and eventually pass into the bulk. A recent experimental and theoretical study explained in detail how adsorption can affect response times.[18, 19, 46, 47] Often maximum or a minimum in potential excursion can occur while the external ion distribution is seeking a new steady state. For example, an iodide-sensing electrode can be initially in a high iodide activity with some bromide ion also present. When the electrolyte is changed to a lower iodide activity (also with a constant bromide level), the potential response passes through a minimum value on its excursion to a steady, more positive final value. This effect is caused by a changing quantity of adsorbed iodide that is removed from the electrode surface with the aid of bromide. The removed iodide is temporarily in the vicinity of the electrode surface, prior to diffusing away into the bulk. The sudden iodide activity surge causes the transient potential to move more negatively before going positive to the steady value corresponding to the lesser bulk iodide concentration. A similar effect occurs when a selective sodium-ion sensor (a glass electrode) is exposed to a mixture of potassium and sodium ions. The potassium ions partition (adsorb) on the glass in competition with the sodium ions, but the potassium ions do not move and give no steady-state response. However temporarily, they replace some sodium ions and cause a temporary surge in the sodium activity at the electrode surface and a sudden maximum in potential.

Whenever surface adsorption can be changed, and the resulting temporary upset in local concentration occurs, then a prolonged response time is expected. This effect is not difficult to show when only one adsorbing, responding ion is involved, but the origin is difficult to prove. Some progress has been made.[18, 19, 46, 47]

Whenever a resistive surface layer is formed on an electrode, and when this layer is more resistive than the bulk membrane, then time responses are increased by an order of magnitude. This type of effect is especially common with plasticized, liquid membranes of all kinds. In order to reach a steady response, ions have to permeate this outer layer. Examples of this behavior for neutral-carrier membranes have been recently demonstrated, and there are many previous studies given in references 48 through 50. When surface layers can be removed, e.g., etching glass electrode surfaces or washing PVC membranes with tetrahydrofuran, the response times can be shortened. Originally these effects were attributed to electrode "poisoning."

Mixtures of salts containing one or more interfering ion may still show prolonged response times, even when no adsorption is involved. This possibility is commonly found for liquid ion-exchanger membranes. It is simply a result of the adjustment of site concentration within the membrane as the steady state is approached. It is a well-known result of the theory of solid and liquid ion-exchangers, bathed in a single responsive salt, that the steady state is reached as soon as the surface of the membrane attains a new concentration after a concentration step. However, for mixtures of ions, the liquid electrode does not reach a steady-state potential distribution until the ions have permeated the entire membrane and reached their steady concentration distribution. This process requires many seconds, perhaps even one minute.

3.8   Temperature Coefficients

Control of the temperature of both the solutions being analyzed and the electrode assembly is essential for accurate potentiometry and amperometry, especially for direct measurements. Increasing or decreasing the temperature has the effect of increasing or decreasing the slope of the response curves in potentiometry as shown in Figure 3.8. The effect

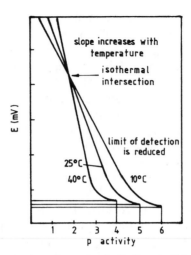

Figure 3.8 Effect of temperature change on electrode potential. Reproduced from Philips, *Guide to the Use of Ion-Selective Electrodes*, Philips Eindhoven, 1975, 9. With permission.)

on the detection limit has been exaggerated in Figure 3.8 for clarity of illustration and it may be noted that there is a small advantage in reducing the temperature of test solutions when working near the limit of detection.[51] Temperature changes also affect the intercepts E_0 or E'. These changes shift the entire calibration curve up or down and have a significant effect on accuracy. Ideally though, all analyses should be performed at the same temperature, but this is not always possible. Values of the various parameters reflecting the basic characteristics of ion-selective membrane electrodes in most of the cases included in instruction manuals are for the standard temperature of 25°C; these must be adjusted for use at other temperatures. In laboratories where there are large temperature fluctuations during the day, the electrodes should be recalibrated frequently in order to correct for any changes in standard potential and calibration slope.

Temperature effects are less important in potentiometric titrations than in direct potentiometry or standard addition (or subtraction), although the sharpness of the inflection corresponding to the end point may be affected. Because of the complexity of gas-sensing membrane electrodes it is difficult to characterize the effect of temperature in this case. The only significant temperature coefficient that can be measured for these gas-sensing systems refers to the usually unrealistic state when the electrode and sample are both in a thermostatically controlled environment for various temperatures; such coefficients were determined by Bailey and Riley[52] for an EIL Model 8000 automatic analyzer.

When a gas-sensing membrane electrode is used under laboratory conditions, with the bulk of the electrode at room temperature and the samples at varying temperatures, the apparent temperature coefficients of, say, an ammonia electrode is an increase of about 2 mV $°C^{-1}$. Thus, care must be taken to ensure that standard and samples are kept at the same temperature and that these electrodes are not subjected to rapid temperature fluctuations, e.g., by exposure to direct sunlight.[52]

3.9 Methodology

3.9.1 Standardization of Membrane Electrodes

For analytical applications, membrane electrodes have to be calibrated, despite the extensive theory presented in Chapter 1. Approximate calibration curves of potential or current vs. log activity or concentration are possible from tabulated data of electrode potentials, junction potentials, and diffusion coefficients. However, accuracy of results demands calibration using solutions with similar composition as the samples. Unlike calibrants for pH-sensitive glass electrodes, a very limited number of ion-buffers are available. These are restricted to heavy-metal ions (such as Ag^+, Cd^{2+}, etc.) that form amine complexes of known formation constants using amines of known basicity. Then, low levels of metal ion activity (to 10^{-20} M) can be reliably made by pH control of amine–metal ion mixtures.[53-57] A limited number of single-salt solutions (e.g., Ca^{2+}, K^+, and Na^+) with known single-ion activities are also available. These values were determined by the same methods and the same types of assumptions used for the U.S. Bureau of Standards pH standards.[58-60]

Saturated solutions of insoluble salts with a common cation or common anion have been used as calibration standards.[61] In addition, serially diluted solutions of dissociated, soluble salts have been used. Neither procedure can be recommended unless the solutions have ionic strength controlled by addition of an inert salt. Solubility products are rarely known accurately enough to serve as a means for calculating an ion activity in a saturated solution with or without inert salts. If inert, ionic strength adjusting salts are not used, a simple saturated solution is an unreliable standard. Similarly, sequential dilutions of soluble, dissociated salts to give concentrations less than 10^{-5} M are also unreliable. Simple dilutions of soluble salts and solutions of a single saturated salt at concentrations less than about 10^{-5} M are prone to ion adsorption on the walls so that concentrations are overestimated.[62] However, saturated solutions in inert electrolytes are not so prone to adsorption of the electroactive ions. Dilutions to 10^{-6} or 10^{-7} are reputed to be possible. The inert salt selected, must not form complexes with the sensed ion. Of course, it must not contain species that cause an interference in the response itself. The selected perchlorate–iodide example in the next paragraph was chosen because perchlorate does not interfere with the silver iodide–based sensor. However, for liquid ion-exchanger types, perchlorate is usually a very serious interference. In those cases, very hydrophilic salts must be used to adjust ionic strength, e.g., salts of acetate, fluoride, hydroxide, sulfate, and phosphate.

It is common practice to standardize potentiometric electrodes using serially diluted solutions of a completely dissociated salt of the ion of interest. Another inert, unresponsive salt is present at constant, high concentration to maintain a constant ionic strength. For example, 0.1 M NaI in 0.9 M NaClO$_4$ could be a base solution. Each dilution, say 10 ml to 100 ml, would be achieved by addition of a diluent solution of 1.0 M NaClO$_4$. Any desired iodide concentration can be made (down to about 10^{-6} M) at constant ionic strength of 1 M. Each dilution can be stored separately to form a family of standard solutions. The activity coefficients are virtually constant and so a calibration plot of millivolts vs. log concentration of iodide can be constructed. The activity scale is merely offset by a constant value that if often practically unimportant. For amperometric sensors, the concentrated standard would be chosen and linearly diluted to give solutions (individual current points) on a linear, rather than a logarithmic, concentration scale.

The avoidance of activity coefficient calculations by using ionic strength control was suggested very early in the development of fluoride-sensitive electrodes.[63] The constant ionic strength diluent contained a known acidity and a known level of complex-forming masking agents, as well as the inert salt. This solution, called total ionic strength adjustment buffer (TISAB), is still important for reliable use of fluoride-sensing electrodes.

The preparation of dilutions from a single, concentrated solution need not require making individual samples of each dilution. A large beaker, say 1 l, can be used to contain the stirred, constant ionic strength diluent solution. Then a concentrated basis solution can be loaded into a burette. Sequential dilutions can be produced in the beaker by adding calculated quantities of the basis solution from the burette. Automatic generation of known, diluted solutions has been suggested by Horvai et al.[64] A flow-mixing system is used, but the method is not practical unless many repetitive calibrations are required. Coulometric generation of low ion activities is ideally suited to ion-selective electrode calibration.

In the non-Nernstian region of potentiometric electrode calibration, or in the nonlinear region of the amperometric electrodes, relationship between E and log C, or between I and C, is no longer linear. Two standards spaced in uniform intervals in the linear region do not suffice to define the curvature. Many standard samples are required to define the curvative regions, especially if a theoretical model will be fit to the response points.

3.9.2 Direct Potentiometry

Direct potentiometry is the simplest method of all for the use of ion-selective electrodes. After preparing a calibration curve with two or three standard activities or concentrations, the ion-selective membrane electrode and reference electrode can be placed in the sample solution (untreated or treated with ionic strength adjuster) and measurements made in much the same way as pH measurements. It is important to remember that high precision is rarely achieved in direct measurements. Quite apart from variations due to electrode drift and problems of the reference electrode and junction potentials, generally laboratory conditions are such that readings to within ± 0.1 mV are difficult. This, in itself, leads to an error of $\pm 0.4\%$ in the activity of a species under Nernstian conditions for a univalent ion and double this for a divalent ion. Of course, if the error is random, it can, in principle, be made as small as desired by making a sufficient number of replicate determinants. The direct potentiometry is suitable for the analysis of samples in which the determinand is present in the "free" state, or in which the determinand may be unbound by appropriate treatment.

3.9.3 Standard Addition and Subtraction Methods

Standard addition and subtraction methods are generally more accurate than direct measurements of concentration, but they involve extra measurements and complicated calculations. They require a change in the unknown concentration of the primary ion A. For standard addition (or known addition) more A is introduced into the unknown, whereas in known subtraction the level of A is lowered by adding a known quantity of a complexing or precipitating agent. Increased accuracy obtained with these methods is possible by the removal of any uncertainty associated with the standard potential E_0. The slope S is determined as usual with standard solutions, and the unknown sample concentration is then calculated from the change in potential obtained with a known increment of the measured ion or complexing agent.

In the method of standard addition to sample, the EMF E_1 is measured for the sample solution of known volume V_1 and the total molar concentration C_1 of the unknown species,

$$E_1 = E' \pm S \log x_1 f_1 C_1 \qquad (3.22)$$

where x_1 is the fraction of free (uncomplexed) primary unknown ion. A new potential E_2 is measured after adding a small volume V_s of a standard solution (concentration C_s) of the species sought; and C_s is generally greater than the unknown so that the added volume must be small compared with the original. Then

$$E_2 = E' \pm S \log x_2 f_2 \frac{V_1 C_1 + V_s C_s}{V_1 + V_s} \qquad (3.23)$$

where x_2 and f_2 correspond to the new free ion fraction and activity coefficient, respectively. Usually one can assume that the fractions and activity coefficients are unchanging. Then, the difference potential is defined

$$\Delta E = E_2 - E_1 \pm S \log \frac{V_1 C_1 + V_s C_s}{C_1(V_1 + V_s)} \qquad (3.24)$$

and the sample concentration C_1 can be calculated from Equation 3.25,

$$C_1 = \frac{C_s}{10^{\pm \Delta E/S}(1 + V_1/V_s) - (V_1/V_s)} \qquad (3.25)$$

In analate addition (sample addition to a standard), the EMF is measured for the electrochemical cell containing V_s cm^3 of standard solution of determined ion at concentration C_s. Then a volume V_a cm^3 of the sample solution is added to the cell, and the new EMF is read. Sample concentration C_1, corresponding to ΔE, has the following value:

$$C_1 = C_s \left[10^{\pm \Delta E/S}(1 + V_s/V_a) - V_s/V_a \right] \qquad (3.26)$$

The known subtraction method depends on lowering the concentration of free ions by addition of a complexing or precipitating agent. Reasoning along lines similar to the known addition method leads to the following version of Equation 3.23:

$$E_2 = E' \pm S \log x_2 f_2 \frac{V_1 C_1 - V_s C_c}{V_1 + V_s} \qquad (3.27)$$

where C_c is the concentration of the complexing (precipitating) agent, with $1:1$ complexation assumed.

3.9.4   Simplification when Standard Addition Volume Is Small

When the standard volume added is very small relative to V_1, Eq. 3.25 simplifies to

$$C_1 = \frac{C_s V_s}{(10^{\pm \Delta E/S} - 1)V_1} \qquad (3.28)$$

and is suited to those standard addition cases when the volume change of sample is not more than about 1%. Equation 3.24 rearranges to the form

$$\Delta E = \pm S \log \frac{C_1 + C_\Delta}{C_1} \qquad (3.29)$$

where C_Δ is the change in concentration after the known addition, and is equal to $V_s C_s / V_1$. Rearrangement gives

$$\frac{C_0}{C_\Delta} = \frac{1}{10^{\pm \Delta E / S} - 1} \qquad (3.30)$$

for which there are extensive tables for given values of S, corresponding both to monovalent and divalent ions. A given change in potential gives a unique value of C_Δ. The preceding equations are derived on the assumption that the activity coefficient of the determinant is identical before and after the addition or subtraction and the degree of complexation of the determinand also remains constant throughout.

Some additional assumptions, such as that no interferents are present in the sample and the ion-selective membrane electrode has a theoretical Nernstian response slope, are made in the derivation of these equations. The last assumption is especially important when using the values already found in various tables. In fact, only a few electrodes have a truly Nernstian value of the calibration slope, but most of the electrodes have a very large range in which the slope is constant. Non-Nernstian slopes can sometimes be controlled by using samples with known concentration and by adding a standard solution.

3.9.5 Potentiometric Titrations

Titrations are controlled chemical reaction procedures that match a standard reagent with an unknown reactant. The purpose is to determine the unknown concentration by finding the equivalent reagent volume, using the known stoichiometry of the reaction. These applications of the laws of equivalence can be applied to acid–base, complex–formation, precipitation–formation, or redox reactions. For the analysis of drugs, the reactions are a special category of complex or precipitation titration because the reaction generally forms an ion pair that may be soluble or insoluble. At the end point, which is close to the equivalence point depending on the sensitivity for detecting species concentrations, an equivalent volume of reagent is defined, V_e. At this point,

$$C_s V_e = C_1 V_1 \qquad (3.31)$$

with C_s the concentration of standard (equivalents per liter), C_1 the unknown concentration (equivalents per liter), V_e the end-point volume of the standard or the titrant, and V_1 the volume of the unknown material.

At higher concentrations, potentiometric titrations are superior to the direct potentiometric methods described previously. In practice, the electrodes are placed in the sample and the reagent is added progressively while the measured potential is observed. The rate of change of potential is slow at first, increases to a maximum as the equivalence point is reached, and reduces again when the equivalence point is passed. A double curve, like an elongated letter S, is obtained when the measured potential is plotted against the added volume of reagent, and the equivalence point is indicated at the steepest point of the curve, where the potential changes most rapidly.

In the occasional cases where a single electrode may sense the ions of the titrant and of the titrand (active species from sample solution)—as in the titration of, say, quinidine with tetraphenylborate ions, where a PVC plastic matrix quinidine membrane electrode would sense the decrease of quinidine ion activity before the equivalence point, and the increase in tetraphenylborate ion activity afterwards—the EMF vs. titrant volume should be symmetrical near the end point.[65] The shape of the curve is less symmetrical if the stoichiometry of the reaction is other than one-to-one. Similar curves are also obtained if the electrode used responds only to the titrant, as, for example, when a biguanide sample is titrated with copper(II) ions with a copper(II) membrane electrode used as sensor (see Section 5.58, in Part II).

One of the major advantages of the titration techniques, both potentiometric titrimetry and Gran's plot methods, is that they enable a large range of species, not directly sensed by ion-selective electrodes, to be measured (many examples may be found in Part II of this book).

3.9.6 Gran Plots

A variation of the standard addition method depends on a modification of the method originated by Gran,[66] in which a potentiometric titration curve is converted to a linear form using a semi-antilog plot. Equation 3.23 is rewritten as

$$(V_1 + V_s) \times 10^{\pm E_2/S} = x_2 f_2 \times 10^{\pm E'/S}(V_1 C_1 + V_s C_s) \quad (3.32)$$

Then, for a well-behaved system, a plot of the left-hand side of the equation (ordinate) vs. the volume of standard added (abscissa) must be linear. The intercept along the negative V_s (where the ordinate is zero) is the volume V_e, the volume of the equivalence point or end point of the titration, with the value determined by,

$$C_1 V_1 = -C_s V_e \qquad (3.33)$$

The sample's initial unknown concentration is calculated from this equation. The determination of the unknown concentration of a species A is thus carried out in a simple manner by adding a standard solution of A.

On the other hand, the response E of an ion-selective electrode, say, to a monovalent cation with response to a suitable reference electrode may be represented by the Nernst equation, without taking into account the liquid junction potential, activity coefficient, complexation association with other ions, and interfering effects:

$$E = E' + S \log C_A \qquad (3.34)$$

where C_A is the concentration of the free cation and E' incorporates the standard and reference potentials; S is the slope of the electrode response. Rearrangement of Equation 3.34 gives

$$C_A = k \, \text{antilog}(E/S) \qquad (3.35)$$

Thus, antilog(E/S) is proportional to C_A and may be plotted as a measure of C_A (along the ordinate) against the volume of titrant added to a titration (along the abscissa) to give a linear plot called Gran's plot. The plot is linear because the decrease of C_A is linearly dependent on the addition of volume of titrant, V. The plot can be extrapolated to $C_A = 0$, i.e., to antilog$(E/S) = 0$, to the intercept of the titrant volume axis and locating point, V_e, which is the equivalence point where the concentration of the A cation is zero.

Because of the relationship between [A] and [B], $[A]^x[B]^y = K_e$, where K_e is an equilibrium constant and x and y are the stoichiometric coefficients, it follows that if Equation 3.35 is valid, then

$$C_B = K' \, \text{antilog}(-E/S') \qquad (3.36)$$

thus, after the equivalence point, by plotting antilog$(-E/S')$ vs. volume added, another straight line is obtained and showing the increase of the titrant concentration after the equivalence point; this second straight line can also be extrapolated to find a $C_B = 0$, where the antilog$(-E'/S) = 0$, to the intercept on the volume of titrant axis locating the same V_e where the level of excess titrant is zero.[67]

The Gran's plot is, therefore, composed of two straight lines: one before the equivalence point, showing the analate concentration decreasing and with negative slope; the other after the equivalence point, showing the titrant concentration increasing and with positive slope.

Semi-antilog Gran's plot paper is supplied by, among others, Orion Research, Inc. This special graph paper is available in two forms, with and without built-in correction for volume changes. The more commonly used type, with volume change correction, automatically corrects the readings for 10% change in sample volume during the titration (or 1% per major division of the axis); thus, if the sample volume is 100 ml, the total volume of added increments of titrant must finally be 10 ml. The ordinate of the paper is antilogarithmic, so that volumes and E, plotted on it, become converted to antilog(E/S). The theoretical value of S is assumed. This avoids the calculation; thus, the measured potentials are plotted directly on the ordinate against the titrant volume along the abscissa. The intercept of the extrapolated line on the abscissa marks the end point (equivalence point). Errors are introduced if the value of S is not theoretical, as is often the case; although, these errors can be compensated for. The effort may make it simpler, especially if accuracy and precision are important, to calculate the correct antilog(E/S) values and to plot them on ordinary graph paper.

As an alternative, it is an easy matter to design a ruler to take in the left-hand side of Equation 3.31 for the direct plotting of the ordinate on ordinary graph paper from the EMF readings. The ruler can be appropriately biased for planned volume changes and a full description of the ruler has been given by Westcott.[68] A major disadvantage of both Orion's Gran plot paper and rulers is their complete dependence on fixed response patterns. This is Nernstian for Orion paper, but rulers can be made for any electrode calibration slope.

Ivaska[69] described a computational method for simultaneous determination of the equivalence volume and the slope of a sensor electrode from standard titration procedures. Treatment of data is simplified by use of equal additions of titrant or standard solution. Equivalence volumes and concentrations were determined with errors of 0.5 to 2% by this method.

3.9.7 Amperometric Titrations

Amperometric titrations are relatively simply related to the potentiometric titrations. The four classes of titration reaction apply here. In advance, the amperometric sensor is selected that applies most conveniently to the reaction. There are always two choices, and sometimes three choices. A sensor for the determinand (the species whose concentration is unknown) is the usual choice, although a sensor for the titrant is equally viable. In redox titrations, the choice of monitoring the reaction product is sometimes selected. In every example, the solution of unknown concentration is stirred and the sensor, the reference and the current-measuring circuit installed. When the determinand is monitored, the amperometric signal decreases linearly to the end point. When the titrant is monitored, the current signal is the very low, background value until the end point. Then the current increases linearly with excess titrant beyond the end point. In case a product component is monitored, the current increases linearly during the titration, but becomes constant at the end point and beyond. No special mathematics are required because the linear responses can be quickly extrapolated to determine the end-point titrant volume. Then Equation 3.31 applies.

References

1. International Union of Pure and Applied Chemistry, *Pure Appl. Chem.*, **48**, 127 (1976). See also IUPAC Information Bulletin No. 169 (1978) and *Ion-Selective Electrode Rev.*, **1**, 139 (1979).

2. International Union of Pure and Applied Chemistry, *Recommendations for Nomenclature of Ion-Selective Electrodes* (Recommendations, 1975), Pergamon Press, Oxford, 1976.

3. D. Midgley, *Analyst*, **104**, 248 (1979).

4. E. Pungor, K. Tóth, and J. Havas, *Mikrochim. Acta*, 689 (1966).

5. J. J. Lingane, *Anal. Chem.*, **39**, 881 (1967).

6. R. P. Buck, *Anal. Chem.*, **40** 1432 (1968).

7. E. Pungor, K. Tóth, and J. Havas, *Acta Chim.*, *Acad. Sci. Hung.*, **58**, 16 (1968).

8. D. C. Muller, P. W. West, and R. H. Muller, *Anal. Chem.*, **41** 2038 (1969).

9. E. W. Baumann, *Anal. Chem.*, **54**, 189 (1971).

10. J. Růžička and C. G. Lamm, *Anal. Chim. Acta.*, **54**, 1 (1971).

11. P. A. Evans, G. J. Moody, and J. D. R. Thomas, *Lab. Pract.*, **20**, 444 (1971).

12. N. Parthasarathy, J. Buffle, and D. Monnier, *Anal. Chim. Acta*, **68**, 185 (1974).

13. J. Buffle, N. Parthasarathy, and W. Haerdi, *Anal. Chim. Acta*, **68**, 253 (1974).

14. D. J. Crombie, G. J. Moody, and J. D. R. Thomas, *Anal. Chim. Acta*, **80**, 1 (1975).

15. J. Kontoyannakos, G. J. Moody, and J. D. R. Thomas, *Anal. Chim. Acta*, **85**, 47 (1976).

16. A. Hulanicki, A. Lewenstam, and M. Maj-Zurawska, *Anal. Chim. Acta*, **107**, 121 (1979).

17. R. A. Durst and B. T. Duhart, *Anal. Chem.*, **42**, 1002 (1970).

18. T. R. Berube, R. P. Buck, E. Lindner, M. Gratzl, and E. Pungor, *Anal. Chem.*, **61**, 453, 1989.

19. T. R. Berube, R. P. Buck, E. Lindner, and E. Pungor, *Anal. Chem.*, **63**, 946 (1991).

20. N. Parthasarathy, J. Buffle, and W. Haerdi, *Anal. Chim. Acta*, **93**, 121 (1977).

21. J. Buffle and N. Parthasarathy, *Anal. Chim. Acta*, **93**, 111 (1977).

22. W. E. Morf, G. Kahr, and W. Simon, *Anal. Chem.*, **46**, 1538 (1974).

23. O. D. Bonner and D. C. Lunney, *J. Phys. Chem.*, **70**, 1140 (1966).

24. N. Ishibashi, H. Kohara, and N. Murakami, *Bunseki Kagaku*, **21**, 1072 (1972).

25. N. Ishibashi and A. Jyo, *Microchem. J.*, **18**, 220 (1973).

26. N. Kamo, H. Hazemoto, and Y. Kobatake, *Talanta*, **24**, 211 (1977).

27. N. Kamo, Y. Kobataki, and K. Tsuda, *Talanta*, **27**, 205 (1980).

28. J. Růžička, E. H. Hansen, and C. J. Tjell, *Anal. Chim. Acta*, **67**, 155 (1973).

29. G. J. Moody and J. D. R. Thomas, *Ion-Selective Electrode Rev.*, **1**, 3 (1979).

30. J. W. Ross, *Science*, **156**, 1378 (1967).

31. H. J. Nielsen and E. H. Hansen, *Anal. Chim. Acta*, **84**, 1 (1976).

32. C. Liteanu, E. Hopîrtean, and I. C. Popescu, *Anal. Chem.*, **48**, 2010 (1976).

33. C. Liteanu, E. Hopîrtean, and I. C. Popescu, *Acta Chim.*, *Acad. Sci. Hung.*, **97**, 265 (1978).

34. W. E. Morf, G. Kahr, and W. Simon, *Anal. Chem.*, **46**, 1538 (1974).

35. R. Scholer and W. Simon, *Helv. Chim. Acta*, **55**, 1801 (1970).

36. W. E. Morf, *The Principles of Ion-Selective Electrodes and of Membrane Transport*, Elsevier Scientific, Amsterdam, The Netherlands, 1981.

37. G. J. Moody and J. D. R. Thomas, *Talanta*, **19**, 623 (1972).

38. E. Pungor and K. Tóth, *Analyst*, **94**, 625 (1970).

39. K. Tóth, I. Gavaller, and E. Pungor, *Anal. Chim. Acta*, **57**, 131 (1971).

40. B. Fleet and H. von Storp, *Anal. Chem.*, **43**, 1575 (1971).

41. D. Ammann, E. Pretsch, and W. Simon, *Anal. Lett.* **7**, 23 (1974).

42. W. J. Blaedel and D. E. Dinwiddie, *Anal. Chem.*, **47**, 1070 (1975).

43. W. E. Morf, E. Lindner, and W. Simon, *Anal. Chem.*, **47**, 1596 (1975).

44. E. Lindner, K. Tóth, and E. Pungor, *Anal. Chem.*, **48**, 1071 (1976).

45. T. R. Berube and R. P. Buck, *Anal. Lett.*, **22**, 1221 (1989).

46. E. Lindner, K. Tóth, and E. Pungor, *Anal. Chem.*, **54**, 202 (1982).

47. M. Gratzl, E. Lindner, and E. Pungor, *Anal. Chem.*, **57**, 1506 (1985).

48. G. Horvai, E. Graf, K. Tóth, E. Pungor, and R. P. Buck, *Anal. Chem.*, **58**, 2735 (1986).

49. K. Tóth, E. Graf, G. Horvai, E. Pungor, and R. P. Buck, *Anal. Chem.*, **58**, 2741 (1986).

50. E. Lindner, Zs. Niegreisz, K. Tóth, E. Pungor, T. R. Berube, and R. P. Buck, *J. Electroanal. Chem.*, **259**, 67 (1989).

51. Philips, *Guide to the Use of Ion-Selective Electrodes*, Philips, Eindhoven, 1975.

52. P. L. Bailey and M. Riley, *Analyst*, **100**, 145 (1975).

53. E. H. Hansen, C. G. Lamm, and J. Růžička, *Anal. Chim. Acta*, **59**, 403 (1972).

54. A. Craggs, G. J. Moody, and J. D. R. Thomas, *Analyst*, **104**, 412 (1979).

55. J. Růžička and E. H. Hansen, *Anal. Chim. Acta*, **63**, 115 (1973).
56. E. H. Hansen and J. Růžička, *Anal. Chim. Acta*, **72**, 365 (1974).
57. J. Havas, M. Kaszas, and M. Varsanyi, *Hung. Sci. Instr.*, **25**, 23 (1972).
58. R. A. Robinson and R. G. Bates, *Anal. Chem.*, **45**, 1666 (1973).
59. R. G. Bates, *Pure Appl. Chem.*, **36**, 407 (1973).
60. R. G. Bates, Further studies of ionic activity scales, in *Ion-Selective Electrodes*, 4 (1984 Matrafured Symposium), Akademiai Kiadó, Budapest, and Elsevier, Amsterdam, 1985, pp. 725–742.
61. J. Vesely, O. J. Jensen, and B. Nicolaisen, *Anal. Chim. Acta*, **62**, 1 (1972).
62. R. Durst and B. T. Duhart, *Anal. Chem.*, **42**, 1002 (1970).
63. M. S. Frant and J. W. Ross, *Anal. Chem.*, **40**, 1169 (1968).
64. G. Horvai, K. Tóth, and E. Pungor, *Anal. Chim. Acta*, **82**, 45 (1976).
65. V. V. Coşofreţ and R. P. Buck, *J. Pharm. Biomed. Anal.*, **3**, 123 (1985).
66. G. Gran, *Analyst*, **77**, 661 (1952).
67. M. Mascini, *Ion-Selective Electrode Rev.*, **2**, 17 (1981).
68. C. C. Westcott, *Anal. Chim. Acta*, **86**, 269 (1976).
69. A. Ivaska, *Talanta*, **27**, 161 (1980).

Part II

ANALYSIS OF PHARMACEUTICALS BY MEMBRANE SENSORS

Chapter 4

INORGANICS

4.1   Aluminum Compounds

There is no aluminum(III)-ion-selective membrane sensor, but this ion in many pharmaceuticals (see Table 4.1) may be easily determined with a fluoride-selective membrane sensor that acts as a potentiometric indicator in the titration with fluoride standard solutions.[1-4] In the presence of sodium ions (from sodium fluoride standard titrant) insoluble cryolite (Na_3AlF_6) or a similar compound is formed, according to

$$Al^{3+} + 6\,NaF \rightarrow Na_3AlF_6 + 3\,Na^+ \qquad (4.1)$$

The ratio of six fluorides to one aluminum favors the determination of small quantities of aluminum. The samples may be titrated with 0.1 M sodium fluoride at pH 5.0 (acetate buffer) in 70% (m/V) ethanol.[2]

Aluminum in antacid formulations is mostly present in non-ionic form, as hydrated aluminum oxide. The sample solution, though, may be regarded as a saturated solution of Al^{3+}, the concentration of which in solution is consequently very small. Each addition of fluoride titrant, before the end point, causes a large and rapid negative shift in potential due to F^- in solution. The potential then shifts back toward the positive direction as fluoride in solution is used up by reaction with Al^{3+}, thus allowing further aluminum to go into solution. This process continues until the end point, when all the aluminum is used up and fluoride is in excess. As the fluoride concentration of the solution rises, steady meter readings are obtained and those data are used in the Gran's plot[5] (see Figure 4.1).

Table 4.1 Aluminum Compounds Assayed by Fluoride Membrane Sensors

Compound	Formula (MM)	%Al	Therapeutic category
Alum:			Precipitates proteins;
Potassium alum	$KAl(SO_4)_2 \cdot 12\,H_2O$ (474.4)	6.69	also a powerful astringent
Ammonia alum	$NH_4Al(SO_4)_2 \cdot 12\,H_2O$ (453.3)	5.95	
Aluminum acetate solution (5%, v/v)	$C_6H_9AlO_6$ (204.1); $Al(OCOCH_3)_3$	13.22	Astringent
Aluminum hydroxide gel	$Al(OH)_3$ (78.0) This is a suspension, each 100 g, of which contains the equivalent of not less than 3.6 g and not more than 4.4 g of Al_2O_3 in the form of aluminum hydroxide and hydrated oxide.	—	Antacid
Aluminum sulfate	This is a hydrated mixture of the normal salt $Al_2(SO_4)_3$ with a small proportion of basic aluminum sulfate. It contains not less than 54.0% and not more than 59.0% of $Al_2(SO_4)_3$. Anhydrous: $Al_2(SO_4)_3$ (343.1)	—	Astringent (more than alum)

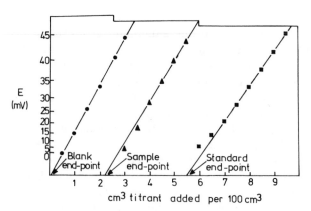

Figure 4.1 Data from titration of aluminum with 0.05 M sodium fluoride solution plotted on 10% volume-corrected Gran's plot paper: ● blank; ■ 10^{-3} M standard aluminum; ▲ sample (50 mg Al^{3+} per 5 cm^3). (Reproduced from Cooper, M. E. A., Ballantine, J., and Woolfson, A. D., *J. Pharm. Pharmacol.*, 31, 403, 1979. With permission.)

Analytical Procedures

i. *Potentiometric titration method:*

An aliquot of the sample solution containing 50 to 100 mg aluminum is accurately weighed and transferred with distilled water into a 250-cm^3 volumetric flask (for aluminum hydroxide gel the sample must first be solubilized with hydrochloric acid), diluted with distilled water to volume, and mixed. 50 cm^3 of this solution is pipetted into a 150-cm^3 beaker containing a magnetic stirring bar. The pH is adjusted to approximately 4.0 to 4.5 with sodium hydroxide solution and then 5.0 cm^3 of acetate buffer (pH 5.0) and approximately 50 cm^3 of ethanol are added. The solution is potentiometrically titrated under stirring, with 0.1 M sodium fluoride standard solution. The EMF readings are recorded vs. titrant volume and the end point is evaluated from the maximum slope of the titration curve (1 cm^3 0.1 M NaF $\equiv$ 0.45 mg Al^{3+}).

ii. *Gran's-plot method:*

Antacid solutions or suspensions are diluted volumetrically 1 in 1000 with distilled water to give an aluminum concentration in the region of 10^{-3} M. For solid formulations, a weight of powder equivalent to one tablet is diluted with distilled water to give, again, an aluminum concentration of about 10^{-3} M. Before the analysis of samples, a blank and a standard solution are titrated with fluoride.

a. *Blank titration*—A volume of 1 cm^3 of the ionic-strength–pH adjustment solution (ISA; 2 M acetate buffer containing sodium acetate [12 g] and glacial acetic acid [57 cm^3] in 1 dm^3) is added to distilled water (100 cm^3) and 5×10^{-2} M fluoride titrant (0.5 cm^3) is added to the stirred solution. The meter is set to read zero with the fluoride/Ag/AgCl double junction electrode couple, and further 0.5-cm^3 fluoride aliquots are added. The electrode potential is plotted against titrant volume on 10% volume-corrected Gran's plot paper scaled so that the 5-cm^3 reading falls near the top of the antilog axis. The resulting straight line is extrapolated back to the horizontal axis to give the blank end point.

b. *Standard titration*—ISA (1 cm^3) is added to 10^{-3} M Al^{3+} standard solution (100 cm^3) and titrated with fluoride as before. When the potential displayed by the meter becomes negative, fluoride ion is in excess. The potential corresponding to each 0.5 cm^3 aliquot of titrant added past this point is recorded and results are plotted as before. Extrapolation of the resulting straight line to the horizontal axis gives the standard end point.

c. *Sample titration*—ISA (1 cm^3) is added to the appropriately diluted sample (100 cm^3) and the solution titrated with fluoride. When fluoride titrant is in excess the potential becomes negative and meter readings stabilize rapidly. Results are treated as under method b, extrapolation back to the horizontal axis giving the sample end point.

The molarity of aluminum in the sample solution is given by Equation 4.2; hence the concentration of aluminum in the original sample could be calculated:

$$Al^{3+}\ (M) = \frac{\text{sample concentration}}{\text{standard concentration}}$$

$$= \frac{(\text{unknown intercept}) - (\text{blank intercept})}{(\text{standard intercept}) - (\text{blank intercept})} \quad (4.2)$$

Note: For alum and aluminum sulfate a supplementary analysis based on sulfate determination is also recommended (see Appendix 3).

4.2 Ammonia Solution and Ammonium Salts

Table 4.2 lists some ammonium compounds of pharmaceutical interest. Their quantitative assay can be accomplished by either ammonia-gas-sensing probes, which are now manufactured by many specialized companies (see Appendix 1), or by an ammonium-selective membrane sensor.

Ammonia-gas-sensing probes use a hydrophobic gas-permeable membrane to separate the alkaline test solution from the internal solution (0.1 M NH_4Cl). A glass pH-electrode and a Ag/AgCl reference electrode are immersed in the internal solution[6]. Because the internal ammonium concentration is very large, the equilibrium equation for the hydrolysis of ammonia,

$$\frac{[NH_4^+][OH^-]}{[NH_3]} = K \sim 2 \times 10^{-5} \quad (4.3)$$

can be approximated by

$$\frac{[OH^-]}{[NH_3]} = K' \quad (4.4)$$

for small values of $[NH_3]$.

Thus, when the probe is immersed in an alkaline test solution, ammonia diffuses through the membrane until the partial pressure of ammonia is the same on both sides of the membrane; this alters the ammonia

Table 4.2 Ammonium Compounds Assayed
by Ammonia-Gas-Sensing Probes

Compound	Formula/ (MM)	Therapeutic category
Strong ammonia solution	This is an aqueous solution of ammonia, NH_3 (17.0), containing 27.0 to 30.0 (m/m) NH_3.	Ammonia, when inhaled, irritates the mucosa of the upper respiratory tract and reflex through the medulla, causes stimulation of the respiration, acceleration of heart, and rise in blood pressure.
Ammonium bicarbonate	NH_4HCO_3 (79.06)	An irritant to mucous membranes; in small doses used as a reflex expectorant
Ammonium bromide	NH_4Br (97.9)	Depressant of the central nervous system; more effective as sedative than as hypnotic
Ammonium chloride	NH_4Cl (53.5)	It is rapidly absorbed from the gastrointestinal tract. The ammonium ion is converted into urea in the liver; the anion thus liberated into the blood stream and extracellular fluid causes metabolic acidosis and decreases the pH of the urine.

concentration in the filling solution, and so causes a pH change that is monitored by the glass electrode, where potential varies in a Nernstian manner with changes in the hydroxide level:

$$E = E_0 - S \log[OH^-] \qquad (4.5)$$

Because the hydroxide concentration is proportional to ammonia concentration (Equation 4.4), the electrode response to ammonia is also Nernstian:

$$E = E_0' - S \log[NH_3] \qquad (4.6)$$

A thorough treatment of the theory of the operation of ammonia electrodes, as well as of other sensors from the same category (SO_2, NO_2, H_2S, and HCN) has been given by Ross et al.[7]

Analytical Procedures

i. *Direct measurement:*

Three standard solutions of ammonium chloride (10^{-2}, 10^{-3}, and 10^{-4} M) are prepared by serial dilution of 0.1 M ammonium chloride stock solution. The ammonia-gas-sensing probe in conjunction with a reference electrode is placed in the 10^{-3} M standard and then 1 cm^3 of 10 M sodium hydroxide solution is added to each 100 cm^3 of standard. The EMF is recorded. The electrode pair is then placed in the 10^{-4} and 10^{-2} M standards, respectively, and the same procedure followed. Finally, the electrode pair is placed in 100 cm^3 of sample solution, and 1 cm^3 of 10 M sodium hydroxide is added. The recorded EMF is compared with the calibration curve constructed by plotting the EMF values vs. log C. In all cases magnetic stirring is used.

ii. *Known addition:*

To measure an unknown sample of ammonium (ammonia), the ammonia-gas-sensing membrane probe (slope 59 mV decade^{-1}) in conjunction with a reference electrode is placed in 100 cm^3 of sample solution. To this solution, 1 cm^3 of 10 M sodium hydroxide is added and the EMF (E_1) is recorded under stirring. A standard solution about 10 times as concentrated as the sample concentration is prepared by diluting 0.1 M ammonium chloride standard. 10 cm^3 of this standard is pipetted into the sample solution. The new EMF value (E_2) is recorded under stirring. The value Q, corresponding to the change in potential, ΔE, is given in Appendix 2 (Table XII). To determine the original sample concentration, Q is multiplied by the concentration of the added standard:

$$C_0 = QC_S \qquad (4.7)$$

where C_0 = the sample concentration, Q = the reading from known addition table, and C_S = the concentration of added standard.

4.3   Calcium Compounds

Calcium is an essential element of tissues and of blood, which contains approximately 10 mg per 100 cm^3—of this, about 5 mg is in the ionized form and the remainder is colloidal, associated with proteins. The average daily requirement of calcium is 500 mg, but larger amounts are necessary during periods of growth, pregnancy, and lactation.

The first two compounds in Table 4.3 are used as antacids, usually given with other antacid substances in mixtures, powders, and tablets; they are also used as a basis for dentifrices.

Table 4.3 Calcium Compounds Assayed by Calcium
Membrane Sensors

Compound	Formula/ (MM)	%Ca
Calcium carbonate	$CaCO_3$ (100.01)	40.08
Calcium chloride	$CaCl_2 \cdot 6\,H_2O$ (219.1)	18.29
Calcium hydroxide	$Ca(OH)_2$ (74.09)	54.10
Calcium gluconate	$C_{12}H_{22}CaO_{14} \cdot H_2O$ (448.4) $[HOCH_2[CH(OH)]_4COO]_2Ca \cdot H_2O$	8.94
Calcium lactate	$C_6H_{10}CaO_6 \cdot 5\,H_2O$ (308.3) $(C_3H_5O_3)_2Ca \cdot 5\,H_2O$	13.00
Calcium levulinate	$C_{10}H_{14}CaO_6 \cdot 2\,H_2O$ (306.3) $[H_3CCOCH_2CH_2COO]_2Ca \cdot 2\,H_2O$	13.09
Calcium pantothenate	$C_{18}H_{32}CaN_2O_{16}$ (476.5) $[HO-CH_2-C(CH_3)_2-\overset{\displaystyle OH}{\underset{\displaystyle H}{C}}-CONH(CH_2)_2COO]_2Ca$	8.56

In most cases calcium is determined after solubilization of the sample with dilute hydrochloric acid, adjustment of the pH of the solution to 10 to 11 with ammonia buffer, and titration with EDTA standard solution in the presence of a calcium-ion-selective membrane sensor. Ross[8] described the first effective membrane sensor selective for calcium ions. It was based on the calcium salt of didecylphosphoric acid (0.1 M) dissolved in di-n-octylphenyl phosphonate as liquid membrane. The reasonable selectivity of this sensor toward typical components of blood serum led to numerous applications in biomedical sciences and also to its use in analytical, inorganic, organic, and physical chemistry. This type of membrane sensor is now produced by many manufacturers[9-11] (see Appendix 1). However, the sensor was subject to pH-interference to an extent depending on the sample calcium levels. There were also problems when using the sensor as indicator in the complexometric determination of calcium in, for example, serum or sea water,[12] where the high sodium level impairs electrode performance. Various attempts have been made to overcome these disadvantages, but a major step forward in electrode design was made by Thomas and co-workers[13-15] when they incorporated the original Orion ion exchanger in a PVC matrix. This prolonged electrode lifetime and allowed the reservoir of ion exchanger within the electrode body to be dispensed with. Different electroactive materials have also been suggested[16-22] and considerable strides have been made

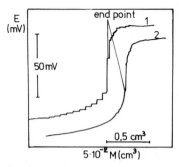

Figure 4.2   Digital titration curves for 5×10^{-2} M Ca^{2+} in borate buffer solution of pH 9.2; 0.05 M EDTA solution was added at a rate of 2.00 cm^3 min^{-1} (curve 1) and 0.50 cm^3 min^{-1} (curve 2). (Reproduced from Pranjić-Anušić, Z., *Acta Pharm. Jugoslav.*, 29, 29, 1979. With permission.)

in calcium-ion sensing with the introduction of a neutral-carrier sensor[22] and sensors of calcium bis-dioctylphenyl phosphonate sensor.[23] Thus, for example, sensors of calcium bis-4-alkylphenyl phosphonates (alkyl = hexyl, octyl, and 1,1,3,3-tetramethyl-butyl) in conjunction with dioctylphenyl phosphonate solvent mediator in PVC matrices give good calcium-ion-selective electrodes with near-Nernstian slopes and detection limits of 1.9×10^{-6} to 2.7×10^{-6} M calcium ions in the absence of ion buffers. A very interesting program of work on the characterization and application of alkylphenyl phosphonate and acyclic neutral-carrier calcium-ion-selective membrane sensors has been carried out by Thomas and co-workers at UWCC in Cardiff and Simon and co-workers at ETH in Zürich[24, 25].

Table 4.4   Results of Calcium Determination in Vital Effervescent Tablets and Vital Granules by Digital Titration with 5×10^{-2} M EDTA Using a Calcium-Selective Membrane Sensor as Potentiometric Indicator[26]

	Vital effervescent tablets	Vital granules
Calcium gluconate expected	400 mg ± 5% in one tablet	400 mg ± 5% in one 4.0-g parcel
No. of measurements	13	9
Calcium gluconate found (mg)	406 ± 3	409 ± 2
Relative standard deviation (%)	±0.7	±0.5

Pranjić-Anušić[26] used a calcium-selective membrane sensor (Philips, IS 561 Ca^{2+}) for determining the calcium content in some pharmaceutical products by potentiometric digital titration with 0.05 M EDTA in borate buffer solution of pH 9.2. The slower titrant delivery (e.g., 0.5 cm^3 min^{-1}) corresponding to curve 2 in Figure 4.2 allows a better stabilization of the sensor signals. The relative standard deviation in this method was 0.3%. The titrations carried out in this way satisfied the requests for the quality control of pharmaceutical products.

In digital titration of samples from Vital effervescent tablets and Vital granules (both containing calcium gluconate as active principle), the total potential change between the beginning of titration and 100% excess of titrant was about 100 mV, which ensured a good precision of the determinations (see Table 4.4).

The analyses performed with a series of various pharmaceutical products containing calcium have shown that the procedure is very rapid and simple and can be successively applied to routine quality control.

Analytical Procedures

i. *Direct potentiometry:*
   Three standard solutions of calcium chloride (10^{-2}, 10^{-3}, and 10^{-4} M) are prepared by serial dilution of 0.1 M calcium chloride stock solution. The ionic strength and the pH are kept constant ($I = 0.1$ M, adjusted with sodium nitrate, pH 9.0, adjusted with ammonia). The calcium membrane sensor and SCE reference electrode are placed in the standard solutions in the order 10^{-3}, 10^{-4}, and 10^{-2} M. The EMF readings (linear axis) are plotted against concentration (log axis). The EMF measurements are made under stirring. The unknown concentration is determined from the calibration graph. Previously, the unknown sample is brought into solution with dilute hydrochloric acid.

ii. *Potentiometric titration:*
   The electrode pair (calcium-ion-selective indicator and SCE reference) is placed in the sample solution (approximately 40 to 50 cm^3 concentration about 10^{-3} M at pH 11, adjusted with ammonia solution) and potentiometrically titrated with 10^{-2} M EDTA. The EMF is recorded as a function of the added titrant volume, and the end point corresponds to the maximum slope on the titration curve (borate buffer medium of pH 9.2 may also be used).

iii. *Pharmaceuticals assay by potentiometric titration:*
   An accurately weighed powder of sample of 400 to 500 mg from tablets, granules, coated pills, etc.) is carefully dissolved in approximately 30 cm^3 of distilled water in a 100-cm^3 beaker covered with a watch glass. After addition of 5 cm^3 of 0.5 M HCl solution, the

acidified solution is boiled for 5 min, then transferred to a 100-cm^3 volumetric flask and diluted with water to the mark. To 5.0 cm^3 of this solution is added 20 cm^3 of borate buffer solution of pH 9.2, and the solution is potentiometrically titrated with 0.05 M EDTA, following the same procedure as described previously.

4.4   Iron Salts

Iron is essential for the formation of hemoglobin and hence for the oxidative processes of living tissues. Compounds of iron, such as those listed in Table 4.5 are used in all forms of mycrocytic and hypochronic anemia of infants, anemia due to excessive or repeated hemorrhage, and anemia associated with infections, infestations, and malignant disease.

Membrane sensors selective to iron(II) and iron(III) are not yet commercially available, but the determination of these ions with ion-selective membrane sensors is not difficult because there are procedures using

Table 4.5   Iron Salts Assayed by Membrane Sensors

Compound	Formula/ (MM)	%Fe
Iron(III) ammonium citrate	This is a complex ammonium iron(III) citrate that contains 20.5–22.5% Fe.	20.5–22.5
Iron(II) fumarate	$C_4H_2FeO_4$ (169.9) $\begin{bmatrix} OOC-CH \\ \| \\ CH-COO \end{bmatrix}Fe$	32.87
Iron(II) gluconate	$C_{12}H_{22}FeO_{14} \cdot 2\,H_2O$ (482.2) $\begin{bmatrix} H\ \ H\ \ OH\ H \\ \|\ \ \ \|\ \ \ \|\ \ \ \| \\ HO-C-C-C-C-COO \\ \|\ \ \ \|\ \ \ \|\ \ \ \| \\ OH\ OH\ H\ \ \ OH \end{bmatrix}_2 Fe \cdot 2\,H_2O$	11.58
Iron(II) succinate	This is a basic salt that may be prepared by the interaction of sodium succinate and iron(II) sulfate in boiling aqueous solution; it contains 34.0–36.0% Fe, calculated on the dried substance.	34.0–36.0
Iron(II) sulfate	$FeSO_4 \cdot 7\,H_2O$ (278.0)	20.09

either $FeCl_4^-$-ion-selective sensors[27, 28] or indirect methods with other M^{z+} ion-selective membrane sensors ($M^{z+} = Ca^{2+} - Mg^{2+}$, Cu^{2+}, Hg^{2+})[29, 30] or a perbromate-ion-selective membrane sensor[31]. The electroactive material for tetrachloroferrate(III)-ion-selective sensor was the ion-association complex tetrachloroferrate–methyltricaprylylammonium in a PVC matrix.

A linear Nernstian response holds for this sensor over the iron(III) concentration range of 10^{-1} to 10^{-4} M for solutions containing a controlled total chloride ion concentration of 6 M.

The interereference of iron(III) with the solid-state copper(II)-ion-selective sensor (Orion, Model 94-29)[32] has been used to determine iron(III) by direct potentiometry.[29] Because copper sulfide from the membrane is continuously leached out, especially in concentrated iron solution, it is not advisable to use the electrode for measurement of high iron concentration. Reproducibility of the copper(II) membrane sensor can be improved by immersing in dilute acid (pH 2.0) or 1 M copper(II) sulfate solution for 1 h before use.

Analytical Procedures

i. *Direct potentiometry:*
   Three standard solutions (10^{-2}, 10^{-3}, and 10^{-4} M) of iron(III) are prepared by serial dilutions of 0.1 M iron(III) standard stock solution. The ionic strength and pH are kept constant ($I = 0.1$ M, adjusted with sodium perchlorate, pH = 2.0, adjusted with perchloric acid solution). The copper(II) membrane sensor (Orion, Model 94-29) and SCE reference electrode are placed in the respective standard solutions and EMF readings (linear axis) plotted against concentration (log axis). The EMF measurements are made under stirring and the unknown concentration is determined from the calibration curve.

ii. *Potentiometric titration:*
   The electrode pair ($FeCl_4^-$-ion-selective membrane indicator and SCE reference) is introduced into a sample solution (30 to 40 cm^3, concentration approximately 5×10^{-3} M containing 6 M total chloride [5 M LiCl + 1 M HCl] at pH 1.0) and titrated with 5×10^{-2} M EDTA solution. The EMF is recorded as a function of the added titrant volume. The end point corresponds to the maximum slope on the titration curve.

4.5 Magnesium Compounds

There is no ion-selective membrane sensor for magnesium, although progress may be expected in the future. There is a divalent cation electrode that consists of an electrode body and a replaceable pretested sensing module. The sensing module contains a gelled internal filling solution—a liquid ion-exchanger or organic amine that can form complexes with all divalent metal ions—and a membrane saturated with liquid ion-exchanger. The membrane separates samples from an electrode filling solution and is selective for divalent cations (e.g., Ca^{2+}, Mg^{2+}). If a complexing agent is used to mask all divalent metal ions, except magnesium, the divalent ion electrode (Orion, Model 93-32, which has replaced the earlier 92-33 divalent cation electrode) will be selective for magnesium ions in the presence of complexing agent. For example, ethyleneglycol-bis(2-aminoethylether)tetracetic acid (EGTA) makes a simple ion-exchanger separation of Mg^{2+} from Ca^{2+} possible.[33] In the presence of EGTA the calibration curve is linear to Mg^{2+} in the range 10^{-1} to 10^{-5} M. A pH of 7.0 ± 0.2 was selected for measuring the activity of magnesium by direct potentiometry.[34]

Table 4.6   Magnesium Compounds Assayed by Membrane Sensors

Compound	Formula/ (MM)	%Mg	Therapeutic category
Heavy magnesium carbonate	A hydrated basic magnesium carbonate of varying composition corresponding approximately to the formula $3\,MgCO_3 \cdot Mg(OH)_2 \cdot 4\,H_2O$	Varies with composition	Antacid and purgative
Light magnesium carbonate	A hydrated basic magnesium carbonate of varying composition corresponding approximately to the formula $3\,MgCO_3 \cdot Mg(OH)_2 \cdot 3\,H_2O$	Varies with composition	Antacid and purgative
Magnesium hydroxide	$Mg(OH)_2$ (58.32)	41.38	An antacid; by the formation of magnesium chloride in the stomach, also a mild saline laxative
Magnesium oxide	MgO (40.3)	60.32	Antacid and laxative
Magnesium sulfate	$MgSO_4 \cdot 7\,H_2O$ (246.5)	9.86	Saline purgative

It is also possible to determine magnesium from many compounds of pharmaceutical interest (see Table 4.6) by potentiometric titration at pH 9.7 adjusted with glycine–NaOH buffer with EDTA solution in the presence of a calcium-ion-selective membrane sensor.[35] Magnesium in the range 0.7 to 5 mg was determined semi-automatically with average errors of about 0.3%.

Except for magnesium sulfate, all the compounds listed in Table 4.6 must first be solubilized with dilute hydrochloric acid.

Analytical Procedures

i. *Direct potentiometry:*

Three standard solutions of magnesium (10^{-2}, 10^{-3}, and 10^{-4} M) are prepared by serial dilution of 0.1 M magnesium chloride solution. The ionic strength and pH are kept constant ($I = 0.1$ M, sodium nitrate, pH = 7.0 adjusted with diethanolamine and/or hydrochloric acid solution). The pair of electrodes (Orion, divalent indicator electrode, Model 92-32 or 93-32, and SCE reference electrode) is immersed in the standards in the order 10^{-3}, 10^{-4}, and 10^{-2} M. The EMF readings (linear axis) are plotted against concentration (log axis). The EMF measurements are made under stirring and the unknown concentration is determined from the calibration curve.

ii. *Potentiometric titration:*

The electrode pair (calcium-ion-selective indicator and SCE reference) is introduced into the sample solution (30 to 40 cm^3, concentration 10^{-3} M, pH 9.7 adjusted with glycine–NaOH buffer) and titrated with 10^{-2} M EDTA solution. The EMF is recorded as a function of the added titrant volume. The end point corresponds to the maximum slope on the titration curve.

4.6   Potassium Compounds

Potassium salts (see Table 4.7) are very important therapeutic agents, but they can be dangerous if they are improperly used. Many potassium-ion-selective membrane sensors were developed during the last two decades[36, 37] but the most successful sensor utilizes valinomycin, a macrocyclic substance from a group of depsipeptides that contains a 36-membered ring.[38] This antibiotic, produced by cultures of *Streptomyces fulfissimus*, forms complexes with alkali metal ions, the stabilities of which decrease in the order $Rb^+ > K^+ \gg Na^+ > Li^+$. The remarkable difference in the stabilities of the potassium and sodium complexes is the reason for the high selectivity of the potassium sensor ($k_{K,Na}^{pot} \sim 1 \times 10^{-4}$) and this is significant in the biomedical sciences.

Table 4.7   Potassium Compounds Assayed by Membrane Sensors

Compound	Formula/ (MM)	%K	Therapeutic category
Potassium acetate	$C_2H_3KO_2$ (98.14) CH_3COOK	39.84	Has been used in cardiac arrhythmias; used as an expectorant and diuretic.
Potassium bicarbonate	$KHCO_3$ (100.1)	31.06	Used for treating gastric hyperactivity
Potassium bromide	KBr (119.0)	32.85	Depressant of the central nervous system; more effective as sedative than as hypnotic
Potassium carbonate	K_2CO_3 (138.20)	56.58	Systemic alkalizer; diuretic
Potassium chloride	KCl (74.56)	52.44	Potassium replenisher
Potassium citrate	$C_6H_5K_3O_7 \cdot H_2O$ (324.4)	38.28	Employed principally to make urine alkaline in treatment of inflammatory conditions of the bladder to prevent crystalluria during treatment with certain sulfonamides
Potassium gluconate	$C_6H_{11}KO_7$ (234.3) $HOCH_2(CHOH)_4COOK$	16.69	Potassium replenisher
Potassium hydroxide	KOH (56.1)	69.69	A powerful caustic; 2.5% solution in glycerin used as a cuticle solvent
Potassium iodide	KI (166.0)	23.55	Antifungal; expectorant; source of iodine

Using the standard addition method, Nobile et al.[39] reported a relative standard deviation of 0.73 to 1.32% for the determination of potassium in commercial pharmaceutical formulations (vials, tablets, liophilized powder) with a potassium-selective membrane sensor (Beckman). The results compared well with those obtained by atomic absorption spectrometry. Direct potentiometric methods[40] as well as potentiometric titration method with sodium tetraphenylborate solution[41] were used with satisfactory results for potassium determination in electrolyte infusion solutions.

Analytical Procedures

i. *Direct measurement:*

Three standards (10^{-2}, 10^{-3}, and 10^{-4} M) are prepared by serial dilution of the 0.1 M potassium chloride stock solution. The ionic strength is kept constant at 0.1 M with calcium chloride solution. The electrode pair (potassium selective membrane with SCE reference electrode connected to the sample solution by a 1 M calcium chloride bridge) is immersed in the standards in the order 10^{-3}, 10^{-4}, and 10^{-2} M. The EMF readings (linear axis) are plotted against concentration (log axis). The EMF measurements are made under stirring, and the unknown concentration is determined from the calibration curve.

ii. *Known addition:*

To measure an unknown sample of potassium, the pair of electrodes (as before) is placed in 100 cm^3 of sample, and 2 cm^3 of 5 M calcium chloride solution (ionic strength adjuster) is added. After recording the stable reading of E_1, 10 cm^3 of standard solution that is about 10 times as concentrated as a sample concentration (prepared by diluting 0.1 M potassium chloride stock solution and with 2 cm^3 of 5 M calcium chloride added to each 100 cm^3 standard) is added to the sample. The solution is thoroughly stirred and the reading E_2 recorded. The value Q, corresponding to the change in potential ΔE ($\Delta E = E_2 - E_1$), is given in Table XII of Appendix 2 (sign " $-$ " is changed for sign " $+$ "). To determine the original sample concentration, Q is multiplied by the concentration of the added standard (Equation 4.7).

4.7   Sodium Compounds

Table 4.8 lists some sodium compounds of pharmaceutical interest that can be assayed by sodium membrane sensors.

Although the discovery of cation-responsive glass electrodes was one of the earliest of the many new developments in ion-selective membrane sensors,[42] only limited progress has been made in improving sodium-ion sensors. Much remains to be achieved, especially with regard to response time and selectivity with respect to H$^+$. Following the notable work of Eisenman and co-workers,[42] who examined the theoretical aspects of glass electrode response, Mattock[43, 44] investigated the behav-

Table 4.8   Sodium Compounds Assayed by Membrane Sensors

Compound	Formula/ (MM)	%Na	Therapeutic category
Sodium bicarbonate	$NaHCO_3$ (84.01)	27.37	Neutralizes acid secretions in the stomach
Sodium carbonate	Na_2CO_3 (106.0)	43.39	Used for the preparation of alkaline baths and of surgical chlorinated solutions
Sodium chloride	NaCl (58.44)	39.34	For maintaining osmotic tension of the blood and tissues
Sodium citrate	$C_6H_5Na_3O_7$ (258.1)	26.73	Used principally to make the urine alkaline in treatment of inflammatory conditions of the bladder to prevent crystalluria during treatment with certain sulfonamides
Sodium fluoride	NaF (42.0)	54.75	Used as dental caries prophylactic in children
Sodium iodide	NaI (149.9)	15.34	Antifungal; expectorant
Sodium lactate	$C_3H_5NaO_3$ (112.06)	20.52	For maintaining osmotic tension of the blood and tissues
Sodium phosphate	$Na_2HPO_4 \cdot 12\,H_2O$ (358.1)	6.42	Saline purgative
Sodium potassium tartrate	$C_4H_4KNaO_6 \cdot 4\,H_2O$ (282.2)	8.15	Saline purgative
Sodium sulfate	$Na_2SO_4 \cdot 10\,H_2O$ (322.2)	7.14	Saline purgative

ior of different sodium-ion-selective membrane electrodes made from EIL BH68, EIL BH104, and Corning NAS 11-18 glasses.

Although it is not possible to claim that the sensor glasses produced so far are the best possible, it is worthwhile to remember that the first conclusions on the role of aluminum oxide[45] were subsequently confirmed. Extensive comparative studies[46-48] revealed that aluminum oxide is by far the most suitable component for making sodium silicate–based glass sensitive to alkali. The optimum ratio of the three oxides in the $Na_2O–Al_2O_3–SiO_2$ three-component system was established by Eisenman et al.[49]

The principal interferents of sodium glass electrodes are hydrogen and silver ions; potassium ions interfere to a lesser extent (typically $k_{Na,K}^{pot} \simeq 10^{-3}$). Consequently, because silver ions are only rarely a sample constituent, sample pretreatment is usually restricted to pH adjustment to ensure pH > pNa + 3.

Neutral-carrier-based sensors have been designed for determining sodium ions.[25] Their discrimination over protons is usually considerably higher than for glass electrodes, whereas selectivities for sodium over potassium are comparable to those of glass electrodes.

Analytical Procedures

i. *Direct measurement:*

For measurements in units of moles per cubic decimeter, 10^{-2}, 10^{-3}, and 10^{-4} M standards are prepared by serial dilution of the 0.1 M sodium chloride stock solution. The ionic strength is kept constant with Orion ionic strength adjuster (ISA) whereby 2 cm^3 ISA are added for each 100 cm^3 standard solution. The electrode pair (sodium-ion-selective indicator electrode and double-junction reference electrode with 0.5% ammonium chloride solution in the outer chamber) is immersed in the standards in the order 10^{-3}, 10^{-4}, and 10^{-2} M. The EMF readings (linear axis) are plotted against concentration (log axis). The EMF measurements are made under stirring and the unknown concentration is determined from the calibration curve.

ii. *Known addition:*

To measure an unknown sample of sodium, the pair of electrodes (see the preceding text) is placed in 100 cm^3 of sample, and 2 cm^3 ISA is added. After potential equilibration, the EMF is recorded (E_1). A standard solution of about 10 times as concentrated as the sample concentration is prepared by diluting 0.1 M sodium chloride stock solution (2 cm^3 ISA is added to each 100 cm^3 standard), and 10.0 cm^3 of this standard is pipetted into the sample. The solution is thoroughly stirred and E_2 (in millivolts) is recorded. The value Q, which corresponds to the change in potential ΔE ($\Delta E = E_2 - E_1$), is given in Table XII of Appendix 2 (sign "$-$" is changed with sign

" + "). To determine the original sample concentration, Q is multiplied by the concentration of the added standard (Equation 4.7.).

References

1. E. W. Baum, *Anal. Chem.*, **42**, 110 (1970).
2. M. S. Frant and J. W. Ross, *Science*, **159**, 1553 (1966).
3. B. Jaselkis and M. K. Bandemer, *Anal. Chem.*, **41**, 855 (1969).
4. A. Hamola and R. O. James, *Anal. Chem.*, **48**, 776 (1976).
5. M. E. A. Cooper, J. Ballantine, and A. D. Woolfson, *J. Pharm. Pharmacol.*, **31**, 403 (1979).
6. A. K. Covington, Ed., *Ion-Selective Electrode Methodology*, CRC Press, Boca Raton, 1982.
7. J. W. Ross, J. H. Riseman, and J. A. Krueger, *Pure Appl. Chem.*, **36**, 473 (1973).
8. J. W. Ross, *Science*, **156**, 1378 (1967).
9. G. J. Moody and J. D. R. Thomas, *Selective Ion-Sensitive Electrodes*, Merrow, Watford, U.K. 1971.
10. V. V. Coşofreţ, *Membrane Electrodes in Drug-Substances Analysis*, Pergamon Press, London, 1982.
11. J. Koryta and K. Stulik, *Ion-Selective Electrodes*, 2nd ed., Cambridge University Press, Cambridge, 1983.
12. M. Mascini and A. Liberti, *Anal. Chim. Acta*, **53**, 202 (1971).
13. G. J. Moody, R. B. Oke, and J. D. R. Thomas, *Analyst*, **25**, 910 (1970).
14. G. H. Griffiths, G. J. Moody, and J. D. R. Thomas, *Analyst*, **97**, 420 (1972).
15. A. Craggs, G. J. Moody, and J. D. R. Thomas, *J. Chem. Educ.*, **51**, 541 (1974).
16. R. W. Cattrall and D. M. Drew, *Anal. Chim. Acta*, **77**, 9 (1975).
17. M. Ghosh, M. R. Dhaneshwar, R. G. Dhaneshwar, and B. Gosh, *Analyst*, **103**, 768 (1978).
18. R. W. Cattrall, D. M. Drew, and J. C. Hamilton, *Anal. Chim. Acta*, **76**, 269 (1975).
19. A. Ansaldi and S. I. Epstein, *Anal. Chem.*, **95**, 595 (1973).
20. D. Ammann, E. Pretsch, and W. Simon, *Tetrahedron Lett.*, **24**, 2473 (1972).
21. D. Ammann, E. Pretsch, and W. Simon, *Anal. Lett.*, **5**, 843 (1972).
22. D. Ammann, M. Gueggi, E. Pretsch, and W. Simon, *Anal. Lett.*, **8**, 709 (1975).
23. J. Růžička, E. H. Hansen, and J. C. Tjell, *Anal. Chim. Acta*, **67**, 155 (1973).
24. H. Freiser, Ed., *Ion-Selective Electrodes in Analytical Chemistry*, Vols. 1 and 2, Plenum Press, New York, 1978 and 1979.
25. D. Ammann, W. E. Morf, P. Anker, P. C. Meier, E. Pretsch, and W. Simon, *Ion-Selective Electrode Rev.*, **5**, 3 (1983).

26. Z. Pranjić-Anušic, *Acta Pharm. Jugoslav.*, **29**, 29 (1979).
27. R. W. Cattrall and C.-P. Pui, *Anal. Chem.*, **47**, 97 (1975).
28. R. W. Cattrall and C.-P. Pui, *Anal. Chim. Acta*, **78**, 463 (1975).
29. Y. S. Fung and K. W. Fung, *Anal. Chem.*, **49**, 497 (1977).
30. E. Hopîrtean, *Rev. Roum. Chim.*, **22**, 1385 (1977).
31. L. A. Lazarou and T. P. Hadjiioannou, *Anal. Lett.*, **11**, 779 (1978).
32. M. J. Smith and S. E. Manahan, *Anal. Chem.*, **46**, 836 (1973).
33. M. Marhol and K.-L. Cheng, *Anal. Chem.*, **42**, 652 (1970).
34. K.-L. Cheng and K. Cheng, *Mikrochim. Acta*, 385 (1974).
35. T. P. Hadjiioannou and D. S. Papastathopoulos, *Talanta*, **17**, 399 (1970).
36. M. S. Ionescu and V. V. Coşofreţ, *Rev. Chim. (Bucharest)*, **31**, 1005 (1980).
37. M. S. Ionescu and V. V. Coşofreţ, *Rev. Chim. (Bucharest)*, **31**, 1088 (1980).
38. L. A. R. Pioda, V. Stankova, and W. Simon, *Anal. Lett.*, **2**, 665 (1969).
39. L. Nobile, L. Benfenati, M. A. Raggi, I. Cavrini, and A. M. Di Pietra, *Pharmazie*, **44**, 66 (1989).
40. S. Zadeczky, D. Küttel, J. Havas, and L. Kecskes, *Acta Pharm. Hung.*, **48**, 131 (1978).
41. G. Peinhardt and J. Siemroth, *Pharmazie*, **38**, 33 (1983).
42. G. Eisenman, Ed., *Glass Electrodes for Hydrogen and Other Cations, Principles and Practice*, M. Dekker, New York, 1967.
43. G. Mattock, *Analyst*, **87**, 930 (1962).
44. G. Mattock, *Chimia*, **21**, 209 (1967).
45. B. Lengyel and E. Blum, *Trans. Faraday Soc.*, **30**, 461 (1934).
46. B. Lengyel and B. Csákváry, *Acta Chim. Acad. Sci. Hung.*, **25**, 370 (1960).
47. Z. Boksay, B. Csákváry, J. Havas, and M. Patkó, *Hung. Sci. Instrum.*, **41**, 41 (1977).
48. Z. Boksay, B. Csákváry, J. Havas, and M. Patkó, in *Ion-Selective Electrodes*, E. Pungor and E. Buzas, Eds., Akad. Kiadó, Budapest, 1978, p. 269.
49. G. Eisenman, D. Rudin, and J. U. Casby, *Science*, **126**, 831 (1957).

Chapter 5

ORGANICS

5.1 Adenosine Phosphate and Adenosine Triphosphate

$C_{10}H_{14}N_5O_7P$ (MM = 347.2)

$C_{10}H_{16}N_5O_{13}P$ (MM = 507.2)

Therapeutic category: vasodilator; co-enzyme

Discussion and Comments

Papastathopoulos and Rechnitz[1] evaluated an electrode for 5′-adenosine monophosphate (5′-AMP) that uses a layer of suspended 5′-adenylic acid deaminase enzyme (AMP deaminase) in conjunction with an ammonia electrode (Orion, Model 95-10). The substrate is selectively deaminated to inosine 5′-monophosphate (5′-IMP) according to Equation 5.1:

$$5'\text{-AMP} \xrightarrow[\text{H}_2\text{O}]{\text{AMP deaminase}} 5'\text{-IMP} + \text{NH}_3 \qquad (5.1)$$

Liberated ammonia gives rise to a constant potential, linearly related to the logarithm of the substrate concentration in the sample solution. This enzyme sensor was assembled by placing the enzyme between an outer circular cellophane dialysis membrane and the gas-permeable membrane of the ammonia electrode. The sensor was preconditioned for at least 3 h in 0.05 M TRIS–HCl buffer (pH 7.5) and was also stored in this buffer when not in use. The sensor showed a linear response to adenosine phosphate over the range 1.5×10^{-2} to 8.0×10^{-5} M (slope = −46 mV decade^{-1}). The time required for a steady-state potential to be reached depended on the substrate concentration: in the 10^{-4} to 10^{-3} M concentration range, response times of the order of 6 min were found, whereas at concentration approaching 10^{-2} M, response times were shortened to about 2 min.

Because sensor response to some possible interferences (5′-ADP, 5′-ATP, 3′,5′-cyclic AMP, adenine, and adenosine) was similar to that in buffer alone (0.05 M TRIS–HCl, pH 7.5) and, moreover, was independent of interferent concentration over the 10^{-2} to 10^{-4} M range, it is clear that the sensor presented a very high selectivity for 5′-AMP over other nucleotides.

A tissue-based membrane sensor consisting of mouse small intestine mucosal cells coupled to the same ammonia-gas-sensing membrane electrode was used as a model system to study the possibility of selectivity enhancement for such biocatalytic membrane sensors.[2] This tissue-based sensor exhibits response to adenosine and also to 5′-AMP and other adenosine-containing nucleotides when a buffer system containing 0.2 M TRIS–HCl and 0.02% sodium azide at pH 8.2 was used. The linear ranges for adenosine and 5′-AMP were similar to each other and were larger than those for the ADP and ATP cases. The limit of detection was lowest for adenosine at 1.9×10^{-5} M, whereas response times for adenosine, 5′-AMP, ADP, and ATP were 11, 11, 18, and 32 min, respectively. The longer response times for the ADP and ATP nucleotides suggest that under these conditions there is less deaminating activity in the biocatalytic layer for these substrates than for the adenosine and 5'-AMP cases.[2] The sensor became adenosine-selective when a buffer system composed of 0.1 M TRIS–HCl, 0.2 M K_2HPO_4, and 0.02% sodium azide at pH 9.0 and an internal electrolyte containing 0.1 M NH_4Cl and 0.6 M NaCl were used in conjunction with the above sensor.

A liquid membrane incorporated with the lipophilic macrocyclic polyamine, 15-hexadecyl-1,4,7,10,13-pentaazacyclohexadecane ([16]ane N_5), for a potentiometric sensor particularly sensitive to ATP^{4-} was prepared by Umezawa et al.[3,4] The sensor is a conventional PVC type using DOP or o-NPOE as a plasticizer.

([16]ane N_5)

$R = — C_{13}H_{33}$

Figure 5.1 shows the calibration curves for the ATP^{4-} ion in different buffer solutions. It is seen that a Nernst-type response holds between 10^{-3} and 10^{-7} M ATP^{4-} with a slope of about 13 mV decade^{-1} in HEPES buffer, which is reasonably close to the theoretical value of 15 mV decade^{-1} based on the Nernst equation.

The calibration curves for ADP^{3-} and AMP^{2-} gave slopes much less than the corresponding theoretical values, which are in accordance with the respective values of association constants for these anions.

The change in potentials in solutions containing a constant amount of ATP^{4-} with varying concentrations of interfering ions such as ADP^{3-}, AMP^{2-}, or some other anions was within 0.1 to 0.5 mV and therefore the interference with these ions was negligible.

Analytical Procedure

Working AMP substrate solutions in 0.05 M TRIS–HCl buffer (pH 7.5) and ATP substrate solutions in 0.01 M HEPES buffer (pH 6.7) are prepared and stored under refrigeration. Three standards are necessary for each substrate determination by direct potentiometry; 10^{-3} and 10^{-4} M solutions are prepared by successive dilutions from the respec-

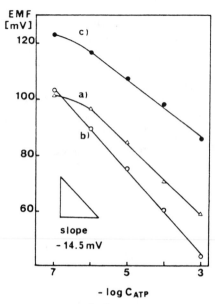

Figure 5.1 Potential response of the [16]ane N_5 liquid membrane electrode for ATP polyanions in different buffer solutions: (a) in 0.01 M MOPSO buffer, pH 6.8; (b) in 0.01 M HEPES buffer, pH 6.7; (c) pH adjusted to 6.9 by adding NaOH solution into the Na_2H_2ATP solution (no buffers added). (Reprinted with permission from Umezawa, Y., Kataoka, M., Takami, W., Kimura, E., Koike, T., and Nada, H., *Anal. Chem.*, 60, 2392, 1988. Copyright 1988 American Chemical Society.)

tive 10^{-2} M stock solution. The standard solutions (10^{-2}, 10^{-3}, and 10^{-4} M) of each substrate are transferred into 150-cm^3 beakers containing Teflon-coated stirring bars. The enzyme sensor sensitive to 5'-adenosine phosphate and the liquid-membrane-type sensitive to adenosine triphosphate, respectively, are immersed in conjunction with a reference electrode successively in the appropriate standards. The graphs E (in millivolts) vs. log[AMP] and E (in millivolts) vs. log[ATP], respectively, are plotted and the substrate concentration is determined from the respective graph.

5.2   Alkaloids

The alkaloids listed in Table 5.1 were determined with various ion-selective membrane sensors. Especially their halide salts are extensively used in medical practice, and their determination in the presence of other compounds by conventional analytical methods is laborious and time-consuming. Their determination in multi-component drugs is even more difficult, because the alkaloids are usually present in smaller concentrations than the other components.

A variety of indicator membrane sensors have been used to determine the alkaloids in Table 5.1. In most cases sodium tetraphenylborate (NaTPB) is used as titrant in respective potentiometric titrations. Vytras[5-7] published three reviews on the use of ion-selective membrane sensors for titration using NaTPB. If either the alkaloid determined or the titrant (or both) has an adequate lipophilic character, the assays involve precipitation of the respective ion pair with low water solubility.

$$Q^+ + X^- \leftrightarrows QX \tag{5.2}$$

Various types of membrane sensors have been suggested for monitoring such titrations.[7] The only commercially available membrane sensors that have been used are the Crytur 19-15 valinomycin electrode[8] and fluoroborate electrode (Orion, Model 93-05).[9] Kálmán et al.[10,11] and Yao et al.[12] determined some alkaloid halides (chlorides and bromides) by potentiometric titration with silver nitrate solution (e.g., 10^{-2} M concentration) and chloride or bromide sensors. The methods showed good precision and accuracy. The presence of various amounts of boric acid in some pharmaceuticals did not affect the determinations.[12]

Coşofreţ et al.[13] used a silver(I)-ion-selective membrane sensor, obtained by impregnating a graphite rod attached to the end of the Teflon tube with silver(I) chelate of 1-(2′,3′,5′-tri-O-benzoyl-β-D-ribofuranozil)-4-thioxo-5-methyl-thio-6-azauracil dissolved in chloroform, for the determination of scopolamine hydrobromide and N-butylscopolammonium bromide in pharmaceutical preparations such as tablets and injectable aqueous solutions. The results were similar to those obtained with a commercial bromide membrane sensor (Radelkis, type OP-Br-7111-D).

Besides commercially available membrane sensors, the coated-wire type of electrode became attractive because of its simple fabrication and manipulation (different materials can serve as conductors: aluminum, platinum, copper, graphite, etc.). During the conditioning time, the membrane solvent (plasticizer) is gradually saturated with an ion pair based on extraction principles; its concentration is given by

$$[QX]_{\mathrm{org}} = K_{\mathrm{ex}}(QX)[Q^+]_{\mathrm{aq}}[X^-]_{\mathrm{aq}} = K_{\mathrm{ex}}(QX)K_{\mathrm{so}}(QX) \tag{5.3}$$

where $K_{\mathrm{ex}}(QX)$ and $K_{\mathrm{so}}(QX)$ are the stoichiometric extraction constants and the solubility product, respectively, of the ion pair (QX) involved. Selectivity of such sensors depends on the extractibility of the ion pair involved and increases with molecular weight (i.e., with increasing lipophilicity) of the ions.

The following salts of alkaloids from various pharmaceuticals have been titrated with 5×10^{-2} M sodium tetraphenylborate solution:

Table 5.1   Alkaloids Assayed by Various Membrane Sensors

Alkaloid	Formula (MM)	Therapeutic category
Anisodamine	$C_{17}H_{23}NO_4$ (305.4)	Anticholinergic
Atropine	$C_{17}H_{23}NO_3$ (289.4)	Anticholinergic
Berberine	$C_{20}H_{18}NO_4^+$ (396.4)	Bitter stomachic; antibacterial; anti-malarial; antipyretic
N-Butylscopolammonium bromide	$C_{21}H_{30}BrNO_4$ (440.4)	Antispasmodic
Caffeine	$C_8H_{10}N_4O_2$ (194.2)	Central stimulant

Table 5.1 Continued

Alkaloid	Formula (MM)	Therapeutic category
Cinchonine	$C_{19}H_{22}N_2O$ (294.4)	Antimalarial

Cocaine	$C_{17}H_{21}NO_4$ (308.4)	Topical anesthetic

Codeine	$C_{18}H_{21}NO_3$ (299.4)	Narcotic; analgesic; antitussive

Ethylmorphine	$C_{19}H_{23}NO_3$ (313.4)	Chemotic; also used as analgesic and antitussive

Homatropine (novatropine)	$C_{16}H_{21}NO_3$ (275.3)	Mydriatic; cycloplegic anticholinergic

Table 5.1 Continued

Alkaloid	Formula (MM)	Therapeutic category
Morphine	$C_{17}H_{19}NO_3$ (285.3)	Narcotic; analgesic

Nicotine	$C_{10}H_{14}N_2$ (162.2)	Ectoparasiticide

Papaverine	$C_{20}H_{21}NO_4$ (339.4)	Smooth muscle relaxant; cerebral vasodilator

Physostigmine	$C_{15}H_{21}N_3O_2$ (275.3)	Cholinergic, miotic

Pilocarpine	$C_{11}H_{16}N_2O_2$ (208.3)	Cholinergic (ophthalmic)

Quinidine	$C_{20}H_{24}N_2O_2$ (324.4)	Cardiac depressant (antiarrhythmic)

Table 5.1 Continued

Alkaloid	Formula (MM)	Therapeutic category
Quinine	$C_{20}H_{24}N_2O_2$ (324.4)	Antimalarial

| Scopolamine | $C_{17}H_{21}NO_4$ (303.4) | Anticholinergic |

| Strychnine | $C_{21}H_{22}N_2O_2$ (334.4) | For destroying rodents and predatory animals and for trapping fur-bearing animals |

| Yohimbine | $C_{21}H_{26}N_2O_3$ (354.4) | Adrenergic blocker |

homatropine bromide (1 and 2%), pilocarpine chloride (1, 2, and 3%), ethylmorphine chloride (1 and 3%), and cinchonine chloride (0.2%), all the salts being dissolved in 2% boric acid. The membrane of the aluminum coated-wire indicator sensor was plasticized by 2,4-dinitrophenyl-n-octyl ether (o-NPOE). The S-shaped potentiometric titration curves with well defined end points were recorded for all samples.

A liquid-membrane sensor that can be used in potentiometric precipitation titrations with NaTPB was described by Christopoulos et al.[14] The liquid ion-exchanger is tetrapentylammonium tetraphenylborate dissolved in 4-nitro-m-xylene. This sensor exhibits near-Nernstian response to the TPB$^-$ anion in the range 5×10^{-6} to 3×10^{-4} M with a slope of about 51 mV decade^{-1}. Simple potentiometric titrations were described for the determination of various alkaloids (usually 5 to 25 μmol) with 10^{-2} M NaTPB. In most of the cases the precision of the method was better than 1%.

Several alkaloids have been analyzed by potentiometric methods using a picrate sensor[15] as indicator electrode. The method is based on the formation of insoluble alkaloid picrates. Both versions (direct potentiometry or potentiometric titrations) are sensitive, rapid, fairly accurate, and simple, and they were employed successfully for the determination of papaverine and quinine in pharmaceutical preparations. They can also be applied to other alkaloids that form insoluble picrate salts. Atropine, codeine, and morphine could not be determined in this way because of the great solubility of the picrate salts.

PVC membrane sensors for anisodamine, N-butyl-scopolammonium bromide and homatropine (all these alkaloids belong to a group of constitutionally related alkaloids occurring in different Solanaceae, especially in *Hyoscyamus niger*) were obtained by using the respective TPB salts as electroactive materials.[16] All sensors showed relatively large linear ranges with near-Nernstian slopes (10^{-2} to 1.3×10^{-5} M for anisodamine, 10^{-2} to 10^{-5} M for N-butylscopolammonium and 1.7×10^{-2} to 6.3×10^{-5} M for homatropine) and short response times (less than 20 s in 10^{-2} to 10^{-4} M solutions).

Generally, inorganic salts did not interfere with any of the three membrane sensors. It was observed that procaine interferes with the anisodamine sensor and it is also a potent interferent for N-butylscopolammonium cation. A similar selectivity pattern was found, as anisodamine and N-butylscopolammonium have similar structures and hence similar electrode behavior. All three sensors have been used for the determination of the respective compounds in injectable solutions or eye drops with standard deviations less than 1%.

Shen et al.[17] found that both anisodamine-TPB and anisodamine-dipicrylamine electroactive materials used for electrode preparation are highly affected by chlorpheniramine ($k_{A,Chl}^{pot} = 40.8$ and 4.1, respectively). The effects of different active materials, solvent and internal reference systems, and other factors on the performances of anisodamine membrane sensors have been discussed in detail. It has been found that the use of di-(2-ethylhexyl)-o-phthalate and dinonyl-o-phthalate as solvents (plasticizers) gave higher electrode slope for a content of 0.5% electroactive material. Both sensors, having linear responses in the range 10^{-1} to 10^{-5} M at pH 3.0 to 7.5 with very short response times (5 s in $> 10^{-2}$ M), were used to assay various anisodamine samples by direct potentiometry as well as for the determination of the pK_a value of anisodamine.

Various ion-pair complexes were proposed as electroactive materials for atropine membrane sensors. These were used both for liquid-membrane sensors (e.g., atropine–reineckate, atropine–5-nitrobarbiturate, atropine–picrolonate, atropine–tetraphenylborate) and PVC types (e.g., atropine–reineckate, atropine–tetraphenylborate, atropine–dipicrylamine, atropine–HgI_4^{2-}).[18-21]

With 5×10^{-3} to 10^{-2} M atropine–reineckate (reineckate = $[Cr(NH_3)_2(SCN)_4]^-$) solution in benzyl alcohol as liquid membrane (an Orion liquid-membrane barrel, Model 92, was used for sensor assembly; reference solution: equal volumes of 0.02 M atropine hydrochloride and 0.1 M KCl), the sensor showed Nernstian response and stable potential readings. Atropine-sensitive membrane prepared by incorporating 10% atropine–reineckate in PVC and plasticized with dioctylphthalate gives similar results (in both cases the linear response range was 10^{-2} to 5×10^{-6} M atropine).[19]

The results obtained with liquid-membrane sensors containing 5-nitrobarbiturate and picrolonate, respectively, as counter-ions, shows that atropine–5-nitrobarbiturate in n-octanol and atropine–picrolonate in p-nitrotoluene also gave fairly stable and sensitive membranes for potentiometric measurements of atropine. The linear response range for these electrodes was 10^{-2} to 10^{-5} M with an average slope of 56.5 $\pm$ 1 mV decade^{-1}. The useful lifetime of both sensors was about two months.

The responses of all these atropine sensors are not noticeably affected by the presence of most common cations and anions; they suffer negligible interference from many basic compounds (aminobenzoic acid, aminopropanol, ethanolamine, diethylamine, piperidine, glycine, urea, etc.) but respond, however, to alkaloids like nicotine, caffeine, strychnine, etc. The responses of membrane sensors are practically unaffected by changes in pH over the range 3.5 to 8.5.

For the determination of atropine, by direct potentiometry, in the range 1 to 200 μg cm^{-3}, an average recovery of 98.7% (standard deviation 1.8%, $n = 12$) was reported.[20] Determination of atropine in some pharmaceutical samples (injections, eye drops, eye ointments) containing 0.1 to 1% atropine was made after a sample extraction or dilution step (average recovery of the nominal values was 98.8%; standard deviation, 1.7%). The results were in good agreement with those obtained by both official USP and BP methods.

Diamandis et al.[18] reported liquid-membrane sensors for atropinium and novatropinium by using as liquid ion-exchangers atropine–tetraphenylborate and novatropine–tetraphenylborate, respectively, dissolved in 2-nitrotoluene. An Orion liquid-membrane electrode body (Model 92) was used as sensor assembly with a Millipore Teflon membrane. The internal reference solution was 10^{-2} M atropine sulfate–0.1 M NaCl for the atropine membrane sensor and 10^{-2} M novatropine sulfate–0.1 M NaCl for the novatropine membrane sensor. Both sensors were conditioned by soaking in 10^{-2} M atropine or novatropine solution, as appropriate, for 24 h before use and were stored in the same solution when not in use. Their operational lifetime was about one month. The atropine membrane sensor showed rapid and near-Nernstian response in the 10^{-2} to 3×10^{-5} M range over the pH range 2 to 8.5, whereas the novatropine membrane sensor showed near-Nernstian response in the 10^{-2} to 3×10^{-6} M range at pH 2 to 10. Other alkaloids interfere in the response of both sensors.

Amounts of atropine in the range 15 to 900 μmol were determined with an average error of about 2%. Both alkaloids were evaluated from some pharmaceutical preparations (capsules and tablets) and there was satisfactory agreement between the results obtained by the proposed methods and the official methods.

Some Chinese papers describe the construction and analytical characterization of berberine membrane sensors. Berberine tetraphenylborate[22-24] and berberine tetra(m-methylphenyl)borate[22] were studied as possible electroactive membranes for berberine–liquid sensor or PVC types. The liquid-membrane sensors showed Nernstian response over the berberine concentration range from 1×10^{-3} to 1×10^{-6} M in the pH range of 2 to 8 with an average slope of 57 ± 1 mV decade^{-1} (limit of detection 5×10^{-7} M). The PVC membrane sensor (dibutylphthalate as plasticizer) also showed near-Nernstian response down to 10^{-6} M over a pH range 6.2 to 11.5. All berberine membrane sensors showed fast response, good reproducibility, and stability. Direct potentiometry as well as potentiometric titrations with NaTPB solution were performed with good results for berberine determination (average recovery 100.5%).

A carbon-coated PVC sensor,[25] preconditioned in berberine solution, presented a large linear response range (10^{-3} to 8×10^{-7} M) with a detection limit of 5×10^{-7} M (slope 57.3 mV decade^{-1} at 23° C). It was also used for berberine assay by direct potentiometry (average recovery 100.3%, standard deviation 1.8%).

Potassium iodomercurate (K_2HgI_4) has been used as titrant in conjunction with a copper-coated wire electrode, for the determination of berberine and other organic bases. A large potential jump near the equivalence point made the electrode useful for determination of berberine in microgram-range amounts.[26]

A berberine ion-selective field effect transistor (ISFET) sensor[23] was obtained by using berberine tetraphenylborate as an electroactive material. The membrane was coated on a platinum wire connected to a MOSFET. The berberine–ISFET showed near-Nernstian response in the range 10^{-2} to 10^{-6} M over pH 3.5 to 11.0 with a slope of 57 ± 1 mV decade^{-1}. It was found that only cetyltrimethylammonium, among the ions tested, interfered in the sensor response.

A liquid-membrane sensor for caffeine, prepared from a solution of caffeine–picrylsulfonate ion-pair complex in 1-octanol, was developed by Hassan et al.[27] An Orion liquid-membrane barrel (Model 92) with an Orion 92-05-04 porous membrane was used for sensor construction. The liquid ion-exchanger was 10^{-2} M caffeine–picrylsulfonate and the reference solution was a mixture of 10^{-2} M caffeine hydrochloride and potassium chloride. The sensor exhibits Nernstian response in the range 10^{-2} to 10^{-6} M caffeine with a cationic slope of 59 mV decade^{-1}. The sensor has a wide pH working range (pH 5.5 to 9.5), fast response time (20 to 90 s), stable response for at least 30 days, and high selectivity for caffeine in the presence of many organic bases and inorganic salts. Some alkaloids (nicotine, ephedrine, quinine, brucine, and strychnine) interfere only when present at concentration levels at least 10 times greater than those for caffeine. Theobromine and theophyllin seriously interfere ($K_{ij}^{pot} > 7$) and should be removed before caffeine assay. Caffeine solutions at the concentration range of 0.6 to 1000 μg cm^{-3} were determined by direct-potentiometry–calibration-curve methods, as well as by the known-addition technique. The results obtained for 15 samples, each in triplicate, showed an average recovery of 99.5% and a mean standard deviation of 1.4% (direct potentiometry) and 99.5% with 1.2% standard deviation (known addition). Caffeine in some pharmaceutical analgesic preparations (tablets, suppositories) was determined by the sensor method as well as by an official USP method. The results obtained showed an average recovery of 100.2% (mean standard deviation of 1.8%).

A cinchonine-selective liquid-membrane sensor containing cinchonine–picrolonate in nitrobenzene as electroactive material was proposed by Yao et al.[28] It was observed that at pH 6 to 7, the sensor responds mainly to the monoprotonated cinchonine, giving a linear response over the concentration range from 10^{-2} to 2×10^{-5} M with a slope of 54 to 57 mV decade^{-1}, whereas at pH 2.2 it responds mainly to the diprotonated species, with a linear response range from 10^{-1} to 5×10^{-4} M and a slope of 28.5 mV decade^{-1} (a similar behavior was observed for a quinidine-selective membrane sensor[29]). Atropine, strychnine, and procaine interfere significantly in the electrode response but Bu$_4$N$^+$, colchicine, caffeine, pilocarpine, theophyllin, and other common inorganic ions do not. Direct potentiometric determination of cinchonine samples down to the parts-per-million level was performed with an average recovery of 99.6% and a standard deviation of 1.8%. Potentiometric titrations were also used for cinchonine assay at the milligram-per-cubic-centimeter level (the average recovery was 100.1% and the standard deviation 1.5%).

A coated-wire cocaine-selective membrane sensor, based on dinonylnaphthalene sulfonic acid, obtained by Cunningham and Freiser[30] has a detection limit of $10^{-5.5}$ M. The calibration curve for this sensor was found to be reasonably reproducible from day to day, provided that the sensor was soaked in 10^{-2} M, pH 4.0 (acetate buffer) between calibrations. After exposure to strong interferences, however, the calibration curve shifted by several millivolts so that any further measurements would result in erroneously high results. The original response was restored by keeping the sensor in 10^{-3} M cocaine cation for several hours, followed by soaking overnight in the buffer solution.[30] A similar phenomenon was observed in the case of quinidine plastic-membrane sensors.[29]

The liquid ion-exchanger for a nicotine-sensitive membrane sensor described by Efstathiou et al.[31] is nicotine hydrogen tetra(m-chlorophenyl)borate $((NicH^+)(TCPB^-))$ dissolved in o-nitrotoluene. This sensor exhibits a near-Nernstian response to $NicH^+$ from 8×10^{-2} to 10^{-5} M, in the pH range 4 to 7.

The construction and analytical applications of nicotine–tetraphenylborate coated graphite membrane and PVC-membrane selective sensors were described by Dai et al.[32] Both sensors exhibit near-Nernstian response to nicotine from 10^{-4} to 10^{-1} M in the pH range of 6 to 7 and are sufficiently selective for the determination of nicotine samples. The coated-membrane sensor potentiometric method is relatively simple and convenient and gives satisfactory results.

A pilocarpine-selective liquid-membrane sensor was constructed for use in pharmaceutical analysis.[33] The active substance for preparation of liquid membranes was obtained by treatment of pilocarpine with NaTPB or ammonium thiocyanatochromate. The Nernstian response was in the range 10^{-1} to 10^{-6} M (pH = 3 to 6) with a slope of 57 mV decade^{-1}, and the response time was < 30 s at 10^{-1} to 10^{-3} M. Procaine and atropine interfered in the sensor response ($k_{ij}^{pot} = 9.57$ and 4.87, respectively), whereas other substances tested did not. The pilocarpine recovery was 97.7 to 102.7% for the direct potentiometric method and 99.5 to 101.7% for the potentiometric titration.

A quinidine plastic-membrane sensor has been reported by Coşofreţ and Buck.[29] The electroactive material of the membrane is quinidine tetraphenylborate salt; 2-nitrophenyloctyl ether has been found to be the best plasticizer for the PVC membrane (the membrane composition was

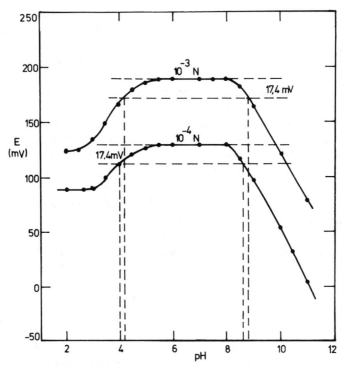

Figure 5.2 Effect of pH on quinidine membrane sensor at two different concentrations of quinidine sulfate solutions. (Reproduced from Coşofret, V. V. and Buck, R. P., *J. Pharm. Biomed. Anal.*, 3, 123, 1985. With permission.)

QdTPB 3.2%, 2-NPOE 64.5%, and PVC 32.3%). The sensor exhibits rapid and near-Nernstian response in the range 2×10^{-2} to 7×10^{-5} N quinidine sulfate over the pH range 6 to 8. In an acidic medium the sensor responds to diprotonated quinidine species (QdH_2^{2+}). From the E(mV)–pH curves (Figure 5.2) basicity constants K_{b_1} and K_{b_2} of quinidine could be evaluated.

Using the pH-dependent equilibria,

$$QdH_2^{2+} \underset{H^+}{\overset{K_{a_2}}{\rightleftharpoons}} QdH^+ \underset{H^+}{\overset{K_{a_1}}{\rightleftharpoons}} Qd \qquad (5.4)$$

$$\textbf{A} \qquad\qquad \textbf{B} \qquad\quad \textbf{C}$$

where pK_{a_1} is the pH at which $[\textbf{B}] = [\textbf{C}]$ and pK_{a_2} is the pH at which $[\textbf{A}] = [\textbf{B}]$. The pK_a values determined are $pK_{a_1} = 8.73$ and $pK_{a_2} = 4.07$ (each represents the mean of two values obtained from both curves in Figure 5.2). The corresponding values of basicity constants are $K_{b_1} = 5.37 \times 10^{-6}$ and $K_{b_2} = 1.18 \times 10^{-10}$; these are in good agreement with previously reported values. The sensor proved useful in potentiometric determinations of quinidine, both by direct potentiometry and by potentiometric titrations. The results of the potentiometric analyses of quinidine sulfate and quinidine gluconate in tablets were obtained with average standard deviations of 0.7 and 1.7% for potentiometric titration method and standard addition method, respectively. The bulk of the excipient did not show any interference in the sensor response.

The fabrication and performance characteristics of PVC-quinine-selective membrane sensors based on the ion-association complex of quinine with tetraphenylborate were described by Anzai et al.[34] Over the range of 10^{-1} to 10^{-5} M, the sensor exhibited a Nernstian-type response with a slope of 56.7 and 56.3 mV decade^{-1} for 10^{-2} M and 10^{-3} M internal solutions, respectively (response time < 10 s for concentrations higher than 10^{-5} M). The authors[34] studied the effect of membrane thickness (dibutylphthalate as plasticizer) on the potentiometric response. The 0.1-mm membrane gave the maximum response of 63.7 mV decade^{-1} (for 10^{-3} M internal solution). A slightly reduced response was observed with the thicker membranes; the sensors with membranes thicker than 0.40 mm showed relatively long response times. Considering both the response and the mechanical strength of the membrane, a 0.21-mm membrane was considered to be suitable for the sensor construction (slope 56.3 mV decade^{-1}). The electrode potentials were little affected by pH in the range of pH 5.3 to 7.3 (for 10^{-3} and 10^{-2} M solutions). The deterioration of the response in the higher and lower pH regions can be explained in terms of the formation of unprotonated and diprotonated species of quinine. The sensors showed good selectivity for quinine relative to several inorganic and organic ions, although the response was interfered with by alkaloids (strychnine, yohimbine, quinidine, and papaverine). Direct potentiometry and potentiometric titrations with 3.3×10^{-3} M NaTPB solution were used for the determination of quinine concentration.

Yao and Liu[35] made a detailed study of variable-valency quinine-selective sensors (quinine–TPB as electroactive material in PVC or nitrobenzene) and the pH effect on sensor linearity slope. They showed that the sensor response toward quinine solutions of different pH values cannot be described simply by using the Nernst equation or the Nicolsky–Eisenman equation. An empirical S–pH relation for a quinine sensor was formulated and acidity constant values, K_{a_1}, for quinine, thiamin, and cinchonine were determined using the sensor. It was found that for homologue membrane solvent the liquid-membrane sensor response slope increases with the dielectric constant of the membrane solvent. The sensor was also proposed for the direct potentiometry and potentiometric titration of quinine.

Liquid-membrane sensors sensitive to scopolamine$^+$ and N-butylscopolammonium$^+$ cations were also described[36]. The following electroactive materials were used: scopolamine–TPB, scopolamine–

dipicrylaminate, scopolamine–picrolonate, and N-butylscopolam-monium–TPB. In all cases nitrobenzene was used as solvent for the respective liquid membrane (10^{-3} M concentration). The basic construction of the sensors has been described elsewhere[37] and consists of impregnating the support material (a graphite rod) with the respective liquid membrane. All sensors exhibited a linear response to the respective organic cation over the range 10^{-1} to 10^{-4} M, and no interference was observed from the common ingredient substances always present in pharmaceutical preparations. The response times of the liquid-membrane sensors were fast, being nearly instantaneous at higher concentrations and requiring less than 3 min with a 10^{-4} M scopolamine hydrobromide or N-butylscopolammonium bromide, respectively. All sensors can be used in the potentiometric determination of the respective alkaloid in pharmaceutical preparations by direct potentiometry and by potentiometric titration. The last method is recommended.

A strychnine membrane sensor, sensitive and selective enough for determining strychnine down to the parts-per-million range was described by Hassan and Elsayes[38]. The sensor is based on a liquid membrane of a strychnine–picrolonate ion-pair complex in nitrobenzene. An Orion liquid-membrane barrel (Model 92) was used for sensor assembly with an Orion 92-81-08 porous membrane to separate the organic phase (5×10^{-3} to 5×10^{-2} M) from the test solution. The membrane sensor has a near-Nernstian response with an average cationic slope of 57 mV decade^{-1} for concentrations ranging from 5×10^{-2} to 5×10^{-5} M in potassium sulfate background and a slope of 52 mV decade^{-1} for 5×10^{-2} to 7×10^{-7} M in water. In the pH range 2 to 7 no great change of membrane potential was observed and the potential difference was only of the order of 10 mV within this range for 10^{-2} to 10^{-7} M concentration. The strychnine base in the aqueous test solution precipitates at pH > 7.

Among possible applications of the sensor are the potentiometric titrations of strychnine with either picrolonic acid or sodium tetraphenylborate (average recovery 99.1% and mean standard deviation 1.2%) and direct potentiometric determination of strychnine (average recovery 101.3% and the mean standard deviation 2.4%) for samples down to 5 ppm.

New strychnine-membrane sensors have been proposed by Chinese scientists.[39, 40] By using strychnine–tetra(m-methylphenyl)borate ion-pair complex as electroactive material in a liquid-membrane type, a sensor with a near Nernstian response down to 5×10^{-6} M strychnine with a slope of 56 mV decade^{-1} and relatively high selectivity over many inorganic and organic cations was obtained.[39] The membrane sensor proved useful for the potentiometric titration of strychnine.

A PVC-membrane sensor containing strychnine–tetraphenylborate as ion-pair complex (DOP as plasticizer) gave a linear response to strychnine in the range 5×10^{-2} to 10^{-5} M with a detection limit of 7×10^{-6} M. No large differences in potential measurements were observed in the range of pH 4 to 6. The sensor, presenting a good selectivity over alkaloids such as atropine, cinchonine, caffeine, nicotine, and berberine, proved useful for strychnine determination by potentiometric titration.[40]

Analytical Procedures

i. *General method, applied to alkaloids for which membrane sensors are available (homemade; see also the preceding text):*
A stock solution of 10^{-2} M alkaloid is prepared by dissolving the respective compound (as chloride, bromide, sulfate, etc.) in distilled water. The pH of the solution is adjusted to the recommended value with NaOH and/or HCl solution or buffer solution (phosphate, citrate, borax, etc.); 10^{-3} and 10^{-4} M alkaloid solutions are obtained from the stock solution by successive dilutions. The sensors (alkaloid-sensitive and SCE) are placed in the standard solutions and EMF readings (linear axis) plotted against concentration (logarithmic axis). The sample concentration is determined from this graph.

ii. *General method, applied to alkaloids that form insoluble ion-pair complexes with sodium tetraphenylborate (see also the preceding text):*
The electrode pair (alkaloid-, tetraphenylborate-, fluoroborate-, or coated-wire-membrane selective as indicator and SCE as reference) is introduced into the sample solution (30 to 40 cm^3; concentration, approximately 5×10^{-3} M, pH adjusted as appropriate—usually between 5 and 6) and titrated with 5×10^{-2} M NaTPB solution. The end points correspond to the maximum slopes on the EMF vs. titrant volume plots.

Note: Picrolonic acid as well as picric acid standard solutions can also be used as titrants for those alkaloids that form insoluble picrolonates and picrates, respectively (see the preceding text).

iii. *Selected analytical methods for alkaloid assay in pharmaceuticals:*
a. *Atropine*—The contents of five ampules of 1% atropine injections (1 cm^3 each) or 5-cm^3 aliquots of 0.5 to 1% atropine eye drops are transferred to a 100-cm^3 measuring flask, diluted to the mark with deionized water, and shaken. Then 2 to 8-cm^3 aliquots of the solutions are transferred to a 100-cm^3 beaker and diluted to 20 cm^3 with 0.1 M potassium nitrate and the pH is adjusted to 4 to 6. Either liquid or PVC–atropine membrane sensor is immersed

in the solution in the conjunction with a reference electrode. The potential readings are recorded when becoming stable and are compared with the calibration graph prepared from pure standard atropine solutions. Atropine in 1% atropine eye ointment is determined after a prior extraction using the USP method. An accurately weighed 10 g of the ointment is dissolved in 20 cm^3 of diethyl ether and treated, in a 100-cm^3 separating funnel, with five successive 10-cm^3 portions of 0.1 N sulfuric acid. The aqueous extract is collected, made alkaline with 20% aqueous ammonia, and extracted with five successive 15-cm^3 portions of chloroform. The chloroform extract is evaporated to dryness and the residue dissolved in 100 cm^3 of 0.05 N sulfuric acid. The atropine content of 2- to 8-cm^3 aliquots of the solution is measured as previously described.[20]

b. *Papaverine*—Thirty papaverine hydrochloride tablets are powdered and extracted with 400 cm^3 of distilled water. After filtering, the volume is adjusted to 500 cm^3 in a volumetric flask. A 15-cm^3 aliquot of this solution is pipetted into the reaction cell containing the picrate membrane sensor and a reference electrode. The sample is titrated with a 10^{-2} M standard sodium picrate solution, using a Gran plot. The region of the equivalence point is reached with three or four large increments of titrant, and then 5 or 6 additional values of the cell potential vs. titrant volume are taken in the range 10 to 100% beyond the equivalence point. The factor $F = (V_0 + V)10^{E/S}$, where V_0 is the initial volume, V is the volume of titrant added, and E is the potential cell, is calculated. The plot F vs. cubic centimeters of titrant added is linear with the x-intercept at the equivalence volume. The line is calculated by a least-squares fit.[15]

c. *Quinidine*—Ten tablets of quinidine (as sulfate or gluconate) are finely powdered and exactly one tenth of the powder is transferred with 25 cm^3 of methanol and TRIS–HCl buffer (pH 7.0) to a 500-cm^3 volumetric flask. The solution is diluted to 500 cm^3 with TRIS–HCl buffer solution (pH 7.0) (solution A). A 10-cm^3 aliquot of solution A is pipetted into a 50-cm^3 volumetric flask containing 2 cm^3 methanol. The solution is diluted to 50 cm^3 with TRIS–HCl buffer (pH 7.0). The contents are shaken and transferred into a 100-cm^3 beaker. The quinidine membrane sensor and a double-junction reference electrode are immersed in the solution. After equilibration the EMF value is recorded and compared with the calibration graph. The value is checked using the standard-addition method. For this purpose 5.0 cm^3 of the standard solution of 5×10^{-3} M quinidine sulfate is added. The change in the millivolt-reading is recorded and used to calculate the concentration of quinidine as sulfate or gluconate per tablet.[29]

5.3   Amantadine

$$C_{10}H_{17}N \ (MM = 151.26)$$

$$NH_2$$

Therapeutic category: antiviral (influenza A); treatment of Parkinsonism

Discussion and Comments

Two amantadine-selective membrane sensors have been described[41]: the membrane sensor based on a 1-adamantanamine–dipicrylamine ion-pair complex were prepared by impregnating the support material (a graphite rod, made water-repellent) attached to the end of a Teflon tube, with a solution of 5×10^{-3} M ion-pair complex in nitrobenzene; the membrane sensor based on dinonylnaphthalene sulfonic acid (DNNS) had the composition 4.0% DNNS, 64.0% 2-nitrophenyloctyl ether, and 32% PVC (m/m). The internal filling solution in the second case was 10^{-3} M 1-adamantanamine hydrochloride of pH 4.6 (acetate buffer solution). The ion-pair complex with DNNS was obtained *in situ* by soaking DNNS–PVC membrane in a solution of 10^{-2} M 1-adamantanamine hydrochloride for 24 h.

Study of the critical response characteristics of both sensors showed that the linear range and consequently the useable linear range and the detection limit are superior for the PVC-sensor type. This is probably because DNNS forms a less soluble ion-pair complex with 1-adamantanamine than that between dipicrylamine and 1-adamantanamine.

The critical response characteristics for both sensors are shown in Table 5.2.

The linearity of E (in millivolts) vs. pH functions depends on the nature of the ion-pair complex used as electroactive membrane. In an acidic medium, only the liquid membrane is strongly affected by pH. This is due to the conversion of the electroactive membrane to its H^+ form at pH values below 4.0. At pH values higher than 8.0, the potentials of both sensors decrease because of the decreased concentration of the protonated adamantanamine that is converted to free base.

Both sensors proved useful in the potentiometric titration of 1-adamantanamine hydrochloride (amantadine, as a drug substance) or tablets. The average recovery of six drug-substance samples, each assayed in duplicate and containing between 4 and 10 mg, was 100.3% and the relative standard deviation was 1.3%. A good recovery was also achieved in the case of four samples of amantadine tablets. The sensor

Table 5.2 Response Characteristics for the Amantadine-Selective
Membrane Sensors[41]

Parameter	Liquid membrane	PVC membrane
Slope [mV $(\log a)^{-1}$]	50.4 ± 0.9^a	55.6 ± 0.8^a
Intercept, E_0 (mV)	319 ± 3.7^b	375 ± 2.9^b
Linear range (M)	10^{-1}–5×10^{-4}	10^{-1}–10^{-5}
Useable range (M)	10^{-1}–10^{-4}	10^{-1}–5×10^{-6}
Detection limit (M)	5×10^{-5}	4×10^{-6}

^{a}Standard deviation of average slope values for multiple calibration in 10^{-2} to
10^{-3} M range.
^{b}Standard deviation of values recorded during one month ($n = 65$).

was also used for the determination of content uniformity of amantadine
tablets (relative standard deviation 2.1%).

Analytical Procedures

i. *Potentiometric titration of 1-adamantanamine hydrochloride
   drug substance:*
   A 10-cm³ aliquot of the sample (containing 4 to 20 mg of 1-adaman-
   tanamine hydrochloride) is pipetted into the reaction cell. About 20
   cm³ of acetate buffer solution of pH 4.6 is added, and with stirring the
   solution is titrated with the standard solution of sodium tetraphenylborate (5×10^{-2} M) using an amantadine–SCE electrode system. The
   inflection point of the potentiometric titration curve is used to determine the end point of the titration.

ii. *Potentiometric assay of amantadine tablets:*
   Amantadine tablets are analyzed by finely powdering 10 tablets from
   the same lot. An accurately weighed portion of the powder equivalent
   to about 25 mg of 1-adamantanamine hydrochloride is transferred to
   the reaction cell and dissolved in about 10 cm³ of distilled water.
   About 20 cm³ of pH 4.6 acetate buffer solution is added and the
   potentiometric titration is carried out as described previously for the
   drug substance.

5.4 Amino Acids

The amino acids listed in Table 5.3, all of pharmacological interest, were
determined with various ion-selective membrane sensors. Some general
analytical methods presented later can be applied for other amino acids
assay, too.

Table 5.3 Amino Acids Assayed by Various Membrane Sensors

Amino acid	Formula (MM)	Therapeutic category	Ref.
Arginine	$C_6H_{14}N_4O_2$ (174.2)	Ammonia detoxicant (hepatic failure)	42–46

$$\underset{H_2N}{\overset{HN}{\diagdown}}C-NH-(CH_2)_3-\underset{\underset{NH_2}{|}}{CH}-COOH$$

Amino acid	Formula (MM)	Therapeutic category	Ref.
Cysteine	$C_3H_7NO_2S$ (121.2)	Has been used as a detoxicant (vet.)	46–54

$$\underset{\underset{SH}{|}}{CH_2}-\underset{\underset{NH_2}{|}}{CH}-COOH$$

Amino acid	Formula (MM)	Therapeutic category	Ref.
L-Dopa (Levodopa)	$C_9H_{11}NO_4$ (197.2)	Anticholinergic; antiparkinsonian	46, 52

$$HO-\underset{HO}{\overset{}{\bigcirc}}-CH_2-\underset{\underset{NH_2}{|}}{CH}-COOH$$

Amino acid	Formula (MM)	Therapeutic category	Ref.
Methyldopa	$C_{10}H_{13}NO_4$ (211.2)	Antihypertensive	52

$$HO-\underset{HO}{\overset{}{\bigcirc}}-CH_2-\underset{\underset{NH_2}{|}}{\overset{\overset{CH_3}{|}}{C}}-COOH$$

Amino acid	Formula (MM)	Therapeutic category	Ref.
Glutamic acid	$C_5H_9NO_4$ (147.1)	Experimental antiepileptic; –hydrochloride as gastric acidifier; —Mg salt hydrochloride as tranquilizer	46, 53, 55, 56

$$HOOC-\underset{\underset{NH_2}{|}}{CH}-(CH_2)_2-COOH$$

Table 5.3 Continued

Amino acid	Formula (MM)	Therapeutic category	Ref.
Glutamine	$C_5H_{10}N_2O_3$ (146.1) $H_2NOC-(CH)_2-CH-COOH$ $\qquad\qquad\quad \underset{\displaystyle NH_2}{\mid}$	For treatment of mental deficiency and alcoholism	57–60
Glycine	$C_2H_5NO_2$ (75.1) H_2N-CH_2-COOH	Nutrient	46, 52, 53
Histidine	$C_6H_9N_3O_2$ (155.2) 	Elevated levels in physiological fluids are responsible for histidemia	61–63
Isoleucine	$C_6H_{13}NO_2$ (131.2) $CH_3-CH_2-CH-CH-COOH$ $\qquad\qquad\quad \underset{\displaystyle CH_3}{\mid}\ \underset{\displaystyle NH_2}{\mid}$	Nutrient	46, 53
Leucine	$C_6H_{13}NO_2$ (131.2) $\begin{array}{l} H_3C \\ \quad\ \ \diagdown \\ \qquad CH-CH_2-CH-COOH \\ \quad\ \ \diagup \qquad\qquad\quad \mid \\ H_3C \qquad\qquad\qquad NH_2 \end{array}$	Nutrient	46, 53, 64
Lysine	$C_6H_{14}N_2O_2$ (146.2) $H_2N(CH_2)_4CH-COOH$ $\qquad\qquad\qquad \underset{\displaystyle NH_2}{\mid}$	Nutrient	46, 52, 53, 65
Methionine	$C_5H_{11}NO_2S$ (149.2) $H_3C-S-CH_2-CH_2-CH-COOH$ $\qquad\qquad\qquad\qquad\quad \underset{\displaystyle NH_2}{\mid}$	Lipotropic	46, 52, 66

Table 5.3 Continued

Amino acid	Formula (MM)	Therapeutic category	Ref.
N-Acetyl-L-Methionine	$C_7H_{13}NO_3S$ (191.3) $H_3C-S-CH_2-CH_2-CH-COOH$ $\quad\quad\quad\quad\quad\quad\quad\quad \mid$ $\quad\quad\quad\quad\quad\quad\quad\quad NHCOCH_3$	Lipotropic	67
Phenylalanine	$C_9H_{11}NO_2$ (165.2) $C_6H_5-CH_2-CH-COOH$ $\quad\quad\quad\quad\quad \mid$ $\quad\quad\quad\quad\quad NH_2$	Nutrient	45, 46, 52, 53, 64, 68–70
Threonine	$C_4H_9NO_3$ (119.1) $H_3C-CH-CH-COOH$ $\quad\quad\quad \mid \quad\quad \mid$ $\quad\quad\quad OH \quad NH_2$	Nutrient	46, 52
Tryptophan	$C_{11}H_{12}N_2O_2$ (204.2)	Nutrient	46, 53
Tyrosine	$C_9H_{11}NO_3$ (181.2) $HO-\bigcirc-CH_2-CH-COOH$ $\quad\quad\quad\quad\quad\quad\quad \mid$ $\quad\quad\quad\quad\quad\quad\quad NH_2$	Nutrient	46, 52 71, 72

   A novel approach to development of bio-selective sensors, exemplified for arginine, has been made by Rechnitz et al.[42] Living micro-organisms, such as bacteria, were employed in place of isolated enzymes at the surface of a gas-sensing membrane sensor. This approach offers several possible advantages over conventional enzyme sensors, e.g., the enzyme extraction and purification steps are eliminated and loss of enzyme activity is often less than in the case of isolated enzymes. As a result, sensor life is prolonged and the biological activity is optimized by the use of a living system.

   The bacterial sensor with response and selectivity to L-arginine was an ammonia-gas-sensing electrode (Orion, Model 95-10) in conjunction with micro-organism *Streptococcus faecium*, which metabolizes L-arginine to produce ammonia according to the following equations:

$$\text{L-arginine} \xrightarrow[\text{deaminase}]{\text{arginine}} \text{citrulline} + NH_3 \qquad (5.5)$$

$$\text{citrulline} + H_3PO_4 \xrightarrow[\text{transcarbamylase}]{\text{ornithine}} \text{ornithine} + \text{carbamoyl-phosphate}$$

$$(5.6)$$

$$\text{carbamoyl-phosphate} + ADP \xrightarrow[\text{kinase}]{\text{carbamate}} \text{carbamic acid} + ATP \quad (5.7)$$

$$\text{carbamic acid} \rightarrow CO_2 + NH_3 \qquad (5.8)$$

Freshly prepared sensors exhibited a linear response to L-arginine in the range 10^{-4} to 6.5×10^{-3} M (0.1 M phosphate buffer, pH 7.4 at 28°C); unfortunately, the response time was approximately 20 min when the sensor was new. The larger response time of the bacterial sensor could be due to the additional diffusion step through the bacterial cell wall.[42]

When *S. lactis* bacterial cells were used in suspension, together with the same ammonia-gas-sensing electrode as a detector, the sensor responded linearly to L-arginine over the concentration range 8.0×10^{-6} to 1.0×10^{-3} M with a slope of 59 mV decade^{-1} and was found to be selective with respect to 18 other L-amino acids.[43] Approximately 3×10^{11} cells were suspended in 5 cm^3 of 0.1 M phosphate buffer of pH 7.8 and they were kept in suspension by magnetic stirring. Use of a smaller number of cells decreases the linear range of the calibration curve, whereas doubling the number stated has no influence on the linear range. The response time of this bacterial sensor system is faster than the sensor described by Rechnitz et al.[42] The same effect has been found with other biological membrane sensors.[73] The preceding sensor system was used with good results to determine L-arginine in the presence of the other L-amino acids and urea.

An arginine sensor based on a coupled enzymatic system consisting of arginase and urease with an ammonia-gas-sensing electrode was constructed by placing a urease–arginase solution between an outer circular dialysis membrane and the gas-permeable membrane of the ammonia electrode and dispersing them uniformly on the surface of the membrane.[44] The lifetime of the sensor was found to be at least three weeks, if stored at 4°C when not in use.

The basis of the sensor is the reactions

$$\text{L-arginine} \xrightarrow{\text{arginase}} \text{urea} + \text{L-ornithine} \qquad (5.9)$$

$$\text{urea} \xrightarrow{\text{urease}} CO_2 + 2\,NH_3 \qquad (5.10)$$

Arginine diffuses into the immobilized enzyme layer of the sensor and produces a stoichiometric quantity of ammonia that gives rise to a constant potential, linearly related to the logarithm of arginine concentration in sample solution (linear range 3×10^{-5} to 3×10^{-3} M, with a response time of 5 min over this range). Arginine in the above range of concentration could be determined with an average error of 1%, and relative standard deviation was 1.0% for a 3.33×10^{-4} M arginine sample.

The development of Ag^+/S^{2-} membrane sensors has made the determination of cysteine by direct potentiometry possible and has extended the possibilities for potentiometric titration. When a Ag^+/S^{2-} sensor is introduced into a solution containing a thiol (e.g., cysteine), silver ions from the membrane and thiolate, $R-S^{y-}$, react:

$$Ag^+ + R-S^{y-} \rightarrow Ag(R-S)_p^{1-py} \qquad (5.11)$$

Tseng and Gutknecht[47] found that the following relation holds for the membrane potential of a silver sulfide membrane sensor introduced into a thiol solution:

$$E_{R-S^{y-}} = E_{Ag^+}^0 + \frac{RT}{F} \ln \frac{\alpha}{K_f} - \frac{pRT}{F} \ln(L - p\alpha) + \frac{RT}{F} \ln \gamma_{Ag^+} \qquad (5.12)$$

where α is the silver ion activity due to available silver ions at the membrane surface, K_f is the apparent formation constant of the silver–thiol complex, L is the total thiol concentration, γ_{Ag^+} is the activity coefficient of the Ag^+. With L large, i.e., $L \gg p\alpha$, a plot of E (in millivolts) vs. $\ln L$ gives a line of slope $-pRT/F$ from which p, the stoichiometric term, can be determined. With E_0 and α known, K_f can also be determined.[47]

The detection limit for cysteine (in 0.1 M NaOH) was 2×10^{-4} M and the slope of the calibration curve was -108 mV decade^{-1}. This compares favorably with the value of -118 mV decade^{-1} predicted by Morf et al.[74] for a 1:2 silver–cysteine complex. The lower value of -108 mV decade^{-1} may be explained by the fact that the sensor used by Tseng and Gutknecht[47] had a sulfide ion response of -27.5 mV decade^{-1}.

Gruen and Harrap[48] used a solid-state silver sulfide membrane sensor (Orion, Model 94-16) and silver nitrate for the potentiometric titration of L-cysteine in an aqueous medium (pH 2.5 to 9) and Selig[49] described the potentiometric determination of cysteine and other thiols with a bromide-ion-selective membrane sensor (Orion 94-35). In the last case, micro-amounts (0.01 to 0.1 mM) were titrated with 5×10^{-3} M mercury(II) perchlorate whereas semi-micro-amounts (0.1 to 0.5 mM) were titrated with 2.5×10^{-2} M mercury(II) perchlorate (in both cases in pyridine buffer).

For the potentiometric titration of cysteine, Liteanu et al.[51] used a mercury(II)-ion-selective membrane sensor (membrane composed of 75% moles silver iodide + 25% moles silver sulfide) and mercury(II) as titrant. Cysteine was determined in the 10^{-5} to 10^{-2} M concentration range. Even for only 0.13 μg cm^{-3} of cysteine the potential change was 80 mV at the equivalence point for 1% error.

Jensen and Rechnitz[50] have constructed a bacterial membrane sensor for L-cysteine by coupling the bacterium *Proteus morganii* with a hydrogen sulfide–sensing electrode. Cysteine is metabolized by *P. morganii* to hydrogen sulfide in the reaction

$$\text{cysteine} \xrightarrow[\text{desulfhydrase}]{\text{cysteine}} \text{pyruvate} + NH_3 + H_2S \qquad (5.13)$$

Approximately 10 mm^3 of the centrifuged bacteria was applied to the gas-permeable membrane of the hydrogen sulfide electrode and held in place with a dialysis membrane. The gas-sensing electrode itself was assembled by using the body and internal pH electrode of the Orion Model 95-10 ammonia-gas-sensing electrode. The freshly prepared bacterial sensors were stored for 12 h in nutrient bath (Difco Labs.) at 30°C before initial use. Henceforth, the sensors were stored in 0.1 M phosphate buffer, pH 6.75. For the 5×10^{-5} to 9×10^{-4} M range of cysteine (in 0.1 M phosphate buffer, pH 6.75 at 37°C), the sensor gave a linear response of slope 25 mV decade^{-1}, the response time being 5 to 8 min.

A biocatalytic L-cysteine sensor, where cucumber leaves were used as biocatalysts in conjunction with an ammonia-gas-sensing electrode (Orion, Model 95-10), was described by Smit and Rechnitz.[54] The substrate L-cysteine is biocatalytically degraded, apparently via L-cysteine desulfhydrase activity to products including NH_3 (see Equation 5.13), which produces the potentiometric response. The ammonia release takes place in 1 : 1 stoichiometry to the substrate. Potentiometric measurements with the leaf disk membrane sensor were carried out at 30°C in pH 7.6, 0.1 M phosphate buffer media. Because the leaf disk was interposed between the sample and the internal ammonia sensor, the L-cysteine response was compared with the ammonia response of the same sensor configuration. Response times were of the order of minutes, with the L-cysteine response 30 to 40% slower than the NH_3 response.

A potentiometric membrane sensor with selective response to L-glutamate over the 2×10^{-4} to 1.3×10^{-2} M range was constructed by immobilizing slices of yellow squash tissue at a carbon dioxide gas sensor.[56] This system represents the first successful use of intact plant

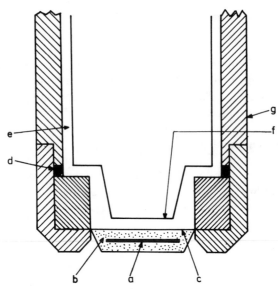

Figure 5.3 Schematic diagram of the squash-tissue-based membrane sensor: (a) slice of yellow squash tissue; (b) BSA conjugate layer; (c) carbon dioxide gas permeable membrane; (d) O-ring; (e) internal electrolyte solution; (f) pH-sensing glass membrane; (g) plastic electrode body. (Components c through g represent the Orion Model 95-02 carbon dioxide gas sensor). (Reproduced from Kuriyama, S. and Rechnitz, G. A., *Anal. Chim. Acta*, 131, 119, 1981, Elsevier Science Publishers, Physical Sciences and Engineering Division. With permission.)

materials as biocatalysts in the construction of bio-selective potentiometric membrane sensors. The biocatalytic activity of this sensor arises from the fact that yellow squash tissue contains glutamate decarboxylase.[56] This enzyme catalyzes the decarboxylation of glutamic acid to produce carbon dioxide and γ-aminobutyric acid according to Equation 5.14:

$$\text{L-glutamic acid} \xrightarrow[\text{pyridoxal-5'-phosphate}]{\text{glutamate decarboxylase}} \gamma\text{-aminobutyric acid} + CO_2 \quad (5.14)$$

Figure 5.3 shows the configuration of the glutamate-selective membrane sensor. A thin slice (about 0.3 mm) of fresh yellow squash, after dipping into a solution containing bovine serum albumin and glutaraldehyde, was immobilized by placing it on the surface of a gas-permeable membrane along with the BSA–glutaraldehyde solution; the assembled sensor was allowed to stand at room temperature for 25 min. Kuriyama and Rechnitz[56] reported that the best results were obtained by using tissue slices from the mesocarp layer of the squash (Figure 5.4); this

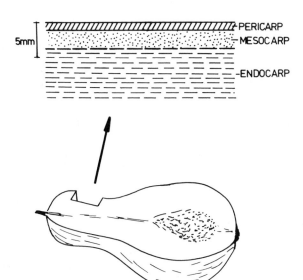

Figure 5.4 Cross section of yellow squash, showing origin of mesocarp biocatalytic layer. (Reproduced from Kuriyama, S. and Rechnitz, G. A., *Anal. Chim. Acta*, 131, 119, 1981, Elsevier Science Publishers, Physical Sciences and Engineering Division. With permission.)

layer appears to contain the highest biocatalytic activity for L-glutamate and is readily separated from the hard pericarp and pulpy endocarp layers of the squash.

The glutamate sensor combines excellent selectivity characteristics over some 25 possible interferences tested with a good reproducibility in a buffer containing 0.1 M phosphate, 40% glycerol, 0.002% chlorhexidine diacetate, and 3×10^{-4} M pyridoxal-5'-phosphate at pH 5.5 (glycerol is added as a proper solvent to stabilize the enzyme and chlorohexidine diacetate is required as a preservative). The useful lifetime of the sensor was only seven days. The sensor was stored in the previously mentioned buffer at room temperature.

An enzyme sensor specific for glutamine has been prepared by entrapping glutaminase on a nylon net between a layer of cellophane and a Beckman 39137 or Thomas 4923-Q10 cation electrode.[57] Glutaminase catalyzes the decomposition of glutamine:

$$H_2NOCCH_2CH_2CH-COO^- \xrightarrow[H_3O^+]{glutaminase} HOOCCH_2CH_2CH-COO^- + NH_4^+$$
$$\qquad\qquad\quad | \qquad\qquad\qquad\qquad\qquad\qquad\qquad\qquad | $$
$$\qquad\qquad\quad NH_3^+ \qquad\qquad\qquad\qquad\qquad\qquad\qquad NH_3^+$$

$$(5.15)$$

Table 5.4 Summary of Response Characteristics of Four Classes of Glutamine-Selective Membrane Sensors[58]

	Enzyme sensor	Mitochondrial sensor	Tissue sensor	Bacterial sensor
Slope (mV decade^{-1})	33.2–48.2	52.5 $\pm$ 2.2	50.0 $\pm$ 2.8	49
Linear range (average) (M)	1.5 $\times$ 10^{-4}– 3.3 $\times$ 10^{-3}	1.1 $\times$ 10^{-4}– 5.5 $\times$ 10^{-3}	6.4 $\times$ 10^{-5}– 5.2 $\times$ 10^{-3}	10^{-4}–10^{-2}
Limit of detection (average) (M)	6 $\times$ 10^{-5}	2.2 $\times$ 10^{-5}	2 $\times$ 10^{-5}	5.6 $\times$ 10^{-5}
Response time^a (min)	4–5	6–7	5–7	5
Useful life-time (days)	1	10^b	30^b	20^b

^{a}Measured at midpoint of linear range.
^{b}Minimum value.

The ammonium ions resulting from Reaction 5.15 are sensed by the sensor. The calibration curve of a freshly prepared glutamine sensor was very steady throughout 8 h continuous operation. The sensor sensitivity was less on the second day even though it was stored at 5°C overnight. This instability can be attributed either to mechanical leakage of the enzyme from the sensor into the solution, or to decomposition of the enzyme. The concentration limit of the glutamine membrane sensor is 10^{-4} M.

Glutamine-selective membrane sensors based upon four classes of biocatalysts immobilized on ammonia-gas sensors have been evaluated by Arnold and Rechnitz.[58] All four classes of biocatalytic membranes depend on the deamination of glutamine and measurement of the ammonia produced using a commercially available ammonia-gas-sensing membrane electrode (Orion, Model 95-10). In all cases the biocatalytic material, e.g., isolated enzyme, bacterial cells, tissue slices, or mitochondrial fraction, is physically supported by an outer dialysis membrane as a layer on the surface of the ammonia electrode.

Each of the four sensor classes have relative advantages and disadvantages arising from the properties and requirements of the individual biocatalysts. Table 5.4 summarizes the most important properties of the four classes of glutamine membrane sensors.

Several workers have reported histidine-selective sensors based on immobilized enzyme or bacterial cell catalyzed reactions.[61-63] Buck and co-workers utilized histidine ammonia lyase (E.C. No. 4.3.1.3) from *Pseudomonas* sp. to develop a histidine sensor that relied on a potentiometric ammonia-gas-sensing electrode as the detector.[62] The response of the sensor was linear between 3×10^{-5} and 10^{-2} M histidine with a slope of 54 mV decade^{-1} (no decline in response over one week of use; response time, 3–8 min). The intact cells were also employed for the construction of the sensor, but selectivity was poor because of the presence of the other deaminating enzymes.[61] A highly selective histidine biomembrane sensor has also been prepared by immobilizing the enzyme histidine decarboxylase (E.C. No. 4.1.1.22) at the surface of a potentiometric carbon dioxide sensor.[63] The enzyme employed was extracted from *Lactobacillus* 30a. The use of concentrated enzyme extract rather than intact bacterial cells is shown to yield biosensors with improved response characteristics. The resulting enzyme-based sensor responds linearly to the logarithm of histidine concentration between 3×10^{-4} and 1×10^{-2} M with a slope typically of 48 to 53 mV decade^{-1} and a useful lifetime of over 30 days.

All histidine membrane sensors found interesting applications for histidine assay in blood or urine samples.

The basis of enzymatic L-lysine membrane sensors depends on the specific enzymatic decarboxylation of L-lysine by immobilized enzyme followed by detection of the released carbon dioxide with a carbon dioxide gas sensing electrode (Radiometer, E-5036).[65]

$$\text{L-lysine} \xrightarrow[\text{decarboxylase}]{\text{L-lysine}} CO_2 \qquad (5.16)$$

The response of the sensor arises from carbon dioxide diffusion to give a change in pH in a thin electrolyte layer trapped between an almost flat pH electrode and the Teflon gas-permeable membrane (Radiometer DG 02).

The lysine-selective enzyme sensor has linear response range of approximately 3×10^{-2} to 10^{-4} M lysine, slope of 0.8 to 0.9 pH decade^{-1} and response times of 5 to 10 min. The lower limit of detection, 6×10^{-5} M, is very near the detection limit of the carbon dioxide gas-sensing electrode itself, which is imposed by the inner electrolyte and atmospheric carbon dioxide. The upper limit is generally between 10^{-1} and 10^{-2} M, which is about the upper practical limit of the carbon dioxide, as governed by the solubility of carbon dioxide.

A methionine-selective membrane sensor can be constructed by using methionine-lyase enzyme (E.C. 4.4.1.11) and an ammonia-gas-sensing electrode (e.g., modified E-5036 electrode, Radiometer) is used to determine the amount of ammonia generated by the enzymatic reaction

$$\text{methionine} \xrightarrow[\text{lyase}]{\text{methionine}} \alpha\text{-ketobutyrate} + CH_3SH + NH_3 \qquad (5.17)$$

The immobilized enzyme is stable for three months, with no loss of activity, and the sensor exhibits a linear response to L-methionine over the 10^{-2} to 10^{-5} M range. The slopes of 42 mV decade^{-1} at 26°C and 49 mV decade^{-1} at 39°C are less than the theoretical values (59 and 62 mV decade^{-1}, respectively) and no interference from the common amino acids was observed.

The N-acetyl-L-methionine sensor, constructed by Nikolelis and Hadjiioannou,[67] can be used for the determination of this substrate in the range 4×10^{-5} to 2×10^{-3} M, by direct potentiometry, with an average error of 1% (1.3% relative standard deviation). The sensor is based on a coupled enzymatic system consisting of acylase and L-amino acid oxidase with an ammonia-gas-sensing electrode. The sensor was constructed by placing an L-amino acid oxidase–acylase solution in 0.1 M TRIS–HCl buffer (pH 8.0), between an outer circular cellophane dialysis membrane and the gas-permeable membrane of the ammonia-gas-sensing electrode (Orion, Model 95-10) and dispersing them uniformly on the surface of the membrane. The sensor was preconditioned by soaking for 2 h in 0.1 M TRIS–HCl buffer (pH 8.0) and was stored in this buffer at 4°C when not in use.

The basis for the sensor is the following reactions:

$$N\text{-acetyl-L-methionine} \xrightarrow[\text{H}_2\text{O}]{\text{acylase}} \text{acetic acid} + \text{L-methionine} \quad (5.18)$$

$$\text{L-methionine} + \text{H}_2\text{O} + \text{O}_2 \xrightarrow[\text{oxidase}]{\text{L-amino acid}} \text{2-oxo-acid} + \text{NH}_3 + \text{H}_2\text{O}_2$$

$$(5.19)$$

N-acetyl-L-methionine diffuses into the immobilized enzyme layer of the sensor and produces a stoichiometric quantity of ammonia, which is monitored by the ammonia electrode. The response time of the sensor is about 5 min in 4×10^{-5} M and 2 min in 2×10^{-3} M. The stability of performance and the working lifetime of the sensor depend on the operational conditions and storage; the lifetime of the sensor was found to be at least 20 days if stored at 4°C when not in use, and only 4 days when stored at room temperature. Unfortunately, the sensor response is highly affected by the presence of various common amino acids.

Three different kinds of enzyme sensor suitable for the determination of phenylalanine have been described by Guilbault and co-workers.[68, 69] The L-phenylalanine membrane sensor based on an iodide-ion-selective electrode uses a dual enzyme reaction layer, namely, L-amino acid oxidase (L-AAO) and horseradish peroxidase (HRP), chemically immobilized in polyacrylic gel. The sensor function is based on

$$\text{L-phenylalanine} \xrightarrow{\text{L-AAO}} H_2O_2 \qquad (5.20)$$

$$H_2O_2 + 2\,H^+ + 2\,I^- \xrightarrow{\text{HRP}} I_2 + 2\,H_2O \qquad (5.21)$$

The iodide ion concentration of the sample solution (added in constant amount) decreases locally at the electrode surface in the presence of L-phenylalanine. The decrease in iodide concentration is sensed by the iodide electrode (optimal parameters are pH = 5.0 and concentration of iodide = 5×10^{-5} M); the sensor response to L-phenylalanine was evaluated as the slope of the potential vs. time curve.

The L-phenylalanine sensor centered on a silicone-rubber-based antibiotic-type ammonium-ion-selective electrode was constructed from a chemically immobilized L-AAO as reaction layer. The L-phenylalanine diffuses into the enzyme layer, where it is converted to ammonium ions. The potential produced is related to the log function of the L-phenylalanine concentration. This sensor gives a linear response to L-phenylalanine in the 10^{-2} to 10^{-4} M range. The response time of the sensor varies from 60 to 180 s, depending on the thickness of the reaction layer.

Guilbault and co-workers[69] constructed a third type of specific enzyme sensor for L-phenylalanine, based on the use of L-phenylalanine ammonia lyase in the air-gap electrode. The ammonium ion, produced from L-phenylalanine (Equation 5.22) in the sample solution is converted to ammonia, which is measured by the change of pH of the air-gap electrode:

$$C_6H_5-CH_2-\underset{\underset{NH_3^+}{|}}{CH}-COO^- \xrightarrow[\text{ammonia lyase}]{\text{L-phenylalanine}}$$

$$NH_4^+ + C_6H_5-CH=CH-COO^- \qquad (5.22)$$

The calibration curve for L-phenylalanine determined with the air-gap electrode is fairly linear from 1×10^{-4} to 6×10^{-4} M L-phenylalanine with a slope of 1.03 pH_e decade^{-1} and a relative standard deviation of 4.67%.

Guilbault and Shu[71] have studied the possibility of using a carbon dioxide gas sensing electrode (Instrumentation Lab., Inc.) as an enzymatic sensor selective to L-tyrosine. The sensor was prepared by dispersing 0.2 cm^3 of enzyme suspension solution (approximately 50 mg of tyrosine decarboxylase per cubic centimeter of 0.1 M citrate buffer at pH 5.5) over the electrode surface covered with a nylon netting. A piece of dialysis film was placed over the enzyme layer to prevent the enzyme from diffusing into the solution.

The enzyme sensor was soaked, with stirring, in buffer solution (0.1 M citrate). When such a sensor is in contact with a sample solution containing L-tyrosine, the carbon dioxide evolved at the electrode surface is stoichiometrically proportional to the concentration of L-tyrosine:

$$\text{tyrosine} \xrightarrow[\text{decarboxylase}]{\text{tyrosine}} \text{tyramine} + CO_2 \qquad (5.23)$$

The best pH was 5.5, when the solubility of tyrosine in water is at its maximum. Between 2.5×10^{-4} and 5×10^{-3} M tyrosine (pH 5.5, citrate buffer), the response of the tyrosine sensor is fast and the calibration slope is 55 mV decade^{-1}. The slope remains constant for 10 h of operation after preparing the sensor.[71]

A bacterial tyrosine-selective potentiometric membrane sensor was proposed[72] in which the desired biocatalytic activity was biochemically induced during growth of the bacterial cells. As a result of this induction a normally ineffective biocatalyst, *Aeromonas phenolegeues* ATCC 29063, could be coupled with an ammonia-gas-sensing electrode in order to produce a useful tyrosine membrane sensor.

The basis of the sensor is the reaction described by

$$\text{L-tyrosine} + H_2O \xrightarrow[\text{pyridoxal-5'-phosphate}]{\text{tyrosinephenol lyase}} \text{pyruvate} + \text{phenol} + NH_3 \qquad (5.24)$$

The sensor showed a linear response in the range 8.3×10^{-5} to 1.0×10^{-3} M tyrosine with a detection limit of 3.3×10^{-5} M for at least eight days; it was applied for the determination of tyrosine in aqueous samples, where tyrosine concentrations ranged from 4.7×10^{-5} to 4.9×10^{-4} M, using the standard-addition method. An average standard deviation of 4% was reported, even at concentrations below the linear range of response of the sensor.

A chemically modified electrode with immobilized enzyme was constructed by covalent attachment of L-amino acid oxidase to a graphite rod via chemical modification of the electrode surface by cyanuric chloride linkage.[64] Logarithmic response with concentration of some L-amino acids (leucine, methionine, phenylalanine) was observed in the 10^{-2} to 10^{-5} M range. Cyanuric chloride was covalently attached to the carbon surface as described in Tse et al.[75] and L-amino acid oxidase (L-AAO) was bound to the cyanuric chloride–modified electrodes by a modified procedure of Wilson et al.[76]

It is well known that L-AAO is a specific catalyst for the decomposition of L-amino acids in the overall reaction

$$R—CH—COO^- + H_2O + O_2 \xrightarrow{\text{L-AAO}} R—C—COO^- + NH_4^+ + H_2O_2$$
$$\underset{NH_3^+}{|} \qquad\qquad\qquad \underset{O}{\|}$$

$$(5.25)$$

Interactions of different surface-functional groups contained in treated graphite, with hydrogen peroxide produced in Reaction 5.25 can be a possible contributor to the potentiometric response of the sensor.[64]

L-amino acids that form soluble copper complexes can be determined by stoichiometric reaction with an excess of copper(II)phosphate suspension and measurement of copper(II) ions produced with a Cu^{2+}-ion-selective membrane sensor.[52] The optimum pH of the reaction (9 to 10) was out of the working pH range of the copper(II) electrode (3 to 7) and thus the stoichiometric reaction of amino acids with an excess of copper(II) phosphate and direct measurement of Cu^{2+} ions after acidifying the solution of the soluble copper–amino acid complexes was preferred.[52]

The stoichiometry of the reaction was studied by measuring the Cu^{2+} brought into solution by the amino acid complex, using a calibration curve from Cu(II) standard solutions. The results obtained for alanine showed a slope of the plot alanine taken (in micromoles) vs. Cu^{2+} found (in micromoles) of 1.994 with intercept 0.051 and $r = 1.000$. Similar results, with slopes of 1.95 to 2.05 were found for glycine, valine, tyrosine, serine, glutamic acid, lysine, aspartic acid, threonine, and proline. The same behavior was shown by the drugs dopa and methyldopa (Aldomet). This group of amino acids gives soluble copper complexes. The other amino acids tested (arginine, glutamine, histidine, tryptophan, cysteine, leucine, phenylalanine, and methionine) give partially or totally insoluble complexes and the method cannot be applied except sometimes in the 5×10^{-4} to 5×10^{-3} M range.[52]

To improve accuracy and avoid errors from any gradual loss of the suspension activity, calibration graphs for individual amino acids were used. The standard deviation for four measurements in the whole concentration range (5×10^{-4} to 5×10^{-2} M) and in all cases is 0.2 to 0.4 mV, which causes a measurement error of about 3%. The proposed method was tested for the determination of dopa and methyldopa in pharmaceutical preparations, the values obtained being in good agreement with those obtained by the USP standard spectrometric methods.

A new method for detection of amino acids in a continuous-flow system was reported with copper wire or copper tubular electrodes as potentiometric detectors.[45] The copper electrodes are shown to respond to the anionic form of amino acids without addition of Cu(II) to the flow system, with a response dependent on flow rate, buffer pH, and the type of amino acid. The potentiometric response of copper-metal electrodes in the presence of amino acids can be attributed to surface reactions at the solid electrode with formation of metal complex species. Copper electrodes tend to oxidize in aqueous solution, and a simplified reaction mechanism can therefore be written as a reaction of Cu^{2+} (from oxidation of the metal) with the amino acid (HA) to form the metal complex of the anion and H^+,[45]

$$Cu \rightarrow Cu^{2+} + 2e \qquad (5.26)$$

$$Cu^{2+} + q\,HA \rightarrow CuA_q^{(2-q)} + q\,H^+ \qquad (5.27)$$

The negative change in potential when the electrode is immersed in an amino acid solution is caused by decrease of Cu^{2+} at the electrode surface due to complex formation.

A trinitrobenzenesulfonate (TNBS) liquid-membrane sensor[77] was used for the determination of some amino acids (cysteine, tryptophan, lysine, phenylalanine, leucine, arginine, glutamic acid, etc.) that react with TNBS:

$$(5.28)$$

both by direct potentiometry or kinetic methods.

The amount of TNBS (in millimoles) that reacts with the amino acid (1 : 1) ratio) was calculated by

$$mmol\ TNBS = V \cdot [TNBS] \cdot (1 - Q) \qquad (5.29)$$

Q being deduced from Equation 5.30,

$$E_f - E_i = \Delta E = -S \log([TNBS]_f/[TNBS]_i) \qquad (5.30)$$

$$\left(Q = 10^{-\Delta E/S} = [TNBS]_f/[TNBS]_i\right)$$

where E_i and E_f correspond, respectively, to the initial $[TNBS]_i$ and final $[TNBS]_f$ of unreacted TNBS in the mixture with amino acid (pH = 12, phosphate buffer).

Measuring ΔE for different concentrations of an amino acid solution and plotting $[TNBS]_i \times 10^{-\Delta E/S}$ vs. [amino acid], a straight line was obtained that was used as a working curve for the assay of respective amino acid.

A kinetic potentiometric method for the determination of amino acids, based on monitoring their reaction with 2,4-dinitrofluorobenzene (DNFB) using a fluoride-selective membrane sensor at pH 9.0 and 25°C was described by Athanasiou-Malaki and Koupparis.[46] The reaction of amino acids with DNFB is a well-known example of nucleophilic aromatic substitution reaction with the formation of an intermediate complex and can be depicted by the following successive reactions:

$$R-NH_2 + (NO_2)_2C_6H_3F \underset{k_{-1}}{\overset{k_1}{\rightleftharpoons}} (NO_2)_2C_6H_3^-(F)N^+H_2R \overset{k_2}{\longrightarrow}$$

$$(NO_2)_2C_6H_3NHR + H^+ + F^- \quad (5.31)$$

Initial-slope and fixed-time (60 s) methods were used to construct calibration graphs, in most instances in the range 1×10^{-4} to $5 \times 10^{-3}\ M$.

The proposed method was applied to the determination of L-dopa and aspartame (L-dipeptide) in pharmaceutical formulations. A recovery study performed on synthetic mixtures of aspartame with various excipients gave a mean recovery of 100.2% (range 98.4 to 102.5%).

Analytical Procedures

i. *General method applied to amino acids for which membrane sensors are available (homemade; see the preceding text):*
A stock solution of $10^{-2}\ M$ amino acid is prepared by dissolving the respective compound in dilute hydrochloric acid. The pH of the solution is adjusted to the recommended value with NaOH solution or buffer solution (phosphate, citrate, etc.). 10^{-3} and $10^{-4}\ M$ amino acid solutions are obtained from the stock solution by successive dilutions. The sensors (amino acid sensitive and SCE) are placed in the standard solutions and the EMF readings (linear axis) plotted against concentration (logarithmic axis). The sample concentration is determined from this graph.

ii. *General method, based on the use of a fluoride-selective membrane sensor (applied to most of the amino acids listed in Table 5.3):*
A volume of 10 cm³ of a working standard or sample solution of the analyte and 5 cm³ of the pH 9.0 mixed borate buffer (it contains $3 \times 10^{-5}\ M$ NaF and $5 \times 10^{-3}\ M$ trans-1,2-diamino-cyclohexane-

N,N,N',N'-tetracetic acid [DCTA]) are pipetted into the thermostated (25°C) reaction cell. After potential had stabilized, under stirring (about 20 s), the recorder pen is adjusted to the higher potential side of the chart recorder and recording is started. The reaction is initiated by injection of 100 μl of DNFB working solution (0.215 M in acetone) with a microsyringe and the reaction curve is recorded for about 2 to 3 min. The content of the cell is evacuated and washed twice with water, and the procedure is continued for the next sample. A blank (H_2O) should be included for each calibration curve. The initial slope $\Delta E/\Delta T$ (in millivolts per second) or the potential change ΔE (in millivolts) is graphically estimated for a 60-s time interval. Using the standard solutions of the analyte, the calibration graph of $\Delta E/\Delta t$ vs. concentration (initial-slope method) or $10^{\Delta E/S} - 1$ vs. concentration (fixed-time method) is constructed. ΔE from the blank should be subtracted from every ΔE measured for the standards or the sample. The slope of the sensor response (S) is periodically determined by successive additions of 100 μl of 1.5×10^{-3} M and 1.5×10^{-2} M NaF standard solutions in 10 cm^3 H_2O mixed with 5 cm^3 of buffer and measuring the E (in millivolts) with the F^--membrane sensor.

[The sample solution should be approximately neutralized, if required (phenolphthalein). For determination of L-dopa in capsules or tablets, not less than 20 tablets or the contents of not less than 20 capsules are weighed and finely powdered or mixed. Then 0.01 to 0.1 mmol of L-dopa is taken for analysis.]

iii. *Selected analytical methods for some amino acids assay:*

    a. L-*Cysteine*—The sample is diluted to 40 cm^3 with distilled water and 2 cm^3 pyridine added. A nitrogen flush is used to prevent air oxidation of the cysteine. The pair of electrodes (bromide-ion-selective or other indicator membrane sensor based on a silver sulfide matrix with SCE reference) is immersed in the sample solution, which is potentiometrically titrated with 5×10^{-3} M mercury(II) perchlorate. The EMF is recorded as a function of titrant volume and the end point corresponds to the maximum slope on the titration curve.[49]

    b. L-*Lysine*—The lysine-selective membrane sensor is immersed into buffered sample, where later the lysine sample is injected. Typically, 2-cm^3 samples are used. The electrode response is noted as pH equilibrium (pH$_e$), P_{CO_2}, or the rate of P_{CO_2} response.[65] Enzyme activity and the effective activity of the enzyme sensor are determined from the rate of P_{CO_2} response. The sensor is equilibrated in 2 cm^3 of 0.5 M acetate buffer (prepared by dissolving the appropriate amount of sodium acetate in distilled water and adjusting the pH at 5.8 with hydrochloric acid) and 0.5 cm^3 of 0.5 M L-lysine in buffer injected. The response, dP_{CO_2}/dt, is determined from the recorder output and the activity calculated as described by Berjonneau et al.[78]

c. L-*Methionine and N-acetyl-L-methionine*—A volume of 5 cm³ of unknown sample solution is pipetted into the thermostated cell where the N-acetyl-L-methionine indicator membrane sensor and the reference electrode are immersed. The EMF is recorded under stirring when it is constant to within ±0.1 mV (in 2 to 5 min). The substrate concentration is calculated from the calibration curve of EMF vs. log(L-methionine or N-acetyl-L-methionine concentration).[67]

5.5 Aminopyrine and Related Drugs

Aminopyrine $C_{13}H_{17}N_3O$ (MM = 231.3)

Antipyrine $C_{11}H_{12}N_2O$ (MM = 188.2)

Analginum (dipyrone) $C_{13}H_{18}N_3NaO_5S$ (MM = 351.4)

Therapeutic category: analgesics; antipyretics

Discussion and Comments

Hopîrtean and Stefănigă[79] studied the response of certain organic cations (alkaloids, vitamins, and amino acids) by a membrane sensor of a platinum wire coated with a plastic membrane.[80] They reported that aminopyrine and antipyrine can be potentiometrically titrated with sodium tetraphenylborate solution using this sensor as indicator. However, the equivalence point is difficult to detect because the potential change at the equivalence point for both aminopyrine and antipyrine is too low even for a ±2% error in the titrant volume.

Both triheptyldodecylammonium iodide (THDA) and trioctyldodecylammonium iodide (TODA) were used to prepare electroactive materials for a PVC–analginum membrane sensor.[81] Experimental results showed that the sensor based on THDA is superior to that based on TODA. The influence of the nature of the plasticizer on the performances of the sensors has been studied. The performances of sensors plasticized with dibutylphthalate (DBP) were compared with those of sensors plasticized with diisooctylphthalate (DIOP) or dinonylphthalate (DNP). The results showed that the sensor plasticized with DBP is lower in detection limit and membrane resistance and is shorter in practical response time. The optimal membrane composition was THDA–analginum ion-pair complex (1.5 to 2.0%), DBP (69%), and PVC (29 to 29.5%). The sensor exhibits a Nernstian-type response to analginum within the concentration range 10^{-1} to 1.8×10^{-5} M with a slope of 60 mV decade^{-1} at 29°C. No significant potential changes in the pH range 4.5 to 6.7 were observed and the response time of the sensor was within 1 min in $\geq 10^{-4}$ M analginum. It was found that among the inorganic and organic ions tested as potential interferents, only ClO_4^- interferes in the sensor response.[81]

Analytical Procedures

i. *Aminopyrine and antipyrine assay:*
   The electrode pair (coated-wire indicator membrane sensor with SCE reference electrode) is introduced into the sample solution (30 to 40 cm^3, approximately 5×10^{-3} M), which is potentiometrically titrated with 5×10^{-2} M standard sodium tetraphenylborate solution. The end points correspond to the maximum slope on the plot of EMF vs. titrant volume.

ii. *Analginum assay:*
   A stock solution of 10^{-2} M analginum is prepared by dissolving the respective amount of the compound in deionized distilled water. The pH of the solution is adjusted with NaOH/HCl solutions to pH 4.5 to 6.7; 10^{-3} and 10^{-4} M analginum solutions are obtained from the stock solution by successive dilutions. The sensors (analginum-selective and SCE) are placed in the standard solutions and the EMF readings (linear axis) plotted against concentration (logarithmic axis). The analginum concentration from the unknown sample is determined from this graph.

5.6   Amitriptyline and Related Drugs

The first four drug substances listed in Table 5.5 were analyzed by a semiautomatic potentiometric technique[82] using a liquid-membrane tetraphenylborate sensor.[14] In all cases, a 10^{-2} M sodium tetraphenylborate solution was used as titrant; optimum pH medium for the first three compounds was 3.3, and for opipramol pH 5.0 is recommended. The end-point jumps were steep and the equivalence point was easily detected. The precision of these compounds in pharmaceutical preparations was better than 1%.

Table 5.5 Amitriptyline and Similar Compounds Assayed by Membrane Sensors

Compound	Formula (MM)	Therapeutic category
Amitriptyline	$C_{20}H_{23}N$ (277.4)	Antidepressant

$$CHCH_2CH_2N(CH_3)_2$$

Clomipramine	$C_{19}H_{23}ClN_2$ (314.9)	Antidepressant

$$(CH_2)_3N(CH_3)_2$$

Imipramine	$C_{19}H_{24}N_2$ (280.4)	Antidepressant

$$(CH_2)_3N(CH_3)_2$$

Opipranol	$C_{23}H_{29}N_2O$ (363.5)	Antidepressant tranquilizer

$$(CH_2)_3-N \quad N-(CH_2)_2OH$$

Protriptyline	$C_{19}H_{21}N$ (262.4)	Antidepressant

$$(CH_2)_3NHCH_3$$

A liquid-membrane sensor for amitriptyline, containing as ion exchanger a mixture of ion-pair complexes of amitriptyline with eosin and tetraphenylborate in *p*-nitrocumene, was described by Mitsana-Papazoglou et al.[83] An Orion liquid-membrane electrode body (Model 92) was used as electrode assembly with a Millipore LCWPO 1300 PTFE membrane. The concentration of ion exchanger was 10^{-2} M with respect to both salts and the internal solution was 10^{-2} M amitriptyline chloride and 0.1 M sodium chloride. The sensor exhibited near-Nernstian response in the range 6×10^{-2} to 6×10^{-5} M with a slope of 60 mV decade^{-1} at 25°C (pH range 1 to 6).

For the amitriptyline membrane sensor, amitriptyline salts with eosin, alizarin red S, tetraphenylborate, anilinonaphthalene-sulfonate, eosin-tetrakis(imidazolyl)borate mixture and eosin–TPB mixture (all dissolved in 2-nitrotoluene) were also tested but the experimental results did not show an improvement in the sensor's behavior.

The amitriptyline membrane sensor was used for the determination of amitriptyline in some pharmaceuticals by direct potentiometry (standard-addition method) or by potentiometric titration method (titrant, sodium picrate; pH 3.3, acetate buffer).

For protriptyline, a coated-wire selective membrane sensor based on dinonylnaphthalene sulfonic acid was prepared.[30] The calibration curve for a set of five protriptyline membrane sensors showed that detection limits of $10^{-6.5}$ M were obtained (linear response range, 10^{-3} to 10^{-5} M). This sensor can be used to determine very low levels of protriptyline in serum and urine samples. Its detection limit compares well to several alternative methods applicable to clinical samples.

Analytical Procedures

i. *Potentiometric titration of antidepressant compounds with standard 10^{-2} M sodium tetraphenylborate solution:*
A volume of 25-cm^3 aliquot of the sample in the range 2×10^{-4} to 1.0×10^{-3} M is pipetted into a 50-cm^3 beaker; 5 cm^3 of the appropriate acetate buffer solution (see the preceding text) is added and the titration is conducted in a usual assay (tetraphenylborate membrane sensor, as indicator, is used). The amount of pharmaceutical compounds present in the sample is calculated, taking into account the maximum slope on the plots of EMF vs. titrant volume.

ii. *Analysis of pharmaceutical preparations:*
For tablet preparation, 20 tablets are weighed and powdered. An appropriate weighed amount of the powder (equivalent to about 0.5 mmol of active principle is transferred to a 500-cm^3 beaker and stirred vigorously with about 400 cm^3 of distilled water for 15 min. The solution is diluted to the mark in a 500-cm^3 volumetric flask and

a 25-cm^3 aliquot is potentiometrically titrated as previously described. For injections (e.g., imipramine) an appropriate volume of the sample (equivalent to 0.2 mmol of active principle) is diluted with water to 200 cm^3 in a volumetric flask and a 25-cm^3 aliquot is potentiometrically titrated as previously described.

iii. *Potentiometric titration of amitriptyline with standard 10^{-2} M sodium picrate solution:*
A volume of 25-cm^3 of the sample in the concentration range of about 10^{-3} *M* is transferred into a 50-cm^3 beaker; 5 cm^3 of the acetate buffer solution (pH 3.3) is added and the titration is conducted in a usual way (amitriptyline membrane sensor, as indicator, is used). The end point corresponds to the maximum slope on the plot of EMF vs. titrant volume.

5.7 Amphetamine and Methamphetamine

$$C_9H_{13}N \text{ (MM = 135.2)} \qquad C_{10}H_{15}N \text{ (MM = 149.2)}$$

$$C_6H_5-CH_2-\underset{\underset{NH_2}{|}}{CH}-CH_3 \qquad C_6H_5-CH_2-\underset{\underset{NH-CH_3}{|}}{CH}-CH_3$$

Therapeutic category: central nervous stimulant agents

Discussion and Comments

An amphetamine-selective membrane sensor described by Luca et al.[84] contains amphetamine–octadecylsulfate ion-pair dissolved in nitrobenzene as an electroactive membrane. The sensor showed a linear response over 10^{-2} to 10^{-5} *M* range and can be used for determination of amphetamine by direct potentiometry. Amphetamine sulfate was also potentiometrically titrated at pH 7.0 with 1% relative standard deviation, using a liquid-membrane tetraphenylborate sensor and 10^{-2} *M* NaTPB solution as titrant.[14]

A coated-wire methamphetamine-selective membrane sensor based on dinonylnaphthalene sulfonic acid[30] presents a near-Nernstian response over the range 10^{-3} to 10^{-5} *M* with a detection limit of 10$^{-5.5}$ *M*. Methadone, cocaine, and protriptyline interfere in its response (k_{ij}^{pot} = 2.41, 0.54, and 2.16, respectively).

Analytical Procedure

10^{-3} to 10^{-5} *M* standards of amphetamine or methamphetamine are prepared by successive dilution from a 10^{-2} *M* stock solution of amphetamine sulfate or methamphetamine hydrochloride. A constant ionic strength ($I = 0.1$ M, adjusted with potassium nitrate) must be used. The standard solutions (10^{-3} to 10^{-5} M) are transferred into 150-cm^3 beakers containing Teflon-coated stirring bars. The respective indicator sensor and the reference electrode (SCE) are successively immersed in the standards. The plots of $-\log$(drug concentration) vs. E (in millivolts) are prepared and the drug concentration in the sample is determined from these calibration plots.

5.8   Antazoline

$C_{17}H_{19}N_3$ (MM = 265.4)

Therapeutic category: antihistaminic

Discussion and Comments

A copper-wire antazoline-selective membrane sensor based on incorporation of antazoline–tetraphenylborate ion-pair complex in a poly (vinylchloride) coating membrane was recently constructed.[85] Four coating membrane compositions were investigated but that containing a mixture of 120 mg PVC, 100 mg DOP, and 30 mg of the ion-pair complex was found to be the best for the sensor preparation. The sensor made by using this membrane composition exhibited a Nernstian-type response to antazoline in the range 1.41×10^{-5} to 0.89×10^{-2} M with a slope of 59 mV decade^{-1} at 25°C and a response time ≤ 10 s. The pH has a negligible effect on the electrode response within the pH range 2.3 to 9.0. The proposed coated-wire sensor is very selective toward antazoline cation with respect to many common inorganic and organic cations, sugars, and amino acids that are frequently present in biological fluids and pharmaceutical preparations. The sensor has been successfully applied for the determination of antazoline in aqueous solutions and pharmaceuticals by using the standard addition method. The recovery, relative error, and standard deviation values were 98.7 to 100.15%, less than 1.5%, and less than 1.2%, respectively.

Analytical Procedure

The standard-addition method is applied, in which small increments of standard antazoline hydrochloride solution (10^{-2} M) are added to 50-cm^3 aliquot samples of various concentrations (3.0×10^{-4} to

1.5×10^{-3} M). The change in the potential reading (at 25°C) is recorded for each increment and used to calculate the concentration of the antazoline hydrochloride sample solution.

For analysis of antazoline formulations, 0.05 to 0.15 g, 0.3 to 1.0 cm³ of tablets, powder, and/or lotion or drops, respectively, is dissolved in 50 cm³ of distilled water and the standard-addition method is applied as previously described.

5.9   Aspirin

$C_9H_8O_4$ (MM = 180.2)

COOH

OCOCH$_3$

Therapeutic category: analgesic; antipyretic; anti-inflammatory

Discussion and Comments

A salicylate-ion-selective membrane sensor was used for the potentiometric determination of aspirin after prior hydrolysis to salicylic acid.[86] Salicylate-membrane sensor contains Aliquot 336 S–salicylate ion-pair as electroactive material dispersed in PVC (di-n-butylphthalate as plasticizer). The response of the sensor was linear in the salicylate concentration range 4×10^{-5} to 1×10^{-1} M with a slope of 56 mV decade^{-1} of concentration; the sensor showed very fast responses, achieved nearly instantaneously or within 15 s. The useful lifetime of the sensor was about three months. Hydrolysis of aspirin tablets was performed by refluxing an equivalent of about 0.5 g acetylsalicylic acid with 0.5 M sodium hydroxide solution for 1 h. An aliquot sample was analyzed by the standard-addition method, with a better precision than that of the BP official method (standard deviation less than 0.9%).

An aspirin-selective membrane sensor, containing aspirin–tricaprylylcetylammonium (N-263) ion-pair complex as electroactive material in PVC membrane (diethylhexylphosphate [DEHP] as plasticizer), shows a near-Nernstian response within 10^{-1} to 10^{-5} M aspirin range (detection limit = 7.3×10^{-6} M) with a slope of 57 ± 1 mV decade^{-1}.[87] When DBP was used as plasticizer the Nernst-type range was shorter (10^{-1} to 10^{-4} M) and the detection limit reached only 5.3×10^{-5} M (slope 58 ± 1 mV decade^{-1}). It was reported that the sensor can work properly within the pH range 6 to 12 with short response times (for 10^{-2} to 10^{-3} M range, less than 10 s; for 10^{-4} to 10^{-5} M, about 30 s) without significant interferences from vitamin C, procaine, caffeine, and sodium isopentylbarbiturate. However, ClO_4^- ions strongly interfere.

Aspirin in pharmaceuticals was assayed with good results by direct potentiometry (recovery 101.3%; standard deviation 4.1%).

Analytical Procedures

i. *Aspirin membrane sensor:*

Ten tablets of aspirin are finely powdered. A portion of the powder equivalent to about 10 mmol of acetylsalicylic acid is quantitatively transferred with distilled water to a 250-cm^3 volumetric flask and the solution is made up with water. An aliquot of 10 cm^3 of this solution is diluted 10 times in a 100-cm^3 volumetric flask with distilled water. The solution is transferred into a 150-cm^3 beaker, where the sensor pair (SCE as reference) is immersed. The EMF is recorded under stirring and compared with the calibration curve.

ii. *Salicylate membrane sensor:*

Ten tablets of aspirin or aspirin/phenacetin/codeine are finely powdered. A portion of the powder equivalent to 400 mg of acetylsalicylic acid is refluxed with 25 cm^3 of 0.5 M sodium hydroxide solution for 1 h. The solution is then filtered and made up to 250 cm^3 in a volumetric flask. An aliquot (10 cm^3) of the solution is neutralized to bromocresol green with dilute sulfuric acid and diluted to 100 cm^3 with water; 25 cm^3 of this solution is mixed with 3 cm^3 of the saturated borax solution and 2 cm^3 of 2 M ammonium sulfate solution. The standard-addition method was employed to evaluate the concentration of salicylate ion in the resulting solution.[86]

5.10   Barbiturates and Thiobarbiturates

Table 5.6 summarizes the barbiturate and thiobarbiturate drugs (compounds with very well known hypnotic activity) that can be assayed by potentiometry with selective membrane sensors.[88-94]

The liquid-membrane sensor obtained by Hopîrtean and Veress[88] displayed a linear response to barbital in the concentration range 10^{-1} to 5.6×10^{-4} M. The sensor was prepared by soaking a hydrophobized G_4 glass frit with a nitrobenzene solution of cetylpyridinium bromide (concentration 0.05% m/v). The organic phase is also the internal reference solution. The electrode function is reproducible with a deviation of less than ± 1 mV and the response time was 2 to 5 min for solutions of concentrations $> 10^{-3}$ M and 8 to 10 min for $< 10^{-4}$ M.

Carmack and Freiser[89] developed a rapid and reliable method for the assay of phenobarbital based on measuring the anion potentiometrically,

Table 5.6 Barbiturates and Thiobarbiturates Assayed by Membrane Sensors

Compound	R_1	R_2	X	Formula (MM)
Barbital	— C_2H_5	— C_2H_5	O	$C_8H_{11}N_2NaO_3$ (206.2)
Phenobarbital	— C_2H_5	— C_6H_5	O	$C_{12}H_{11}N_2NaO_3$ (254.2)
Inactin	— C_2H_5	—CH—CH$_3$ $\vert$ C_2H_5	S	$C_{10}H_{15}N_2NaO_2S$ (250.3)
Thiopental	— C_2H_5	—CH—CH$_3$ $\vert$ C_3H_7	S	$C_{11}H_{17}N_2NaO_2S$ (264.3)

with a coated-wire selective membrane sensor. This sensor is based on the ion-pair complex between phenobarbital and tricaprylylmethylammonium from Aliquat 336 S (chloride salt). The construction of coated-wire membrane sensors is described by Freiser and co-workers.[95–97]

Weighed amounts of the electroactive material are either dissolved in a 5% (m/v) solution of PVC in tetrahydrofuran or mixed with the epoxy mixture (containing equal masses of resin and curing agent). The end of a platinum wire is repeatedly dipped into the mixture until a uniform coating is obtained. This usually requires about three dippings. The sensors are cured and air-dried overnight; the exposed portion of the wire is then wrapped tightly with Parafilm. All sensors are initially conditioned by soaking in a 0.1 M phenobarbital solution for 1 h. Immediately before use the membrane sensors are soaked in a dilute (approximately 10^{-4} M) phenobarbital for about 15 min.

The optimum membrane composition is 70% (m/m) of the Aliquat salt in the PVC films and 50% (m/m) for the epoxy films. The PVC coated-wire sensor gives a linear response from 10^{-1} to 10^{-4} M with a slope of 55 ± 2 mV decade^{-1}.

Response times of these coated-wire sensors are fast, being nearly instantaneous at higher concentrations and requiring less than 1 min for 10^{-4} M phenobarbital. EMF readings are reproducible to better than ± 1 mV over the entire concentration range. However, the absolute EMF varies daily by 5 to 15 mV, indicating a one-point standardization before each run. The useful lifetime of these sensors is at least three months.

Results of the potentiometric analysis of phenobarbital tablets using an epoxy coated-wire sensor showed a good agreement with the official

USP method. In contrast to the 4 h required for assay by USP method, an electrode assay can be accomplished within 20 min.[89]

A field effect transistor sensitive to phenobarbital anion was described by Covington et al.[90]. The electroactive material of the membrane was tricaprylmethylammonium–phenobarbital ion-pair. A solution containing electroactive material and PVC in tetrahydrofuran was applied by three solution castings to the gate of the chemically sensitive field effect transistor (CHEMFET); the device was pre-conditioned for 24 h by soaking it in 10^{-1} M sodium phenobarbital.

The linear response range was between 10^{-1} and 10^{-3} M sodium phenobarbital (with added hydroxide to the calibration solutions) and between 10^{-1} and 10^{-4} M when the calibration solutions did not contain sodium hydroxide. The sensitivity of the device decreased rapidly below 10^{-4} M sodium phenobarbital, because of hydroxide interference.[90] Other anions (Cl^-, NO_3^-, SO_4^{2-} and HPO_4^{2-}) interfered, too. Use of the device in whole blood plasma, therefore, would not give accurate results. On the other hand, this device has a significantly superior performance over the coated-wire phenobarbital sensor, regarding its stability and response time ($\tau_{95} = 200$ ms for a stepped change in phenobarbital concentration).

A PVC membrane sensor with inner graphite contact has been proposed for construction of a selective electrode for phenobarbital.[91] Two site carriers were used to prepare the electroactive materials for the membranes, hexadecyltrioctylammonium iodide (HTOA) and hexadecyltriphenylphosphonium iodide (HTPP), respectively; DOP was used as plasticizer.

The main characteristics of those phenobarbital membrane sensors are presented in Table 5.7.

Amobarbital, phenytoin, ethacrynic acid, and salicylic acid interfere in the sensor response.

Phenobarbital was assayed with an average recovery of 98.5% ($n = 10$, coefficient of variation = 0.86%) in aqueous samples containing 0.5 to 2 mg cm^{-3} phenobarbital.

Thiopental membrane sensors containing thiopental–HTOA or thiopental–HTPP as electroactive materials in PVC matrix were constructed and characterized.[91,93] Table 5.8 summarizes the main characteristics of thiopental membrane sensors, containing different concentrations of electroactive material in the membrane (in all cases, DBP used as solvent mediator).

In solutions with thiopental concentrations $\geq 10^{-4}$ M the response times were within 10 s, whereas in more dilute solutions the response times were about 1 min. Barbital and phenobarbital do not interfere in the thiopental electrode response, but ethacrynic acid showed a slight interference ($K_{Th, Et}^{pot} = 0.23$).

Table 5.7 Potentiometric Response Characteristics of Phenobarbital Membrane Sensors[91]

Site carrier and its concentration in the membrane (%)	Linear range (M)	Detection limit (M)	Slope (mV decade^{-1})
HTPP			
5	$2.2 \times 10^{-1}\text{--}5 \times 10^{-4}$	2.2×10^{-4}	57.1
3	$10^{-1}\text{--}10^{-4}$	7.9×10^{-5}	57.3
1	$10^{-1}\text{--}5.6 \times 10^{-4}$	1.8×10^{-4}	54.5
0.5	$10^{-1}\text{--}4.9 \times 10^{-4}$	2.0×10^{-4}	54.5
HTOA			
5	$10^{-1}\text{--}7.9 \times 10^{-4}$	2.5×10^{-4}	56.6
3	$10^{-1}\text{--}3.2 \times 10^{-4}$	7.9×10^{-4}	57.3
1	$10^{-2}\text{--}2.5 \times 10^{-4}$	1.7×10^{-4}	42.9
0.5	$10^{-2}\text{--}10^{-3}$	5.0×10^{-4}	26.7

Table 5.8 Comparison of Thiopental Membrane Sensor Performances[93]

Site carrier and its concentration in the membrane (%)		Linear range (M)	Detection limit (M)	Slope (MV decade^{-1})
HTOA	3	10^{-1}–7.9×10^{-5}	1.2×10^{-5}	56.0
	2	10^{-1}–1.0×10^{-4}	2.2×10^{-5}	55.0
	1.5	10^{-1}–8.9×10^{-5}	2.1×10^{-5}	55.5
	1	10^{-1}–1.0×10^{-4}	2.4×10^{-5}	57.0
	0.5	10^{-1}–7.9×10^{-5}	1.9×10^{-5}	50.0
HTPP	3	10^{-1}–6.3×10^{-5}	2.2×10^{-5}	54.5
	2	10^{-1}–1.0×10^{-4}	1.9×10^{-5}	54.0
	1.5	10^{-1}–1.0×10^{-4}	1.9×10^{-5}	53.5
	1	10^{-1}–1.0×10^{-4}	2.0×10^{-5}	54.0
	0.5	10^{-1}–6.0×10^{-5}	1.6×10^{-5}	52.5

Sodium hydroxide (0.01 M) was selected as measuring medium for thiopental assay by potentiometric method. The sensor based on thiopental–HTOA membrane was used for determination of thiopental in pure drug solutions (average recovery 100.1%; coefficient of variation = 1.0%, $n = 10$) as well as in some injection preparations.

Coşofreţ and Bunaciu[92] investigated the potentiometric response of the silver sulfide crystal membrane sensor (Orion, Model 94-16) to thiobarbiturate anion. If the sensor is introduced into a solution containing Ag^+ ions on both sides of the membrane, a potential difference is developed, given by

$$E - E_0 = \frac{RT}{F} \ln a_{Ag^+} \qquad (5.32)$$

where a_{Ag^+} is the activity of the silver ion at the sample solution/membrane interface. This sensor also responds to inorganic S^{2-} ions:

$$E - E_0' = - \frac{RT}{2F} \ln a_{S^{2-}} \qquad (5.33)$$

The response to S^{2-} is due to an equilibrium change in a_{Ag^+} at the membrane surface.[98] In the direct potentiometric determination of sulfur-containing organic compounds with this sensor[47, 99, 100] it is assumed that the electrode function is the result of interactions between sulfur-containing functional groups and Ag^+ ions from the membrane.

In the presence of thiobarbiturate ions (5, 5-di-substituted thiobarbiturates are present in two predominant forms in alkaline solutions), the sensor responds according to

$$E = E_0'' - S \log a_{\text{thiob.}}$$ (5.34)

Where S is the slope of the membrane sensor, which in this case should be -59.1 mV per decade of concentration, by taking into account the reaction

thione form   thiolic form

(5.35)

where $a_{\text{thiob.}}$ is the activity of thiobarbiturate ions (e.g., inactin, thiopental).

The linear range of the silver sulfide membrane sensor in 0.1 M NaOH solutions of inactin and thiopental, respectively, is about 10^{-3} to 10^{-5} M thiobarbiturate, with slopes in agreement with the corresponding value for a $1:1$ silver–thiobarbiturate compound (54 mV decade^{-1} for thiopental and 51 mV decade^{-1} for inactin, respectively).

In dilute solutions (e.g., 10^{-5} M), response times of about 15 min were observed. These response times are rather too long and, therefore, the potentiometric titration of thiobarbiturates with silver nitrate solution has also been studied. A silver (I)-ion-selective liquid-membrane sensor[101] (membrane, silver diethyldithiophosphate in CCl_4, 5×10^{-4} M; linear response to Ag^+, 10^{-1} to 10^{-5} M; slope 59.5 mV decade^{-1}; response times of a few seconds in concentrated solutions and 3 to 4 min in dilute solutions) was used in potentiometric titrations of thiobarbiturates and the results were similar to those obtained by a commercial silver sulfide membrane sensor (Orion, Model 94-16).

The potentiometric titrations were carried out in distilled water, with good results down to 5×10^{-4} M thiobarbiturate. The absolute error was $\pm 0.5\%$ for 5 to 25 mg of thiopental. The potentiometric titrations cannot be carried out in alkali medium (10^{-1} or 10^{-2} M sodium hydroxide) because of co-precipitation of silver oxide.

Hopkala[94] also used a Ag(I)-membrane sensor as well as a Cu(II)-membrane sensor as indicator electrode in potentiometric titrations of thiopental with silver nitrate and copper sulfate standard solutions, respectively.

The applicability of liquid-membrane sensors for estimation of partition coefficients was demonstrated by Staroscik and Blaskiewicz.[102] The correlation of selectivity coefficients of sensors based on tricaprylmeth-ylammonium-ion pairs with the octanol–water partition coefficients was demonstrated for some barbiturates and thiobarbiturates.

Analytical Procedures

i. *Direct potentiometry, applied to compounds listed in Table 5.6 (except for inactin) for which membrane sensors are available (homemade); see also the preceding text:*
A sample containing 40 to 50 mg barbiturate or thiobarbiturate (as sodium salt) is taken in a 100-cm^3 volumetric flask and diluted to volume with water. The pH is adjusted to approximately 9.0 to 9.5 with 10^{-2} M sodium hydroxide, and the EMF of the solution is measured by the electrode pair (indicator and SCE as reference) and compared with standards.

ii. *Potentiometric titration, applied to thiobarbiturates:*
The pair of sensors (indicator, silver(I) or silver sulfide membrane; reference, SCE) is introduced into the sample solution (30 to 40 cm^3, approximately 10^{-3} M), which is titrated under stirring with 10^{-2} M silver nitrate solution. The end point corresponds to the maximum slope on the plot of EMF vs. titrant volume.

5.11   Benzoic Acid

$$C_7H_6O_2 \ (MM = 122.1)$$

COOH

Therapeutic category: antibacterial; antifungal

Discussion and Comments

Many benzoate-membrane sensors were constructed and applied to the potentiometric determination of benzoate anion.[88, 96, 103–110] The sensor described by Benignetti et al.[103] consists of a 2-mm-thick liquid membrane (10^{-2} M trimethylhexadecylammonium benzoate in 1-decoanol) set between Millipore cellulose acetate disks. The sensor is rapid (< 30 s) and reproducible in response, having linearity between 5×10^{-1} and 5×10^{-4} M benzoate with 59 ± 1 mV decade^{-1} slope. The sensor is affected in its response by some inorganic and/or organic anions, such as carbonate, sulfate, phosphate, halides, nicotinate, oxalate, etc.

The sensor has been used for the direct determination of sodium benzoate in some commercial expectorant medicinal syrups of varying composition (substituted diamines, citric acid, sugar, alcohols, glycerol, methol, ammonium chloride, etc.), but in some cases pre-extraction was necessary, always with pH adjustment.[103]

The benzoate sensors of Freiser and co-workers[96, 104, 105] gave linear responses between 10^{-1} and 10^{-3} M benzoate with slopes of 58.5 mV decade^{-1} (liquid membrane of Aliquot 336 S–benzoate in 1-decanol) and 53 mV decade^{-1} (coated wire with the same electroactive material).

A membrane sensor sensitive to benzoate (0.05% hexadecylpyridinium bromide in nitrobenzene) has been described by Hopîrtean and Veress.[88] The liquid membrane is prepared by soaking a G_4 glass frit with the liquid ion exchanger. The organic solution is also used as internal reference solution. The electrode response is near-Nernstian (slope 54 mV decade^{-1}) to benzoate over the 4.5×10^{-2} to 3.2×10^{-4} M concentration range.

Shigematsu et al.[106] used a liquid-membrane sensor where the organic phase was a 10% (v/v) solution of methyltrioctylammonium benzoate. The internal aqueous reference phase of 0.01 M potassium chloride also made to 0.01 M in potassium or sodium benzoate was used. The sensor is sensitive to benzoate in the 10^{-1} to 10^{-4} M range with a slope of 58 mV decade^{-1} in the 6 to 10 pH range.

A benzoate-selective membrane sensor containing 5×10^{-2} M solution of tr-n-octylmethylammonium benzoate in o-dichlorobenzene as ion exchanger showed a linear response to benzoate anion over the range 10^{-1} to 10^{-4} M.[107] The selectivity of the sensor was enhanced by the addition of p-t-octylphenol (a strong proton donor) to the liquid ion-exchanger solution. A linear relationship was found between the change in the logarithmic selectivity coefficient and the pK_a value of the aliphatic monocarboxylic acid interference (see Figure 5.5). Large decreases in the selectivity coefficients were observed for ions having a small proton-acceptor ability such as perchlorate and trifluoromethane-sulfonate. An improvement in selectivity of a liquid-membrane benzoate sensor containing decanol was also reported.[103] However, the improvement in selectivity was less than that reported by Hara et al.,[107] because the phenol is a much stronger proton donor than decanol.

The behavior of a benzoate-ion-selective membrane sensor containing tributylcetylphosphonium benzoate (TBCPB) (prepared by reacting TBCP–bromide with silver benzoate in 1 : 1 water–ethanol solution) in 1-decanol or nitrobenzene (concentration 10^{-2} M) showed similar characteristics in respects to slope, response time, reproducibility and accuracy.[108] The Nerstian-type linear ranges were 1 to 10^{-4} M (1-decanol) and 1 to 5×10^{-4} M (nitrobenzene) (in both cases, slope = 59 ± 1 mV decade^{-1}). Both sensors were affected in their response by perchlorate ($k_{ij}^{pot} = 31$), trichloroacetate ($k_{ij}^{pot} = 39$), nitrate ($k_{ij}^{pot} = 19$), and iodide

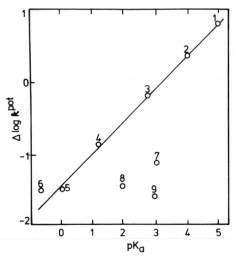

Figure 5.5 Relationship between $\Delta \log k^{\text{pot}}$ and pK_a for some carboxylates: (○) benzoate; (1) acetate; (2) formate; (3) monochloroacetate; (4) dichloroacetate; (5) trichloroacetate; (6) trifluoroacetate; (7) salicylate; (8) hydrogen maleate; (9) hydrogen phthalate. For (7) through (9) the first dissociation constant, pK_{a1}, is taken. (Reproduced from Hara, H., Okazaki, S., and Fujinaga, T., *Anal. Chim. Acta*, 121, 119, 1980, Elsevier Science Publishers, Physical Sciences and Engineering Division. With permission.)

($k_{ij}^{\text{pot}} = 7$). Even so, they were successfully employed in direct determinations, in analysis by standard-addition methods, and also in titrations of benzoate with acids; in this latter case, the accuracy was a little lower (4 to 6%) than in the others. The results, compared with those previously reported[103, 109] show no great influence of the sensor electroactive material, but a marked influence of the membrane solvent.

When hexadecyltrioctylammonium benzoate in nitrotoluene was used as a liquid membrane, a benzoate sensor with good performance was obtained.[110] The Nernst-type response range within 2.5×10^{-4} to 5×10^{-1} M benzoate and its good selectivity over citrate, malate, lactate, vitamin C, acetate, oxalate, and many other inorganic anions (except I^- and NO_3^-) makes it practical for benzoate determination in various pharmaceuticals, in pH range 6.5 to 10.

Analytical Procedure

Standard solutions of 10^{-2} and 10^{-3} M are prepared by successive dilutions from a 10^{-1} M sodium benzoate stock solution. All solutions are adjusted to constant ionic strength ($I = 0.1$ M) with a salt of an anion that does not interfere in the sensor response (e.g., sodium sulfate, sodium citrate, etc.). The standard solutions (10^{-1} to 10^{-3} M) and the sample solution are transferred into 150-cm^3 beakers containing

Teflon-coated stirring bars. The benzoate membrane and the reference sensors are immersed successively in the standards, and the EMF is measured. The benzoate concentration in the sample is determined from the calibration graph.

5.12 Bisquaternary Compounds

The construction, characterization, and analytical evaluation of selective membrane sensors highly sensitive to bisquaternary compounds of pharmaceutical interest, such as those listed in Table 5.9, have been reported by Coşofreţ and Buck.[111] It is well known that the number of methylene groups in the central chain between the two cationic heads in α, ω-bis (trimethylammonium)alkanes has a significant effect on pharmacological activity.

Triphenylstilbenylborate (TPSB) was used as counter-ion in preparing the ion-association complexes. These were prepared *in situ*, by soaking the TPSB (potassium salt)–PVC membranes in the appropriate bisquaternary-drug solution. Of the plasticizers tested, 2-nitrophenyloctyl ether σ-NPOE showed the best behavior in terms of response time and reproducibility. The membrane compositions were 3.2% TPSB, 64.5% σ-NPOE, and 32.3% PVC (m/m).

Table 5.9 Bisquaternary Compounds Assayed by Membrane Sensors

Bisquaternary compound	Formula (MM)	Therapeutic category
Decamethonium bromide	$C_{16}H_{38}Br_2N_2$ (418.4)	Skeletal muscle relaxant
	$[(CH_3)_3 \overset{+}{N}(CH_2)_{10} \overset{+}{N}(CH_3)_3]\, 2\, Br^-$	
Hexamethonium chloride	$C_{12}H_{30}Cl_2N_2$ (273.3)	Skeletal muscle relaxant; antihypertensive
	$[(CH_3)_3 \overset{+}{N}(CH_2)_6 \overset{+}{N}(CH_3)_3]\, 2\, Cl^-$	
Succinylcholine bromide	$C_{14}H_{30}Br_2N_2O_4$ (450.2)	Skeletal muscle relaxant
	$\begin{array}{l} CH_2-COOCH_2\overset{+}{N}(CH_3)_3 \\[2pt] \mid \qquad\qquad\qquad 2\,Br^- \\[2pt] CH_2-COOCH_2\overset{+}{N}(CH_3)_3 \end{array}$	

Table 5.10   Response Characteristics for Bisquaternary-Drug Membrane Sensors[111]

Parameter	Decamethonium sensor	Hexamethonium sensor	Succinylcholine sensor
Slope^a (mV $(\log a)^{-1}$)	29.45 ± 0.29	28.03 ± 0.44	29.05 ± 0.35
Intercept (mV)	157 ± 1.8^b	147 ± 2.2	152 ± 2.1
Linear range (M)	10^{-2}–2.5×10^{-7}	10^{-2}–2.5×10^{-7}	10^{-2}–10^{-6}
Usable range (M)	10^{-2}–10^{-7}	10^{-2}–5×10^{-8}	10^{-2}–10^{-7}
Detection limit			
(M)	1.58×10^{-7}	3.16×10^{-8}	1.12×10^{-7}
(ng cm^{-3})	64	6.4	29

^{a}Average values calculated for 10^{-3} to 10^{-5} M range with standard deviation of average slope value for multiple calibration.$^{(n=5-7)}$
^{b}Standard deviation of values recorded during one month.

The critical response characteristics of the TPSB-based bisquaternary-drug membrane sensors are summarized in Table 5.10. The highest concentration of bisquaternary drugs used for calibrations was 10^{-2} M. All sensors can be used in the concentration range 10^{-2} to 10^{-7} M, a linear E (in millivolts) vs. pC plot being obtained down to 10^{-6} M (pH range 2.0 to 10.5; stable EMF readings within 15 s in the linear ranges of the respective calibration curves).

All the TPSB-based sensors exhibit negligible interference from common inorganic cations or from some amino acids and neurotransmitters. Choline, acetylcholine, and quaternary ammonium compounds interfere in the sensor responses; as expected by inspection of the bisquaternary-drug structures, the greatest interference was observed in the case of the succinylcholine sensor.

Results for measurements of the pure drug solutions at < 1.0 μg cm^{-3} showed a good recovery, standard deviation being $< 2.0\%$. Succinylcholine was also determined in injectable solutions with good precision and an average recovery of 100.5%.

Analytical Procedures

i. *Direct potentiometric measurement of bisquaternary drugs in the microgram range:*
   The appropriate bisquaternary-drug membrane sensor and the SCE reference electrode are immersed in the aqueous sample solution (25.0 cm^3) at pH 7.0 (0.1 M TRIS–HCl buffer). After equilibration by stirring, the EMF value is recorded. The figure obtained is checked by

the standard-addition method. For this purpose, a volume of 1.0 cm^3 of a 10^{-4} M standard solution of the bisquaternary drug is added and the change in millivolt reading (accuracy ± 0.1 mV) is recorded and used to calculate the concentration of the drug.

ii. *Direct potentiometric assay of succinylcholine in injectable solutions (e.g., 10 mg cm^{-3}):*
A volume of 1.0 cm^3 of injectable solution from the mixture solution of about 10 vials is pipetted into a 250-cm^3 volumetric flask; 12.5 cm^3 of 1% (m/v) methyl-p-hydroxybenzoate as preservative is added and the solution is diluted to the mark with 0.1 M TRIS–HCl buffer solution of pH 7.0. A 50-cm^3 aliquot is pipetted into a 100-cm^3 beaker. The electrode pair is immersed into solution and after equilibration the EMF value is recorded; 5.0 cm^3 of 10^{-2} M standard solution of succinylcholine chloride is added under stirring, and the change in millivolt reading is recorded and used to calculate the concentration of succinylcholine. The result for succinylcholine chloride is obtained from the equation

$$C \left(\text{mg cm}^{-3} \text{ of vial solution} \right)$$

$$= C_S \times 9032.5 \times \left(1.10 \times 10^{\Delta E/S} - 1 \right)^{-1}$$

where C_S is the concentration of the standard solution, ΔE is the change in millivolt reading caused by the standard addition, and S is the electrode slope, which must be evaluated very accurately before each series of measurements.

5.13   Bromisoval

$$C_6H_{11}BrN_2O_2 \ (MM = 223.1)$$

Therapeutic category: hypnotic, sedative

Discussion and Comments

For halogen liberation from a compound like bromisoval (bromine linked to a saturated carbon atom), hydrolysis of the sample with an aqueous solution of 20% NaOH with a reflux of about 30 min was frequently used[37, 112, 113] (for details, see Coşofreţ,[98] paragraph 52, p. 74). Goina et al.[112] determined bromine in bromisoval by potentiometric titration with silver nitrate with an absolute error of 0.4%. They also determined bromisoval (after alkaline mineralization) by direct potentiometry at 0.1 M ionic strength (1% absolute error for 10 mg bromisoval).

Analytical Procedure

The weighted sample (approximately 50 mg) is dissolved in approximately 10 cm^3 of methanol in a 250-cm^3 round-bottomed flask, and 20 cm^3 of 20% sodium hydroxide solution is added. The solution is brought to reflux for 30 min. After cooling the contents are quantitatively transferred to a 200-cm^3 volumetric flask and made to volume with distilled water. A 50-cm^3 aliquot is transferred to a 150-cm^3 beaker, neutralized with 2 N nitric acid solution (methyl red), and titrated potentiometrically with 10^{-2} M silver nitrate solution (bromide- or other halide-selective membrane sensor or silver(I)-membrane sensor as indicator). The end point corresponds to the maximum slope of the titration curve (E (in millivolts) vs. volume).

5.14   Bromoform and Other Halogenated Volatile Anesthetics

Bromoform   Chloroform   Ethyl chloride
$CHBr_3$ (MM = 252.7) $CHCl_3$ (MM = 119.4) C_2H_5Cl (MM = 64.5)

Therapeutic category: volatile anesthetics

Discussion and Comments

Results for the microdetermination of chlorine and bromine in highly volatile halogenated organic compounds by the oxygen flask method according to Schöniger are in many instances too low because of incomplete combustion.[114, 115] Several types of container for liquids have been used in Schöniger's method.[116, 117]

Milligram samples of highly volatile halogenated compounds can be assayed by the combustion tube method,[115, 118, 119] where the combustion and final analysis are carried out separately.

In the method of Potman and Dahmen[120] the sample of volatile halogenated compound is introduced into a combustion system by injection. After oxygen combustion over quartz and platinum at 1000°C, the liberated halide is absorbed in 80% acetic acid containing 1.5% nitric acid, 12% hydrogen peroxide, and 17.3% water (v/v) and mercury(II), chloride or bromide (0.140 mg/100 cm^3 and potentiometrically titrated with mercury(II). If the concentration of hydrogen peroxide is greater than 1.2% (v/v), oxidation of sulfide from the silver sulfide crystal membrane sensor (used for titration) occurs. Chloroform and bromoform were determined by Potman and Dahmen's procedure[120] with relative standard deviations of 1.29 and 0.82%, respectively.

Analytical Procedure

The combustion and titration apparatus described in Potman and Dahmen[120] is used. Mercury(II) is added from the burette at such a rate that during the titration the potential is held to within ± 10 mV of the set point. The titration is complete when no more than 0.1 mm^3 of mercury(II) has to be introduced per minute.

5.15   Bupivacaine and Other Related Anesthetic Drugs

Table 5.11 lists some local anesthetic drugs for which sensitive membrane sensors have been constructed and characterized.[121–126]

Dinonylnapthalenesulfonic acid, in a PVC membrane (composition 4.0% DNNS, 64.0% plasticizer [o-NPOE], and 32% PVC) was found to be an adequate site carrier for bupivacaine- and mepivacaine-selective membrane sensors.[121] The sensors displayed linear response within 5×10^{-6} to 10^{-2} M for bupivacaine (slope 55.4 mV decade^{-1}) and 3×10^{-5} to 10^{-2} M for mepivacaine (slope 52.5 mV decade^{-1}) with fast response times. The functions E (in millivolts) vs. pH were linear in the ranges 2 to 7 and 5 to 7 for bupivacaine- and mepivacaine-membrane sensors, respectively. Among many organic compounds tested as potent interferents (amino acids, vitamin C, nicotinamide, vitamin B$_1$, procaine, scopolamine, etc.) no one showed significant interference. Both sensors were used for determination of bupivacaine and mepivacaine (as hydrochlorides), respectively, from pharmaceutical preparations (injection solutions) with a recovery of 99.6% (bupivacaine) and 98.5% (mepivacaine) from the label amount.

Shirahama et al.[122] have described ion-selective crown-ether membrane sensors for dibucaine, hexylcaine, and other local anesthetics. The principle of the sensors is based on selective transport of cationic local anesthetics, in the form of a complex with a neutral carrier, through a hydrophobic PVC gel membrane.

Dibenzo-24-crown-8 (DB 24) and dibenzo-18-crown-6 (DB 18) ethers are known to form complexes with both inorganic and organic cations, and they function as neutral carriers in electrode membranes. Therefore, these crown ethers were tested for the feasibility of use as specific cation-sensitive sensors for the charged species of aromatic amine local anesthetics. Dioctylphthalate was used as plasticizer in the PVC membranes.

Table 5.11   Bupivacaine and Related Anesthetic Drugs Assayed by Membrane Sensors

Drug substance	Formula (MM)	Ref.
Bupivacaine	$C_{18}H_{28}N_2O$ (288.4)	121, 122
Mepivacaine	$C_{15}H_{22}N_2O$ (246.3)	121
Dibucaine	$C_{20}H_{29}N_3O_3$ (343.4)	122
Hexylcaine	$C_{16}H_{23}NO_2$ (261.3)	122
Lidocaine	$C_{14}H_{22}N_2O$ (234.3)	123–125
Tetracaine	$C_{15}H_{24}N_2O_2$ (264.3)	122, 126

Bupivacaine structure:

C_4H_9 on N; piperidine ring — COHN — aryl with H_3C (two methyl groups, 2,6-dimethyl)

Mepivacaine structure:

CH_3 on N; piperidine ring — COHN — aryl with H_3C (two methyl groups)

Dibucaine structure:

quinoline N — $O(CH_2)_3CH_3$; $CONH(CH_2)_2N(C_2H_5)_2$

Hexylcaine structure:

C_6H_5—COO—CH(CH_3)—CH_2—NH—(cyclohexyl)

Lidocaine structure:

aryl with CH_3, CH_3 — $NHCOCH_2N(C_2H_5)_2$

Tetracaine structure:

$CH_3(CH_2)_3NH$—(phenyl)—$COOCH_2CH_2N(CH_3)_2$

The DB 24 membrane sensor (1% crown ether, 20% PVC, and 79% DOP) was used to measure dibucaine response. It is sensitive to dibucaine down to 0.3 mM solution with near-Nernstian slope. A selectivity coefficient $k_{\text{Dib}^+,\text{Na}^+}^{\text{pot}} = 1.2 \times 10^{-3}$ shows that the sensor is more selective to dibucaine than to Na$^+$. This sensor also responds to hexylcaine (1.5 times more than to dibucaine). Benzocaine, bupivacaine, cocaine, dibucaine, procaine, lidocaine, and tetracaine were also tested. Among these anesthetics, dibucaine and hexylcaine were ideally sensed by the DB 24 sensor.

A PVC-gel membrane containing DB 18 also senses dibucaine and hexylcaine, but less ideally. The ring size of DB 18 may be too small to form a well-fitted stable complex, leading to a less sensitive membrane sensor.

The performance characteristics have been published for three lidocaine ion-selective membrane sensors containing as electroactive materials either lidocaine–dipicrylamine[123] and lidocaine–reineckate,[124] ion-pair complexes in nitrobenzene or a lidocaine–dinonylnaphthalenesulfonic acid ion-pair complex[123] in a PVC matrix.

Nitrobenzene was found to be a suitable solvent for the liquid membranes of the sensor constructed by impregnating a support material (a graphite rod, made water-repellent[123] or an Orion 92-05-04 porous membrane[124]), because of its high partition coefficient and its high dielectric constant, which imparts a high conductance to the membrane. In the first case, a concentration of 5×10^{-3} M electroactive material was used, whereas in the second case a concentration of 0.5% (m/v) was found to be adequate with respect to main electrode characteristics.

The lidocaine–dinonylnaphthalenesulfonic acid ion-pair complex was obtained *in situ*, by soaking the dinonylnaphthalenesulfonic acid–PVC membrane in 10^{-2} M lidocaine hydrochloride. The membrane composition was 4.0% DNNS, 64.0% o-NPOE, and 32.0% PVC. The sensors displayed Nernst-type ranges within 10^{-1} to 10^{-4} M, 10^{-1} to 3×10^{-5} M, and 10^{-1} to 3.2×10^{-5} M when dipicrylamine, reineckate, and DNNS, respectively, were used as site carriers in the respective membranes. For dipicrylamine- and DNNS-based sensors the respective slopes were 58.2 ± 0.6 and 57.3 ± 0.7 mV decade^{-1}. Hassan and Ahmed[124] have found a calibration slope of only 29 mV decade^{-1} by supposing that lidocaine contains two basic centers (divalent cation), which, unfortunately, is not true.

The determination of lidocaine hydrochloride in injection solutions[123, 124] and gels, ointments, and sprays[124] show an average recovery of minimum 99% and a standard deviation of better than 1.7%. Lidocaine has been also assayed in a pharmaceutical product used in dentistry (which also contains razebyl and voltaren and other inorganic and organic excipients) by both membrane electrode technique and densitometry.[125] The results obtained compared well and met the product specification.

PVC- and liquid-membrane tetracaine-selective sensors containing te-tracaine–tetraphenylborate, tetracaine–picrate, and tetracaine–rein-eckate, respectively, as electroactive materials were constructed and characterized by Shen et al.[126] The best sensor was that containing tetracaine–TPB in PVC, which shows near-Nernstian response in the range 10^{-1} to 10^{-5} M (slope 57 mV decade^{-1}) with a detection limit of 1.6×10^{-6} M. The membrane sensor is not affected by pH change in the range 3 to 7, gives fast response (less than 10 s in 10^{-2} to 10^{-4} M and about 1 min in solutions $< 10^{-4}$ M) and has good reproducibility and stability. Except for procaine and for tetrabutylammonium ion, no inter-ference was observed with common inorganic and organic substances tested. The sensor proved useful for the determination of tetracaine samples: direct potentiometry gives an average recovery of 99.9% (rela-tive standard deviation 1.15%), whereas potentiometric titration with NaTPB solution shows an average recovery of 99.5% (relative standard deviation 0.66%).

Analytical Procedures

i. *Standard-addition method for bupivacaine and mepivacaine (applied both to the pure drug substance and to injectable solu-tions):*
   The sample to be assayed (50 cm^3, pH 5.0, citrate buffer) is trans-ferred into a 100-cm^3 beaker. The electrode pair is immersed into solution and after equilibration the EMF value is recorded; 1.0 or 2.0 cm^3 standard solution of 10^{-2} M bupivacaine hydrochloride or mepivacaine hydrochloride is added under stirring, and the change in millivolts is recorded and used to calculate the concentration of the respective anesthetic.

ii. *General method, applied to anesthetics for which membrane sensors are available (homemade; see also the preceding text):*
   A stock solution of 10^{-1} M anesthetic drug is prepared by dissolving the respective compound (usually as chloride) in distilled water. The pH of the solution is adjusted to the recommended value with NaOH and/or HCl solution or buffer solution. 10^{-2} and 10^{-3} M anesthetic solutions are obtained from the stock solution by successive dilu-tions. The sensors (anesthetic-sensitive and SCE) are placed in the standard solutions, and EMF readings (linear axis) are plotted against concentration (logarithmic axis). The sample concentration is deter-mined from this graph.

iii. *Potentiometric assay of lidocaine hydrochloride in injectable solutions:*
   A volume equivalent to the content of an ampoule (concentration 1 to 2%) is transferred with distilled water into a 25-cm^3 volumetric flask

and diluted to volume. Volumes of 3 to 8 cm^3 of this solution are pipetted into a 50-cm^3 beaker. About 20 cm^3 of potassium hydrogen phthalate buffer solution of pH 9.5 are added and the sample is potentiometrically titrated with standard 10^{-2} M sodium tetraphenylborate solution. The volume of titrant at the equivalence point is obtained in the usual way.

5.16   Carbetapentane

$$C_{20}H_{31}NO_3 \ (MM = 333.5)$$

$$H_5C_6\diagdown \diagup COOCH_2CH_2OCH_2CH_2N(C_2H_5)_2$$

Therapeutic category: antitussive

Discussion and Comments

Both PVC-coated copper plate and coated copper-wire sensors sensitive to carbetapentane were constructed by using carbetapentane–phosphotungstate ion-pair complex as electroactive material and dibutylphthalate as plasticizer.[127] The membrane sensors, activated only for 45 min in 5×10^{-3} M carbetapentane solution, showed a linear response in the range 10^{-2} to 2×10^{-6} M with a slope of 59 mV decade^{-1} (pH range 2.2 to 7.0).

From Table 5.12, which shows the effect of various plasticizers on the electrode function, can be seen that many other plasticizers were suitable for sensor preparation.

From the many inorganic and organic ions tested as potent interferents, only $(C_4H_9)_4N^+$ and cloperastine showed positive log $k_{A,B}^{pot}$ values (0.15 and 0.57, respectively).

Table 5.12   Effect of Various Plasticizers on the Carbetapentane Membrane Sensor Response[127]

Phthalate	Dimethyl	Diethyl	Dibutyl	Diisooctyl	Dinonyl	Didecyl
Slope [mV (log C)$^{-1}$]	55	58	59	59	59	59
Linear range, $1 \times 10^{-2} \sim$ (M)	3.3×10^{-4}	2.3×10^{-4}	2.0×10^{-6}	2.6×10^{-6}	8.0×10^{-6}	5.6×10^{-6}

The membrane sensors were successfully used for the determination of carbetapentane citrate (Toclace) in aqueous solutions, urine, blood, and tablets, both by direct potentiometry and potentiometric titration (3×10^{-2} M NaTPB as titrant).

Analytical Procedure

The electrode pair (carbetapentane-selective membrane as indicator and SCE as reference) is introduced into the sample solution (30 to 40 cm^3, concentration approximately 3×10^{-3} M, and pH adjusted to approximately 5.0) and potentiometrically titrated with 3×10^{-2} M NaTPB solution. The end point corresponds to the maximum slope on the plot of EMF vs. titrant volume.

5.17   Cephalosporins

Table 5.13 summarizes the cephalosporins (antibiotic agents) that can be determined by a potentiometric method using a silver sulfide ion-selective membrane sensor and 10^{-2} M lead(II) nitrate solution as titrant. The method is based on alkaline degradation of respective cephalosporin and conversion of the resulting sulfide into lead(II) sulfide.[128] Different cephalosporins were shown to give different but reproducible yields of sulfide, otherwise already confirmed by a spectrophotometric method.[129]

The degradation of cephalosporins were carried out in 1 M NaOH solution containing 2% ascorbic acid, for a 5×10^{-3} M level at 100°C for the length specified in the last column of Table 5.13.

Aliquots of these degraded solutions were potentiometrically titrated in 1 M NaOH solution medium containing 2% ascorbic acid and 10% (v/v) of glycerol. Lead nitrate was used instead of silver nitrate for the titration of sulfide, mainly because there is less adsorption of sulfide ion on lead sulfide than on silver sulfide and this gave more accurate results.[128, 130] The addition of glycerol increased the stability of the potential readings.

Matsumoto et al.[131] used a bacterial membrane sensor for the determination of cephalosporins. *Citrobacter freundii* produces cephalosporinase and catalyzes the following reaction of cephalosporin, which liberates hydrogen ions:

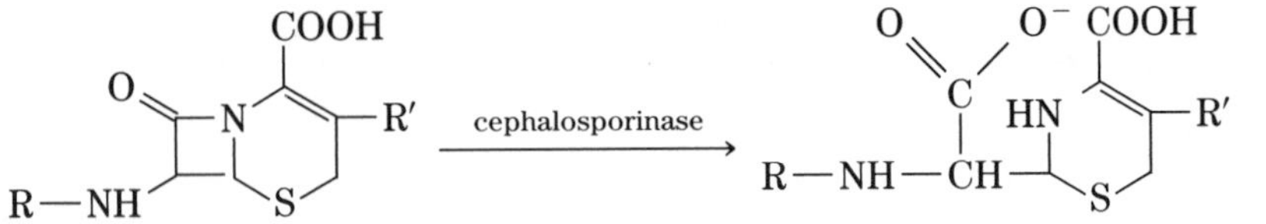

$$(5.36)$$

Table 5.13 Cephalosporins Assayed by Silver Sulfide Membrane Sensors

$$R-NH-\overset{S}{\underset{O}{\Big|}}\overset{}{\underset{N}{\Big|}}R'$$
COOH

Cephalosporin (MM)	R	R'	Molar yield of H_2S formed in 1.0 M NaOH solution at 100°C[a]	Time required (min)
7-ACA 7-Aminocephalosporanic acid $C_{10}H_{12}N_2O_5S$ (272.3)	H —	$- CH_2OCOCH_3$	48.1	40
7-ADC 7-Aminodeacetoxy-cephalosporanic acid $C_8H_{10}N_2O_3S$ (214.3)	H —	$- CH_3$	57.3	35
Cefaclor $C_{16}H_{18}N_3O_4SCl$ (383.9)	$C_6H_5-\underset{NH_2}{\underset{\mid}{CH}}-CO-$	$- CH_2Cl$	43.5	40
Cefazolin $C_{14}H_{14}N_8O_4S_3$ (454.4)	$\underset{N}{\overset{N=N}{\diagup}}\underset{\diagdown}{N}-CH_2-CO-$	$\overset{N-N}{\underset{\underset{CH_2}{\mid}}{S}}\diagdown_{S}\diagdown_{CH_3}$	25.1	40
Cefuroxime $C_{16}H_{16}N_4O_8S$ (424.4)	(furan)$-\underset{N-OCH_3}{\overset{\parallel}{C}}-CO-$	$- CH_3$	14.2	50
Cephalexim $C_{16}H_{17}N_3O_4S$ (347.4)	$C_6H_5-\underset{NH_2}{\underset{\mid}{CH}}-CO-$	$- CH_3$	65.8	30
Cephaloglycin $C_{18}H_{19}N_3O_6S$ (405.4)	$C_6H_5-\underset{NH_2}{\underset{\mid}{CH}}-CO-$	$- CH_2OCOCH_3$	19.5	40

Table 5.13 Continued

Cephalosporin (MM)	R	R'	Molar yield of H_2S formed in 1.0 M NaOH solution at 100°C[a]	Time required (min)
Cephalonium $C_{19}H_{17}N_4O_5S_2$ (445.5)	[furan ring]-CH_2-CO-	$-{}^+N$[pyridine]$-CO-NH_2$	17.4	60
Cephaloridine $C_{19}H_{17}N_3O_4S_2$ (415.5)	[furan ring]-CH_2-CO-	$-CH_2-{}^+N$[pyridine]	18.1	60
Cephalothin $C_{16}H_{16}N_2O_6S_2$ (396.4)	[furan ring]-CH_2-CO-	$-CH_2OCOCH_3$	19.7	45
Cephoxazole $C_{21}H_{18}N_3O_7SCl$ (491.9)	[Cl-substituted benzisoxazole ring with $N-O$ and CH_3]-CO-	$-CH_2OCOCH_3$	14.8	60
Cephradine $C_{16}H_{19}N_3O_4S$ (349.4)	[cyclohexadiene ring]-$CH-CO$- with NH_2	$-CH_3$	61.9	30

[a] Mean of two or three determinations; all within 1% error.

Cephalosporin may therefore be determined from the hydrogen ions generated into the medium by immobilized cephalosporinase. The immobilization of the enzyme was difficult because its molecular mass is only 30,000 and the enzyme is unstable. Therefore whole cells of *C. freundii* were immobilized in collagen membrane and set up as a bacteria collagen membrane reactor that, used with a combination glass electrode, can be used in the determination of cephalosporins (for details, see also Satoh et al.[132] and Karube et al.[133]).

Analytical Procedure

Approximately 5×10^{-3} M solutions of cephalosporins in 1 M sodium hydroxide solution, containing 20 g dm^{-3} of ascorbic acid, are degraded at 100°C for the length of time recommended as mentioned in the last column of Table 5.13. Aliquots (25 cm^3) of these degraded solutions are diluted in a water-jacketed titration vessel held at 25°C, with 100 cm^3 of 1 M sodium hydroxide solution containing 20 g dm^{-3} of ascorbic acid and 10% (v/v) of glycerol. The solutions are potentiometrically titrated with 10^{-2} M lead nitrate solution using the silver sulfide membrane sensor as indicator and SCE connected to the solution by a 0.1 M potassium nitrate salt bridge as reference. The concentration of the respective cephalosporin is calculated from the volume at the equivalence point, evaluated in a usual way.

5.18   Chloral Hydrate

$$C_2H_3Cl_3O_2 \ (MM = 165.4)$$

$$\underset{Cl}{\overset{Cl}{\underset{\textstyle |}{\overset{\textstyle |}{Cl}}}}{C}{-}CH{\overset{OH}{\underset{OH}{<}}}$$

Therapeutic category: hypnotic; sedative

Discussion and Comments

A simple and sensitive micromethod for chloral hydrate determination based on oxidation with iodine in chloroform solution was described by Zaki.[134] The iodide ion produced in the aqueous extract is assayed by an iodide-selective membrane sensor (Orion, Model 94-53) by either a direct measurement, standard-addition technique, or potentiometric titration with standard silver nitrate solution. Two equivalents of iodide ion are produced per mole of chloral hydrate:

$$CCl_3CHO \cdot 2\,H_2O + I_2 \rightarrow CCl_3COOH + 2\,H^+ + 2\,I^- \qquad (5.37)$$

The reaction is quantitative in basic medium of ammonium borate. Elimination of the excess of OH$^-$ ions before measurements with an iodide-selective membrane sensor is not necessary because these ions do not interfere. However, the pH of the sample solution was adjusted between 6 and 7 with nitric acid before potentiometric titration with Ag$^+$, in order to avoid the reaction between Ag$^+$ and OH$^-$ after the equivalence point.

The application of the standard-addition technique and potentiometric titration results in an average recovery of 99.9% with a standard deviation not exceeding 0.1% ($n = 15$). An average recovery of 98.1% (standard deviation less than 1.8%, $n = 15$) was obtained by the direct measurement technique. Down to 0.1 mg of chloral hydrate, samples can be successfully determined with a negligible blank.[134]

Analytical Procedures

An aqueous sample containing 0.1 to 4.0 mg chloral hydrate is mixed in a 50-cm^3 separatory funnel with 10 cm^3 0.1 N chloroform iodine solution and 5 cm^3 of aqueous 1.0 N ammonium borate. The mixture is shaken for 10 min. A volume of 10 cm^3 of double-distilled water is added and the mixture is shaken again for 1 min. The two phases are separated and any traces of iodine in the aqueous solution are removed by extraction with chloroform. The aqueous solution is transferred to a 200-cm^3 beaker and analyzed by one of the following techniques. A blank is performed under similar conditions.[134]

   i. *Direct measurement method:*
      A standard iodide ion solution having a concentration near the expected sample concentration is prepared; 2 cm^3 of 5 N potassium nitrate solution is added per 100 cm^3 standard and stirred. The electrode pair is placed in the standard solution and the EMF is recorded. The volume of the extracted sample is completed to 100 cm^3 with double-distilled water and 2 cm^3 of 5 N potassium nitrate solution. The mixture is stirred and the potential (in millivolts) is measured as before. The concentration of chloral hydrate equivalent to the produced iodide ions is calculated from the relation

$$pI^- = \frac{E_{obs} - K}{0.057} \qquad (5.38)$$

      The value of the constant K in Equation 5.38 is determined experimentally from the potential reading of the standard iodide solution.
  ii. *Standard addition method:*
      The extract is diluted to 100 cm^3 with double-distilled water; 2 cm^3 of 5 N potassium nitrate solution is added and the standard-addition technique is applied. The directly readable iodide ion concentration is used to calculate the chloral hydrate concentration.
 iii. *Potentiometric titration method:*
      The volume of the extract solution is completed to approximately 50 cm^3 with double-distilled water, stirred, and pH-adjusted between 6 and 7 with 1 N nitric acid solution. The indicator and reference electrodes are immersed into the solution and the titration is performed with 5×10^{-3} M silver nitrate solution. As the end point is approached, the titrant is added in 0.02-cm^3 increments. The equivalence point is determined from the first derivative, $\Delta E / \Delta V$.

5.19   Chlorambucil

$$C_{14}H_{19}Cl_2NO_2 \text{ (MM = 304.2)}$$

$$CH_2-CH_2-COOH$$

$$N(CH_2CH_2Cl)_2$$

Therapeutic category: antineoplastic agent

Discussion, Comments, and Procedure

See Section 5.13 and Dessouky et al.[135]

5.20   Chloramine-T

$$C_7H_7ClNNaO_2S \text{ (MM = 227.7)}$$

$$\underset{\text{O}}{\overset{\text{O}}{\parallel}}$$

$$Cl-N=S-O-Na$$

$$CH_3$$

Therapeutic category: bactericide

Discussion and Comments

Koupparis and Hadjiioannou[136] have described a chloramine-T–selective membrane sensor obtained by modifying an Orion nitrate-selective sensor by converting its liquid ion-exchanger, consisting of [(bathophen-anthroline)$_3$Ni]nitrate in 2-nitro-p-cymene, to the chloramine-T (CAT) form. The CAT-selective membrane sensor has been used successfully in direct and indirect potentiometry for determining chloramine-T, ascorbic acid, arsenic(III), and iodide.[136-138]

The electrode function is linear to chloramine-T at pH 7.0 (phosphate buffer) in the 10^{-2} to 10^{-4} M range. During the operative life of the sensor, which is about one month, the slope of the calibration curve was

59 to 62 mV decade^{-1} at 25°C. The potential is practically independent of pH in the range 5 to 9 for 10^{-2} and 10^{-3} M CAT solutions and in the range 5 to 8 for 10^{-4} M CAT solutions. At lower pH values there is an increase in potential because of the decrease in the CAT concentration caused by a shift in the equilibrium between CAT-anion (RNCl$^-$) and the free acid (RNHCl):

$$RNCl^- + H^+ \rightarrow RNHCl \qquad (5.39)$$

The free acid disproportionates to give 4-methylbenzene sulfonamide and sparingly soluble dischloramine-T:

$$2\,RNHCl \rightleftharpoons RNH_2 + RNCl_2 \quad (R = CH_3C_6H_4SO_2^-) \qquad (5.40)$$

Accordingly, a white precipitate appears in CAT solutions below pH 4.8.

Analytical Procedures

i. *Direct measurements:*
   Standard solutions of 10^{-2} to 10^{-4} M are prepared by successive dilutions from 10^{-1} M chloramine-T stock solution. All standard solutions need to be buffered at pH 7.0 with 0.1 M phosphate buffer solution, so that the ionic strength and hence the activity coefficients are constant. The standard solutions are transferred into 150-cm^3 beakers containing Teflon-coating stirring bars and the electrode pair. The chloramine-T concentration in the sample is determined from the calibration graph.

ii. *Potentiometric titration:*
   The pair of electrodes (indicator, CAT-selective membrane) is introduced into the sample solution (30 to 40 cm^3, approximately 10^{-3} M) and titrated with 10^{-2} M ascorbic acid solution. The electrode potential is recorded as a function of the added titrant volume and the end point determined in a usual way.

5.21   Chloramphenicol and Its Esters

Table 5.14 lists the chloramphenicol and its esters (antibacterials; antibiotics with a wide range of antimicrobial activity) that can be determined by a method described by Hassan and Eldessouki.[139] The method is based on an easy reduction procedure for chloramphenicol and its esters in various pharmaceutical preparations, without prior extraction, by the use of cadmium metal. The reduction is followed by the determination of released Cd^{2+} ions by potentiometry with a cadmium-selective membrane sensor (Orion, Model 94-48).

Table 5.14   Chloramphenicol and its Esters Assayed by Membrane Sensors

$$R-O-CH_2-CH-CH-OH$$

with NH, CO, $CHCl_2$ substituents and the aromatic ring bearing NO_2

Compound	R	Formula (MM)
Chloramphenicol	$-H$	$C_{11}H_{12}Cl_2N_2O_5$ (323.1)
Chloramphenicol cinnamate	$-CO-CH \\ \parallel \\ CH \\ \mid \\ C_6H_5$	$C_{20}H_{18}Cl_2N_2O_6$ (435.3)
Chloramphenicol palmitate	$-CO-(CH_2)_{14}-CH_3$	$C_{27}H_{42}Cl_2N_2O_6$ (561.6)
Chloramphenicol sodium succinate	$-CO-(CH_2)_2-COONa$	$C_{15}H_{15}Cl_2N_2NaO_8$ (445.2)

The time required for the quantitative reduction of chloramphenicol with cadmium metal in the presence of 0.05 M hydrochloric acid is not more than 15 min under boiling:

$$O_2N-\text{(ring)}-CH-CH-CH_2-OH + 3\,Cd + 3\,HCl \rightarrow$$
$$\qquad\qquad\quad \underset{OH}{|}\ \ \underset{NHCOCHCl_2}{|}$$

$$H_2N-\text{(ring)}-CH-CH-CH_2-OH + 3\,CdCl_2 + 2\,H_2O \quad (5.41)$$
$$\qquad\qquad\quad \underset{OH}{|}\ \ \underset{NHCOCHCl_2}{|}$$

For chloramphenicol esters, a prior hydrolysis (by treatment with 1 M alcoholic potassium hydroxide solution, at room temperature) is necessary.

Cadmium ions released (six equivalents for one mole of drug) in Reaction 5.41 are potentiometrically titrated with 10^{-2} M EDTA (pH 10, adjusted with ammonia buffer). The results of Hassan and Eldessouki[139] show an average recovery of 99.3% and a mean standard deviation of

0.5% (for pharmaceutical preparations, the standard deviation was 1.5%). (For details see Abdalla et al.[129] and Coşofreţ,[98] pp. 239–242).

A dichloroacetate selective sensor with a liquid membrane of tetraoctylammonium dichloroacetate, dissolved in o-nitrotoluene, was employed for the potentiometric determination of chloramphenicol in pharmaceutical preparations.[140] The method is based on measurements of the dichloroacetate anion released by alkaline hydrolysis of the drug:

$$O_2N\!-\!\langle\text{benzene}\rangle\!-\!\underset{\underset{OH}{|}}{CH}\!-\!\underset{\underset{NHCOCHCl_2}{|}}{CH}\!-\!CH_2\!-\!OH \xrightarrow[-CHCl_2COO^-]{OH^-}$$

$$O_2N\!-\!\langle\text{benzene}\rangle\!-\!\underset{\underset{OH}{|}}{CH}\!-\!\underset{\underset{NH_2}{|}}{CH}\!-\!CH_2OH \quad (5.42)$$

The dichloroacetate membrane sensor exhibits near-Nernstian response to dichloroacetate over the concentration from 10^{-2} to 2×10^{-5} M in the pH range 3.5 to 10 and could be used in 10% (v/v) propyleneglycol solution (an adequate solvent for chloramphenicol) with a very slight decrease in the linear range.

The hydrolysis of chloramphenicol is a general acid–base catalyzed reaction. The drug is very unstable in the alkaline medium. When known amounts of pure chloramphenicol were used, it proved impossible to obtain quantitative release of dichloroacetate by hydrolysis at elevated temperature in 0.1 M NaOH (Figure 5.6). Decreasing the temperature to 30°C gave almost quantitative release of dichloroacetate within 70 to 80 min.[140] Dichloroacetate proved to be stable under these conditions. The dichloroacetate anion is hydrolyzed in alkaline solutions at elevated temperatures, yielding glyoxylate—which may then undergo a Cannizzaro reaction, yielding glycolate and oxalate:

$$Cl_2CHCOO^- \xrightarrow[-Cl^-]{2\,OH^-} (HO)_2CHCOO^- \rightleftharpoons CH(O)COO^- + H_2O \quad (5.43)$$

$$2\,CH(O)COO^- \xrightarrow{OH^-} HOCH_2COO^- + (COO)_2^{2-} \quad (5.44)$$

The average error found for the determination of pure chloramphenicol samples (> 5 mg) by the standard-addition method was about 2.1%. Chloramphenicol was also determined in capsules, suppositories, and eye drops by the proposed method and the polarographic method, and

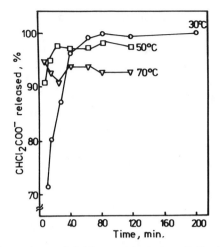

Figure 5.6 Percentage release of dichloroacetate during alkaline hydrolysis of chloramphenicol (in 0.1 M NaOH) at three different temperatures. (Reproduced from Pentari, J. G. and Efstathiou, C. E., *Anal. Chim. Acta*, 153, 161, 1983, Elsevier Science Publishers, Physical Sciences and Engineering Division. With permission.)

the results agreed well. The accuracy of the proposed method was further checked by means of recovery experiments with the capsules. The average recovery found was 104% (range 96 to 110%).[140]

The potentiometric method for the determination of chloramphenicol has the advantage that it eliminates prior separation steps and may be applied directly to colored and turbid solutions. For their spectrophotometric determination, a prior extraction of chloramphenicol is absolutely necessary.

Chloramphenicol as well as its esters may also be determined with a chloride-selective membrane sensor[135] or a silver sulfide membrane sensor[141] following mineralization by Schöniger's procedure. The sample is burnt in an oxygen flask, and the products are absorbed in an alkaline solution of hydrazine. Chloride is then potentiometrically titrated, in a medium of water–acetic acid–isopropyl alcohol (27:3:5, v/v), with 0.01 N silver nitrate solution.

Analytical Procedures

i. *Use of cadmium(II)-selective membrane sensor:*
   A 1- or 2-cm^3 aliquot of the sample solution is transferred to a 100-cm^3 conical flask with a ground-glass neck and a side arm with bubbler; 10 cm^3 of 0.05 M HCl is added and the flask fitted with a

water condenser. The contents are heated on a sand bath while passing carbon dioxide (about 50 bubbles per minute) through the side arm. When the solution starts boiling, 50 to 100 mg of cadmium metal turnings, previously washed with 6 M HCl followed (thoroughly) with double-distilled water, are introduced and boiling continued for 15 to 20 min in a carbon dioxide atmosphere. The flask and its contents are cooled and the reaction solution transferred to a 50-cm^3 volumetric flask and diluted to volume with double-distilled water. This test solution is used for the potentiometric determination. A blank is prepared under similar conditions.

Pure powders or pharmaceutical preparations containing chloramphenicol esters should be hydrolyzed prior to reduction with cadmium metal. An aliquot containing 10 to 15 mg of the esters is transferred to a test tube (20 × 2 cm), 1 cm^3 of 1 M alcoholic potassium hydroxide solution is added, and the mixture shaken for 2 min and transferred to the reaction vessel; 10 cm^3 of 0.05 M HCl is added and the assay completed as before.

An aliquot of the test solution of 25.0 cm^3 (3 to 7 mg of chloramphenicol) is transferred to a 100-cm^3 beaker. The pH is adjusted to 10 with aqueous ammonia (approximately 1 cm^3 of 25% solution). The electrode pair is placed into the sample solution and titrated with 10^{-2} M EDTA. The titrant volume, evaluated in a usual way, is used to calculate the content of the drug.

ii. *Use of dichloroacetate-selective membrane sensor:*
The samples (capsules or suppositories) containing chloramphenicol are homogenized if necessary. The appropriate amounts of homogenized samples, containing 5 to 15 mg of chloramphenicol are weighed in vessels calibrated at the 20-cm^3 mark, with well-fitting stoppers; 5.0 cm^3 of 0.1 M sodium hydroxide is pipetted into each vessel. After shaking, the vessels are immersed in a water bath thermostated at 30 ± 1°C for 70 to 80 min with frequent shaking. Each solution is neutralized with 0.7 M phosphoric acid solution. The contents are diluted to 20.0 cm^3 with distilled water (solution A); 15.0 cm^3 of solution A is transferred to a 50-cm^3 beaker, where the electrode pair is immersed. Under stirring the corresponding potential E_1 is recorded. A volume of 50.0 μl of 0.2 M sodium dichloroacetate solutions is added and the new potential, E_2, is recorded. The amount of chloramphenicol in the weighed sample is calculated from the equation

$$\text{Chloramphenicol (mg)} = \frac{4.309}{1.0033 \times 10^{(E_1 - E_2)/S} - 1}$$

5.22   Chloroquine

$$C_{18}H_{26}ClN_3 \ (MM = 319.9)$$

Cl

N

NHCH(CH$_2$)$_3$N(C$_2$H$_5$)$_2$

CH$_3$

Therapeutic category: antimalarial; antiamebic; lupus erythematosus suppressant

Discussion and Comments

A sensitive and selective PVC-membrane sensor, based on chloroquine–dinonylnaphthalenesulfonate ion-pair complex as the electroactive material and o-nitrophenyloctyl ether (o-NPOE) as plasticizer, was useful for the determination of chloroquine in ranges down to 1 μg cm^{-3} with good precision.[142] The membrane composition was 4.0% DNNS, 32.0% PVC, and 64.0% o-NPOE (m/m). The DNNS in the polymer membrane was converted to the chloroquine form by soaking the electrode in a solution of 10^{-2} M chloroquine diphosphate for 24 h after electrode preparation. The membrane sensor, containing 10^{-3} M chloroquine diphosphate solution at pH 4.2 (hydrogenphthalate buffer) as internal solution, was stored between measurements in the same solution.

The sensor has a near-Nernstian response within 10^{-2} to 10^{-6} M (slope 28.6 mV decade^{-1}) with a detection limit of 2×10^{-7} M (64 ng cm^{-3}) and exhibits negligible interference from various drugs (isoniazid, dopamine, choline, acetylcholine, carbachol, succinylcholine, etc.), amino acids, and lower quaternary ammonium compounds. There is essentially no response to the inorganic cations commonly present in pharmaceutical and biological samples. Many lipophilic amines (alkaloids, phenothiazines, phencyclidine, amitriptyline and similar compounds, higher quaternary ammonium compounds, etc.) interfere. As far as is known, chloroquine is never present together with such compounds, either in pharmaceuticals or in clinical samples; therefore, relevant selectivity coefficients for these compounds were not evaluated. Stable EMF readings were obtained within 30 s in 10^{-2} to 10^{-5} M range. In very dilute solutions ($\leq 10^{-6}$ M) response times were 1 to 2 min. Reproducibilities of EMF readings were acceptable in both these ranges.

The chloroquine membrane sensor proved useful for quantifying chloroquine (chloride or phosphate form) by the standard-addition method. The analysis of pure chloroquine diphosphate solutions (< 2.0 μg cm^{-3}) provided an average recovery of 100.2% with a relative standard deviation of 1.7% ($n = 10$). The determination of chloroquine in pharmaceutical preparations (tablets and injectable solutions) also gave good results (relative standard deviation < 1.6%).

When a fluoroborate-membrane sensor was used as indicator electrode in the potentiometric titration of chloroquine diphosphate, a relatively high potential jump was obtained (about 200 mV), which made location of the equivalence point easy.[143]

Analytical Procedures

i. *Direct potentiometric measurement of chloroquine in the microgram range:*
   The electrode pair (double-junction silver–silver chloride or SCE as reference) is placed in the aqueous solution of chloroquine (25.0 cm^3) buffered at pH 4.2. After equilibration by stirring, the EMF is recorded and compared with the calibration graph. The value obtained is checked by a standard addition. For this purpose, 1.0 cm^3 of standard 10^{-4} M chloroquine diphosphate is added.

ii. *Potentiometric assay of chloroquine hydrochloride in injectable solutions (e.g., 50 mg ml^{-1}):*
   The contents of five ampules of chloroquine hydrochloride injections are mixed in a 100-ml beaker. An aliquot (1.0 cm^3) is diluted to the mark with distilled water in a 100-cm^3 volumetric flask (solution A); 1.0 cm^3 of this solution is pipetted into a 250-ml volumetric flask, and the volume is adjusted to the mark with the hydrogenphthalate pH 4.2 buffer (solution B). An aliquot (25 cm^3) of solution B is pipetted into a 100-cm^3 beaker and both indicator and reference electrodes are immersed. After electrode equilibration by stirring and after recording the EMF, 1.0 cm^3 of standard 10^{-3} M chloroquine diphosphate at pH 4.2 is added, and the change in EMF is recorded and used to calculate the chloroquine hydrochloride content of the ampules.

iii. *Potentiometric assay of chloroquine diphosphate in tablets (e.g., 500 mg per tablet):*
   Ten tablets from the same lot are finely powdered. A portion of the powder, equivalent to about 100 mg of chloroquine diphosphate, is dissolved in distilled water in a 100-cm^3 volumetric flask and the solution is diluted to the mark (solution A). An aliquot (1.0 cm^3) of

solution A is pipetted into a 250-cm^3 volumetric flask and the volume is adjusted to the mark with the hydrogenphthalate pH 4.2 buffer (solution B). An aliquot (25.0 cm^3) of solution B is used for the chloroquine assay as previously described.

5.23 Chlorpheniramine and Diphenhydramine

$C_{16}H_{19}ClNO_2$ (MM = 274.8) $C_{17}H_{21}NO$ (MM = 255.4)

Cl — ... CH(CH$_2$)$_2$N(CH$_3$)$_2$ ... CHO(CH$_2$)$_2$N(CH$_3$)$_2$

Therapeutic category: antihistaminic agents

Discussion and Comments

Sensitive membrane sensors both for chlorpheniramine and diphenhydramine were reported since 1973, when Kina et al.[144] published the construction and analytical evaluation of a diphenhydramine membrane sensor.

PVC-matrix membrane sensors sensitive to chlorpheniramine have been developed by Fukamaki and Ishibashi[145] and by Liang and Huang.[146] In the first case, a PVC–tetrahydrofuran solution (12.5% PVC), 1.2-dichloroethane solution containing chlorpheniramine–tetraphenylborate salt (18.75%), and dioctylphthalate were mixed in a mass ratio of 8 : 4 : 2. The mixture was spread on a glass plate and left over 48 h for the membrane to form. Membrane disks were attached to the end of PVC tubes in order to construct chlorpheniramine sensors. The sensor exhibits near-Nernstian response to chlorpheniramine cation between $10^{-1.5}$ and $10^{-4.5}$ M, and the membrane potential is independent of pH from 4.5 to 8.0. Interferences by sodium, potassium, ammonium, and calcium ions were extremely low ($k_{ij}^{pot} < 10^{-4}$). Ephedrine, methylephedrine, caffeine antipyrine, etc., usually contained in common drugs, did not interfere between pH 5 and 7, but diphenhydramine ion interfered ($k_{CP, Diph}^{pot} = 1.13$).

A chlorpheniramine membrane sensor proposed by Liang and Huang also contains chlorpheniramine–tetraphenylborate ion-pair complex in a PVC matrix, but dibutylphthalate was used as plasticizer (membrane composition, chlorpheniramine–tetraphenylborate : DBP : PVC, 1 : 69 : 30 [m/m]).[146] The sensor shows near-Nernstian response within 10^{-2} to

Table 5.15 Some Performance Characteristics
of Diphenhydramine-Selective Membrane Sensors[147]

Electroactive material	Linear range (M)	Detection limit (M)	Slope (mV decade^{-1})
Diphenhydramine–TPB	10^{-1}–3.2×10^{-6}	1.6×10^{-6}	60
Diphenhydramine–reineckate	10^{-1}–10^{-5}	6.3×10^{-6}	53
Diphenhydramine–HgI$_4^{2-}$	10^{-1}–2.0×10^{-5}	1.2×10^{-5}	55
Diphenhydramine–picrolonate	10^{-1}–5.0×10^{-5}	3.2×10^{-5}	51

10^{-5} M range (pH 5 to 7) with an average slope of 56 ± 0.1 mV decade^{-1} and a detection limit of 2.29×10^{-6} M. When the concentration of electroactive material in the membrane was very high (e.g., 20%), the near-Nernstian range reduced to 10^{-2} to 10^{-4} M, but the slope remained unchanged.

The sensor's response was unaffected by many inorganic and organic compounds (e.g., benzoate, paracetamol, aminopyrine, moroxidine, promethazine, etc.) and had a lifetime of at least six months.

Both chlorpheniramine membrane sensors were used with good results for chlorpheniramine determination in pure solutions as well as tablets.

Various counter-ions (dipicrylamine, tetraphenylborate, reineckate, [HgI$_4$]$^{2-}$, and picrolonate) were used to prepare electroactive materials for both liquid- and PVC-membrane diphenhydramine-selective sensors.[144,147,148] The liquid-membrane sensor selective to diphenhydramine cation constructed by using diphenhydramine–dipicrylamine ion-pair in nitrobenzene (10^{-4} M concentration) and an Orion liquid-membrane barrel gave a Nernstian response (slope 59 mV decade^{-1}) in the 10^{-1} to 10^{-5} M diphenhydramine range (pH from 3 to 5), without significant interferences from common inorganic and organic cations.

The effects of various anion-associating agents (see Table 5.15), plasticizers, and other factors on the performances of diphenhydramine sensors were discussed in detail by Shen and Li.[147] The best plasticizers were DBP and di-2-ethylhexylphosphate (DEHP) and diphenhydramine–tetraphenylborate was found as the best electroactive material with respect to detection limit, electrode slope, and reproducibility of potential measurements. These sensors show Nernstian response over the diphenhydramine concentration range from 10^{-1} to 3.2×10^{-6} M, in the pH range from 2.5 to 5.5. From many inorganic and organic cations tested as potent interferents, only dibazol, levamisole, and quinine showed serious interference.

The membrane sensor proved useful in the determination of diphenhy-dramine in both pure solutions and injections, by direct potentiometry or potentiometric titrations. For potentiometric titration of diphenhydra-mine, a tetraphenylborate liquid-membrane sensor was also used.[149]

Shoukry et al.[148] also constructed diphenhydramine-selective mem-brane sensors of both the coated-wire and more conventional polymer-membrane types. They are based on incorporating diphenhydramine–tetraphenylborate ion pair in plasticized film. The membrane composi-tion (m/m) was 4.7% diphenhydramine–TPB, 47.6% DOP, and 47.7% PVC. For construction of coated-wire sensors, spectroscopically pure copper wires of 3-mm diameter were used. The membrane sensors showed a Nernstian response to diphenhydramine over a relatively wide range of concentrations, but in all cases the linear ranges were shorter than those reported by Shen and Li.[147]

The membrane sensors were successfully applied in the potentiomet-ric determination of diphenhydramine in pure solutions and in pharma-ceutical preparations (e.g., syrups) by direct potentiometry using the standard-addition and potentiometric titration methods.

When sodium tetraphenylborate standard solution was used for the potentiometric titration of both chlorpheniramine and diphenhydramine (as maleate and hydrochloride, respectively) a fluoroborate-ion-selective indicator sensor was used.[150] Other commercially available liquid-mem-brane sensors such as the perchlorate and nitrate sensors as well as a solid-state cyanide membrane sensor also respond to investigated cations. Relative standard deviations of 0.49 and 0.53% were reported for chlor-pheniramine and diphenhydramine, respectively.

Analytical Procedures

i. *Direct measurement (chlorpheniramine and diphenhydramine membrane sensor, respectively):*
Standard solutions from 10^{-2} to 10^{-4} M of the respective drug are prepared by successive dilutions of 10^{-1} M stock solution of chlor-pheniramine maleate and diphenhydramine hydrochloride, respec-tively. A constant ionic strength (e.g., $I = 0.1$ M, adjusted with sodium nitrate) and pH between 5 and 7 (for chlorpheniramine) and between 3 and 5 (for diphenhydramine) must be used. The EMF of the standard solutions are measured under stirring, using a SCE as reference electrode. The unknown concentration of the sample is determined from the calibration graph, E (in millivolts) vs. log C.

ii. *Potentiometric titration (the respective drug sensor, TPB mem-brane sensor or BF_4^--membrane sensor):*
The electrode pair (SCE as reference) is introduced into the sample solution (30 to 40 cm^3, approximately 5×10^{-3} M) and potentiomet-rically titrated with standard sodium tetraphenylborate solution

$(5 \times 10^{-2}\ M)$. The end point is determined from the maximum slope of the titration curve of EMF vs. titrant volume.

5.24   Chlorzoxazone

$$C_7H_4ClNO_2\ (MM = 169.6)$$

Therapeutic category: skeletal-muscle relaxant

Discussion and Comments

A simple potentiometric method for the determination of chlorzoxazone, based on the use of a carbon dioxide gas sensing electrode (Orion, Model 95-02) was described by Tagami and Muramoto.[151] Chlorzoxazone decomposes into an aminophenol compound and sodium carbonate on refluxing with 3 M sodium hydroxide solution for 2 h:

$$\text{(5.45)}$$

After acidifying with 6 M hydrochloric acid and adjusting the pH to 8.5, the carbon dioxide was determined with a CO_2-gas-sensing electrode.

In order to expel carbon dioxide in an apparatus equipped with a reflux condenser fitted with a soda lime tube, a stream of nitrogen was passed through the apparatus to displace the air. The electrode potential reached the maximum at a boiling time of 80 min. After 20, 40, and 60 min, the recoveries were 94, 98, and 99%, respectively.

The concentration of drug was first determined in the pure drug powder. The amount of chlorzoxazone was estimated with an average error of 0.1% and the standard deviation was 0.39%. In order to apply the method to tablet assay, a previous extraction step was required. The extraction was carried out five times with acetone (approximately 4 cm^3), and the resultant extract was then evaporated. The residue was also boiled for 2 h with 3 M sodium hydroxide solution, and the procedure continued as previously described. The mean recovery was 99.6% and the standard deviation was 0.24%.[151]

Analytical Procedure

i. *Decomposition of chlorzoxazone:*

A mixture of 423.93 mg chlorzoxazone and 50 cm^3 of carbonate-free 3 *M* sodium hydroxide is placed in a 200-cm^3 three-necked flask in a stream of nitrogen. A reflux condenser equipped with a soda lime tube to prevent entrance of carbon dioxide from the air is attached, and the other neck of the flask is closed with a stopper. The mixture is boiled gently in an oil bath for 2 h. The solution is then cooled and diluted with about 50 cm^3 of water. A drop of phenolphthalein is added, and the alkaline solution is continuously neutralized with 6 *M* hydrochloric acid until the indicator turns to pink. The solution is then adjusted to pH 8.5 with dilute hydrochloric acid solution using a pH-meter. The procedure is carried out in a stream of nitrogen. The solution is transferred into a 250-cm^3 volumetric flask and diluted to volume with water. The concentration of the final drug solution is 1×10^{-2} *M*, corresponding to 1×10^{-2} *M* carbon dioxide.

Sample solutions for measurements are obtained by dilution of this stock solution with water.

ii. *Measurement:*

A 50-cm^3 aliquot of the samples and standard solutions used are placed in the cell and the air in the cell is replaced with nitrogen. The cell is then closed with a rubber stopper and incubated at 20°C. To obtain more analytical accuracy, a calibration curve is prepared for every set of determinations, because the slope of the linear calibration plot is not constant. The standard procedure is as follows: The sensor with 0.01 *M* sodium bicarbonate internal filling solution is washed with water and further immersed for about 3 min in fresh 0.1 *M* sodium chloride solution in order to complete the washing; then the old internal filling solution is replaced with fresh internal filling solution. After acidifying of the first standard with 5 cm^3 of 0.1 *M* citrate buffer (pH 4.5), the electrode is placed in the first standard solution, and the EMF is recorded. After electrode washing as previously described, the electrode is placed in the sample solution acidified with buffer, and the EMF is recorded. After another washing, the electrode is placed in the second standard solution acidified with buffer, and the EMF is recorded. The sample concentration is determined from the calibration curve.

iii. *Assay of tablets:*

Twenty tablets are weighed and then finely powdered. A portion of the powder, equivalent to 423.93 mg chlorzoxazone is accurately weighed out. To this sample, 20 cm^3 of acetone is added. The solution is stirred to extract the drug and then centrifuged. The supernatant acetone solution is removed and 20 cm^3 of acetone is added to the residue. The extraction procedure is carried out four times. The collected acetone fractions are evaporated to dryness. The residue is then refluxed for 2 h with 50 ml of 3 N sodium hydroxide solution and the same procedure as described before is followed.[151]

5.25   Cholesterol

$$C_{27}H_{46}O \ (MM = 386.6)$$

Therapeutic category: pharmaceutic aid (emulsifying agent)

Discussion and Comments

The determination of total cholesterol in blood serum plays an important role in the clinical diagnosis of disease states. A potentiometric analysis method for cholesterol using a double-enzyme procedure in an automated analysis system using the reaction sequence

$$\qquad (5.46)$$

$$+ \ H_2O_2 \quad (5.47)$$

$$H_2O_2 + 2\,I^- + 2\,H^+ \xrightarrow{\text{Mo(VI)}} I_2 + 2\,H_2O \qquad (5.48)$$

was described.[152] The enzymatic Reactions 5.46 and 5.47 were carried out in an automated analysis system under controlled conditions for a fixed time interval. A specially constructed flow-through membrane electrode was used to monitor the change in iodide concentration produced

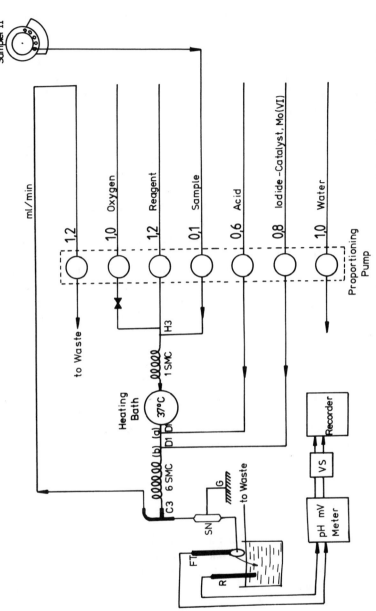

Figure 5.7 Schematic of autoanalysis system for cholesterol determination: H_3 and D_1—fittings; SMC—mixing coil; SN—stainless steel ground contact; C—recorder ground input; FT—sensing electrode; R—reference electrode; VS—voltage suppressor. (Reprinted with permission from Papastathopoulos, D. S. and Rechnitz, G. A., *Anal. Chem.*, 47, 1792, 197. Copyright 1975 American Chemical Society.)

by the Mo(VI)-catalyzed indicator reaction. Figure 5.7 shows a schematic diagram of the cholesterol analysis system used. The optimum conditions for the analysis are also indicated in Figure 5.7.

Working solutions for the enzymatic analysis were prepared in two steps. First, a stock solution (pH 6.8) of 5×10^{-2} M in Na_2HPO_4, 5×10^{-2} M in NaH_2PO_4, 6×10^{-3} M in sodium cholate, 4×10^{-3} M in sodium azide, and containing 5 cm^3 dm^{-3} of Triton X-100 surfactant was prepared, just before starting the analysis, cholesterol esterase and cholesterol oxidase were added to the stock solution to yield enzyme concentration of 100 to 120 units per liter, respectively. For the indicator reaction, solutions 5×10^{-4} M in KI containing 1 g dm^{-3} of the $(NH_4)_6Mo_7O_{24} \cdot 4\,H_2O$ catalyst and 1.6 M $HClO_4$ were employed. Within 13 min at 37°C both enzymatic reactions proceed simultaneously and are terminated (point (a), Figure 5.7) in the stream by the aspiration of 1.6 M $HClO_4$. The acid treatment also precipitates serum proteins and adjusts the pH to the optimum value for the indicator reaction. Reagents for the later reaction are introduced at point (b) of Figure 5.7. The indicator reaction is permitted to proceed for a fixed time interval (about 4 min) in six delay coils at room temperature before the debubbled stream reaches the sensing electrode. In the enzymatic reactions, sodium cholate acts as an emulsifier whereas the presence of Triton X-100 nonionic surfactant serves to keep the free cholesterol and cholesterol oxidase enzyme in solution; sodium azide acts as an enzyme preservative and minimizes catalase interference.[152] In the previously described optimized conditions, the calibration curve was linear over the total cholesterol concentration range of 80 to 420 mg dm^{-3}. The precision and accuracy of the method are attractive for routine determinations of total cholesterol in clinical serum samples.

5.26   Cholic Acids

An ion-selective membrane sensor based on a liquid membrane containing tributylcetylphosphonium benzoate in nitrobenzene was used, for the first time, to determine cholic acids listed in Table 5.16, in different pharmaceutical products, using direct potentiometry and standard-addition methods.[153]

Each one of the well-known methods for the determination of cholic acids presents some difficulties, such as toxicity of reagents and the high cost and complexity of the equipment. The potentiometric method using a benzoate-ion-selective membrane sensor is simple, rapid, and sufficiently accurate. The sensor presents a linear response to benzoate over the range 10^0 to 10^{-4} M (slope 59 $\pm$ 1 mV per decade of concentration) with a very short response time (1 to 20 s) and good reproducibility

Table 5.16   Cholic Acids Assayed by Membrane Sensors

Cholic acid	Formula (MM)	Therapeutic category
Cholic	$C_{24}H_{40}O_5$ (408.6)	Choleretic

Chenodeoxycholic	$C_{24}H_{40}O_4$ (392.6)	Experimental, in prevention and dissolution of gallstones

Deoxycholic	$C_{24}H_{40}O_4$ (392.6)	Choleretic

Lithocholic	$C_{24}H_{40}O_3$ (376.6)	Choleretic

Ursodeoxycholic	$C_{24}H_{40}O_4$ (392.6)	Cholagogic

(± 1 mV) and accuracy. The sensor responds also to all cholic acids listed in Table 5.16, at pH 11, with selectivity coefficients varying from 0.11 (deoxycholic) to 2.46 (chenodeoxycholic). The slopes of the electrode response toward the mentioned cholic acids varied from 59 ± 1 (cholic) to 180 ± 4 (lithocholic) mV decade^{-1}.

Two pharmaceutical products containing chenodeoxycholic acid and ursodeoxycholic acid, respectively, were analyzed both by calibration-curve method and standard-addition method. Both pharmaceutical products contain various percentages of cornstarch/starch, magnesium stearate, and aerosol/precipitated silica. Only stearate is a potential interferent, but its low concentration due to low solubility does not affect the measurements.

The same authors[154] reported a cholate liquid-membrane sensor employing benzyldimethylcetylammonium cholate (BDMCACh) as electroactive material. The sensor assembly characteristics are as follows: electrode body, PTFE; electroactive material, BDMCACh ($C_{49}H_{85}O_5N \cdot H_2O$); membrane solvent, decan-1-ol (dielectric constant 8.1 relative to vacuum); membrane solution concentration, 10^{-2} M BDMCACh; internal solution, sodium cholate 10^{-2} M, potassium chloride 10^{-2} M; internal reference electrode, Ag/AgCl; PTFE disks and Millipore supports, diameter 1.3×10^{-2} M, thickness 1×10^{-4} M, and pore size 2×10^{-7} M. According to their results, several conclusions were drawn evidencing the superiority of cholate sensors over benzoate sensors. In general, faster response times (Figure 5.8) are obtained using a cholate membrane sensor (approximately 10 s was the maximum time obtained). The precision is higher for a cholate membrane sensor than for a benzoate sensor, and the linearity range is wider and the detection limits are lower; also, the values of the selectivity coefficients for different anions are low enough to prevent any common interference when using the cholate membrane sensor. The authors[154] found that the slopes of the calibration graphs for the aforementioned cholic acids increase as the number of hydroxyl groups in the steroid ring decreases and when one of these hydroxyl groups is in the β-position toward the plane of the ring.

Various drugs, containing either chenodeoxycholic acid or ursodeoxycholic acid were assayed by an enzymatic–spectrophotometric[155] and by a potentiometric method with a cholate sensor and a benzoate sensor. The better results for precision and accuracy were obtained by standard-addition and Gran's-plot methods, using the cholate sensor.

The cholate-selective liquid-membrane sensor has been applied also to the determination of the pool of cholic acids in human bile. Pretreatment of the sample (deproteinization with ethanol; treatment with carbon black for removal of the bile pigments; strong alkaline hydrolysis to deconjugate bile acids; precipitation of cholic acids in acid medium, followed by their extraction with an organic solvent or, alternatively, by filtration of the precipitated acids on a Gooch filter, to free the acids

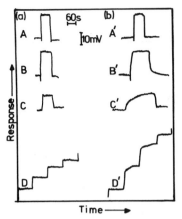

Figure 5.8 Comparison between the response of (a) the cholate membrane sensor with that of (b) the benzoate membrane sensor in standard solutions of sodium cholate with changing cholate concentration (C): curves A and A'—C(initial) = 2.0 × 10^{-3} M, C(after addition) = 4.5 × 10^{-3} M, and C(after dilution) = 1.9 × 10^{-3} M; curves B and B'—C(initial) = 3.8 × 10^{-4} M, C(after addition) = 9.0 × 10^{-4} M, and C(after dilution) = 3.8 × 10^{-4} M; curves C and C'—C(initial) = 4.0 × 10^{-5} M, C(after addition) = 6.8 × 10^{-5} M, C(after dilution) = 4.4 × 10^{-5} M; curves D and D'—responses of the membrane sensors for successive increases of cholate concentration; D, C = 6.7 × 10^{-4} to 2.2 × 10^{-3} M; D', C = 8.0 × 10^{-4} to 2.5 × 10^{-3} M. (Reproduced from Campanella, L., Sorrentino, L., and Tomassetti, M., *Analyst*, 108, 1490, 1983. With permission.)

from the micelles and to remove inorganic anions that can interfere, and washing the cholate aqueous solution with ether to eliminate cholesterol and lipids still present) was necessary for reliable potentiometric determination.[156]

A polymeric membrane sensor based on PVC, 2-ethylhexylsebacate as plasticizing agent, and the same ion exchanger (5% m/m) and containing 10^{-2} M sodium cholate–KCl mixture solution as internal solution displayed linear responses within 5.3 × 10^{-3} to 8.0 × 10^{-5} M, 2.0 × 10^{-4} to 1.6 × 10^{-5}, and 3.1 × 10^{-4} to 1.6 × 10^{-5} M for cholic, chenodeoxycholic, and ursodeoxycholic acids, respectively.[157] The electrode slopes were 56.2, 58.2, and 56.0 mV decade^{-1}, respectively. The sensor, exhibiting a very short response time ($\leq$ 10 s), was in use for several months with a good repeatability of measurements in the linear range (1.3%, as pooled standard deviation). The membrane sensor was also successfully used for the determination of chenodeoxycholic and ursodeoxycholic acids in some antilithogenic commercial drugs.

Analytical Procedure

Standard solutions of the respective cholic acid are prepared by dissolving the drug substance at pH 11 with distilled water. The sensors (benzoate or cholate membrane as indicator and SCE as reference) are

Table 5.17   Choline and Its Esters Assayed by Membrane Sensors

Compound	Formula (MM)	Therapeutic category
Choline chloride	$C_5H_{14}ClNO$ (139.6) $HOCH_2CH_2\overset{+}{N}(CH_3)_3Cl^-$	Parasympathomimetic
Acetylcholine bromide	$C_7H_{16}BrNO_2$ (226.1) $CH_3COOCH_2CH_2\overset{+}{N}(CH_3)_3Br^-$	Parasympathomimetic
Methacholine chloride	$C_8H_{18}ClNO_2$ (195.7) $CH_3COOCHCH_2\overset{+}{N}(CH_3)_3Cl^-$ $\quad\quad\quad\mid$ $\quad\quad\; CH_3$	Has the muscarinic action as acetyl-choline but is more stable

placed in these solutions and EMF readings (linear axis) are plotted against concentration (logarithmic axis). The sample concentration is determined from this graph.

For pharmaceuticals assay, the respective drug is dissolved in distilled water at pH 11 and after sedimentation of a little insoluble matter the solution is filtered and appropriately diluted. The same procedure is followed.

5.27   Choline and Its Esters

Choline and its esters (see Table 5.17) may be determined with ion-selective membrane sensors.[158-167].

The Corning 476200 acetylcholine-ion-selective membrane sensor used by Baum[158] had a high selectivity for acetylcholine relative to choline (15 : 1) and common inorganic cations such as Na^+, NH_4^+, and K^+. A linear response was obtained in the concentration ranges 10^{-5} to 10^{-1} M for choline and 10^{-6} to 10^{-1} M for acetylcholine, and the calibration slope was almost Nernstian in the first case and Nernstian in the second. The response of the membrane sensor is rapid even at low organic cation concentration (about 10 to 20 s).

The solid-state membrane sensor of Baum et al.[160] consists of an electroactive PVC membrane prepared from a solution of acetylcholine tetra-4-chlorophenylborate in a phthalate ester that serves as a plasticizer. (In the initial efforts to prepare the sensor, 3-nitro-1,2-dimethyl-benzene was used as the solvent in the liquid membrane; the phthalate esters, e.g., dibutylphthalate and dioctylphthalate, used in PVC membranes are better solvents for the choline esters and thus exhibit a "leveling effect.") The polymer-membrane sensor is markedly superior in performance to the liquid-membrane sensor, the first having a very short response time between determinations and a very short uptime of only about 10 min after overnight storage.

A liquid-membrane sensor selective to acetylcholine may be prepared from acetylcholine dipicrylaminate in nitrobenzene (0.1% v/v) soaked in a hydrophobized G_4 frit.[161] The membrane sensor presents a near-Nernstian response to acetylcholine in the 10^{-1} to 7×10^{-6} M range (slope 57.4 mV decade^{-1}) and has a very good selectivity over inorganic cations (Na^+, K^+, NH_4^+, and Ca^{2+}).

An interesting membrane sensor selective to acetylcholine is based on immobilized acetylcholinesterase on the active surface of a pH glass electrode.[162] By the hydrolysis reaction of acetylcholine, catalyzed by acetylcholinesterase (AChE), acetic acid is obtained according to

$$CH_3COOCH_2CH_2(CH_3)_3\overset{+}{N} + H_2O \xrightarrow[-CH_3COO^-]{AChE} HOCH_2CH_2(CH_3)_3\overset{+}{N} + H^+$$

$$(5.49)$$

The sensitive surface of the glass electrode was coated with a homogeneous solution containing 1 mg AChE, 2 mm^3 17.5% albumin (pH 8.5), and 30 mm^3 12.5% glutaraldehyde (pH 8.5). The enzyme membrane sensor was initially placed in a 0.01 M buffer (pH 7.0) and between measurements was kept at 4°C. This sensor gave a linear response to acetylcholine only in the 10^{-2} to 10^{-4} M range (optimum pH 8.3 to 8.6).

An enzyme micro-electrode for acetylcholine determination has been designed[165] by immobilization of acetylcholinesterase on the active surface of a glass micro-electrode. When phosphate buffer of pH 8.0 (10 mM) was used for potential measurement, a detection limit of 10^{-5} M was achieved. In a biological medium such as cerebrospinal fluid, a detection limit of 10^{-4} M has been estimated.

The construction and performance characteristics of ion-selective-membrane micro-electrodes for acetylcholine and its metabolite, choline, based on their complexes with dipicrylamine in various solvents were described by Jaramillo et al.[164] For all solvents with nitro groups (Table 5.18) slopes approach the theoretical value.

Solvents with relatively high dielectric constants, such as 2-nitro-toluene, 3-nitrotoluene, 3-nitro-o-xylene, o-nitroanisole, 2-nitro-p-cymene, and nitrobenzene, are adequate for dissolution of acetylcholine–dipicrylaminate ion-pair complex and showed near-Nernstian slopes.

Table 5.18　Effect of Solvent on Slope and Detection Limit
of Acetylcholine–Dipicrylaminate Membrane Sensors[164]

Solvent	n	Slope ($\pm$s.d.) [mV (pACh)$^{-1}$]	Detection limit (10^{-5} M)
Nitrobenzene	22	59.4 $\pm$ 2.0	2
3-Nitro-o-xylene	6	58.4 $\pm$ 1.1	2
2-Nitrotoluene	6	57.7 $\pm$ 0.5	3
3-Nitrotoluene	5	57.8 $\pm$ 0.6	2
o-Nitroanisole	15	58.8 $\pm$ 1.0	2
2-Nitro-p-cymene	5	54.4 $\pm$ 1.4	3
m-Dimethoxybenzene	5	50.2[a]	6
Toluene	5	50.6[a]	2
o-Nitroanisole[b]	5	51.0 $\pm$ 1.6	3
2-Nitro-p-cymene[b]	5	55.6 $\pm$ 1.6	5
o-Nitroanisole[c]	11	58.5 $\pm$ 0.7	2

[a] Data from Sullivan.[168]

[b] The complex used was acetylcholine–tetraphenylborate. The dipicrylaminate gives better performance (selectivity and detection limit) to the liquid membrane in most of the solvents studied.

[c] The sample solution was dissolved in phosphate buffer saline solution at pH 7.4.

In general, the response time of the acetylcholine micro-electrode was 2 to 3 min even at low concentrations. The response time is influenced mainly by the solvent, the hydrophobicity of the micro-electrode tip, and the presence of interfering ions.[164]

Naturally occurring components of neural tissue did not interfere strongly, although many cholinergic drugs (arecholine, carbachol, atropine, scopolamine, apomorphine, etc.) interfered. For *in vivo* measurements in rat striatum, o-nitroanisole (slope 58.8 $\pm$ 1 mV decade^{-1}) was preferred to the other solvents for liquid membrane because of its lower toxicity and faster equilibration of the liquid-membrane constituents. The choline membrane sensor may have more applicability than the acetylcholine membrane sensor for *in vivo* measurements because acetylcholine is removed rapidly from extracellular fluid, being hydrolyzed to choline and acetic acid.[164]

Tor and Freeman[166] described a new method for the construction of a self-mounted, uniform enzyme membrane, directly attached to the primary sensor's surface, without the need for a physical support, such as a dialysis membrane or rubber O-ring. The method is based on the controlled chemical cross-linking of prepolymerized linear chains of polyamide-hydrazide by dialdehydes, such as glyoxal.

The construction of an enzyme membrane on the surface of the primary sensor (pH glass electrode) is based on the following three-step procedure:[166]

1. An enzyme solution, in a concentrated buffered aqueous solution of an appropriate prepolymer, is prepared; also the sensor surface is cleaned and dried.
2. The sensor is dipped in the enzyme–prepolymer solution for a short time and removed; the absorbed viscous layer is carefully drained and the remaining absorbed film is air dried.
3. The condensed enzyme polymer film is chemically cross-linked by dipping the coated sensor into an ice-cold glyoxal solution. Following appropriate washing to remove non-bound glyoxal, the enzyme sensor is ready for use without need for any additional physical supports.

For an acetylcholinesterase membrane sensor, three buffers (phosphate, MOPS, and HEPES, where MOPS = 3-N-(morpholino) propanesulfonic acid and HEPES = N-(2-hydroxyethyl)-piperazine-N'-2-ethanesulfonic acid) were tested for the concentration range 1 to 100 mM at pH 8.0. The HEPES buffer exhibited the best results: at 5 mM concentration, pH 8.0, 25°C, a linear calibration graph within the concentration range 2×10^{-5} to 1×10^{-3} M was obtained. The response curves of the acetylcholinesterase sensor to different concentrations of acetylcholine show that response times were short: 0.5 to 1 min for low concentrations (e.g., 4×10^{-5} M) and 2 to 3 min for high concentrations. During a period of 6 months and 180 measurements there was no change in the response and sensitivity of this enzyme membrane sensor.[166]

A new type of acetylcholine sensor was made with an ion-selective field effect transistor (ISFET) and acetylcholine receptor.[167] The acetylcholine receptor was fixed on a polyvinylbutyral membrane that covered the ISFET gate. When acetylcholine was injected into this system, the differential gate output voltage gradually shifted to the positive side and reached a constant value. This response was due to the positive charge of acetylcholine. A linear relationship was obtained between the initial rate of the differential gate output voltage change and the logarithmic value of the acetylcholine concentration.

Kina et al.[163] obtained a very sensitive membrane methacholine sensor of useful response range to 10^{-6} M. Dipicrylamine anion was used as the ion-exchanger site in the liquid membrane with nitrobenzene as solvent. Either a U-shaped glass tube or an Orion liquid-membrane barrel was used for constructing the sensor.[169] The sensor was fast in response, giving steady EMF within 1 s, although the actual response time depends on the efficiency of mixing. Such a rapid response is useful for the

potentiometric titration of methacholine chloride with sodium tetraphenylborate solution.

Analytical Procedures

i. *For all compounds listed in Table 5.17, using a liquid- or poly-
    mer-membrane indicator sensor:*
    For measurements in units of moles per cubic decimeter, 10^{-2}, 10^{-3},
    and 10^{-4} standards are prepared by serial dilution of 0.1 M solution
    of the appropriate compound. The ionic strength is kept constant
    (e.g., to 0.1 M) with sodium nitrate solution. The indicator membrane
    and reference electrodes are placed in the standards, and the millivolt
    readings (linear axis) are plotted against concentration (logarithmic
    axis). The measurements are made under stirring. The unknown
    concentration is determined from the calibration curve.

ii. *For methacholine chloride:*
    The electrode pair (methacholine-selective and SCE) is introduced
    into the sample solution (30 to 40 cm^3, approximately 5×10^{-3} M)
    and titrated with sodium tetraphenylborate solution (5×10^{-2} M)
    The end point corresponds to the maximum slope of the titration
    curve of EMF vs. titrant volume.

5.28   Cimetidine and Ranitidine

$C_{10}H_{16}N_6S$ (MM = 252.3)

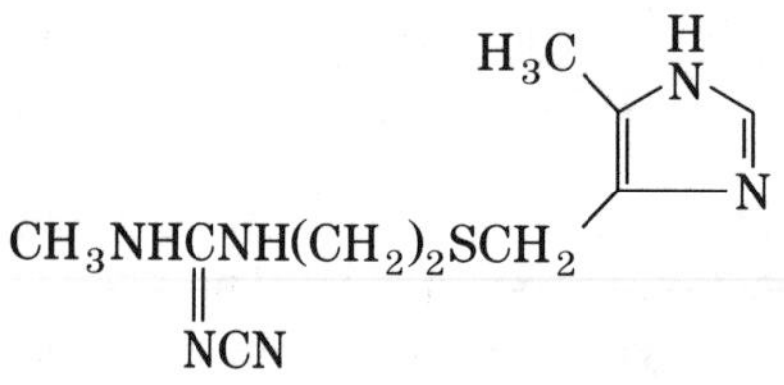

$$CH_3NHCNH(CH_2)_2SCH_2$$
$$\|$$
$$NCN$$

$C_{13}H_{22}N_4O_3S$ (MM = 314.4)

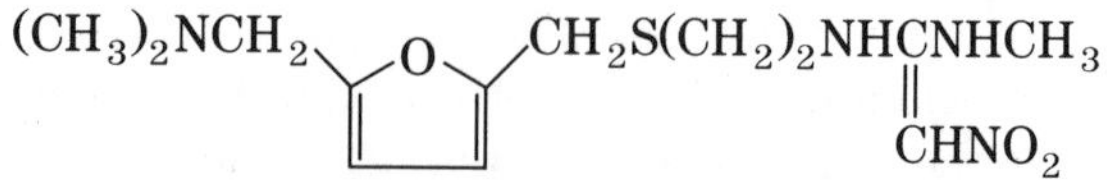

$$(CH_3)_2NCH_2 \quad O \quad CH_2S(CH_2)_2NHCNHCH_3$$
$$\|$$
$$CHNO_2$$

Therapeutic category: antagonists (to histamine H$_2$ receptors),
especially in treatment of peptic ulcer (C) and duodenal ulcer (R)

Discussion and Comments

Liquid-membrane and PVC-matrix ion-selective sensors that respond to
the cationic forms of cimetidine and ranitidine have been described.[170]
The ion exchangers used were cimetidine hydrogen tetra(*m*-chloro-

phenyl)borate $((CimH^+)(TCPB^-))$ and ranitidine hydrogen tetra(m-chlorophenyl)borate $((RanH^+)(TCPB^-))$, dissolved either in p-nitrocumene or entrapped in PVC polymer in the presence of 2-nitrophenyloctyl ether as plasticizer.

An Orion liquid-membrane electrode body (Model 92) with a Millipore LCWPO 1300 PTFE membrane, were used for construction of liquid-type membrane sensors. The liquid ion-exchangers were approximately 10^{-2} M $(CimH^+)(TCPB^+)$ or $(RanH^+)(TCPB^-)$.

For ranitidine sensors, the response in water was Nernstian down to a concentration of 10^{-6} M. In 5×10^{-3} M acetate buffer (pH 4.6) the response was Nernstian down to a concentration of 2.5×10^{-6} M due to a slight interference by Na^+ ions. The limit of linear response in 20% acetonitrile solution in 5×10^{-3} M phosphate buffer of pH 6.5 was 5×10^{-5} M. The slope of calibration curve was (in millivolts per decade) 58.4 for water, 58.6 for acetate buffer, and 57.3 for 20% acetonitrile (20% acetonitrile medium is used for measurements in urine samples). The limit of linear Nernstian-type response of the cimetidine sensor was 2×10^{-5} M in water and 4×10^{-5} M in 5×10^{-3} M acetate buffer of pH 4.6 (slopes of 50.7 and 50.8 mV decade^{-1}, respectively).

The plots of EMF vs. pH showed that the potential is practically unaffected by changes in pH over the ranges 3 to 7 for ranitidine and 3 to 5.5 for cimetidine. These plots were used to calculate the dissociation constant K_a of the cationic acid, which for ranitidine is equal to

$$K_a = [\text{Ran}][\text{H}^+]/[\text{RanH}^+] \tag{5.50}$$

The plot of $[\text{Ran}]/[\text{RanH}^+]$ vs. $1/[\text{H}^+]$ is a straight line that passes through the origin and has a slope of K_a. The ratio $[\text{Ran}]/[\text{RanH}^+]$ can be calculated at each pH value from

$$[\text{Ran}]/[\text{RanH}^+] = \text{antilog}(\Delta E/S) - 1 \tag{5.51}$$

where ΔE is the potential difference $E_1 - E_2$ between potential E_1 at the plateau of the E vs. pH plot and the potential E_2, which corresponds to a certain pH value. The pK_a values calculated for both ranitidine and cimetidine (8.37 and 6.98, respectively) were in good agreement with the literature data (8.2 and 6.8, respectively).

Both membrane sensors, being sufficiently selective, were used for the assay of the active compounds in pharmaceutical preparations (injection solutions and tablets).

Poly(vinylchloride–matrix) micro-electrodes sensitive to ranitidine cation were suitable for potentiometric measurements in 250 μl of stirred solutions. New methods were devised for the selective extraction of ranitidine from urine and serum samples by using reversed-phase octadecylsilane-bonded silica. Very sensitive potentiometric methods were proposed for the determination of ranitidine in urine and serum samples. The methods have been applied successfully for the determination of ranitidine in urine in the range 2.5×10^{-5} to 5×10^{-4} M and in serum in the range 1×10^{-6} to 1.5×10^{-5} M, in a preliminary pharmacokinetic experiment.[170]

Analytical Procedures

i. *Cimetidine for injection:*
A 2.0-cm^3 aliquot of the commercial product was diluted with 0.1 M acetate buffer (pH 5.2) to a final volume of 500 cm^3; 20 cm^3 of the resulting solution (V_x) was used for analysis. A first potential reading was recorded for this solution. Subsequently, a second potential reading was obtained after the addition of a small volume (V_s) of concentrated drug-solution of concentration C_s. The initial concentration C_x, of the sample is calculated from

$$C_x = C_s V_s / \left[10^{\Delta E / S} (V_x + V_s) - V_x \right] \qquad (5.52)$$

where ΔE is the change in potential and S is the slope of the electrode response.

ii. *Cimetidine tablets:*
At least five tablets are made into a powder. An appropriate amount of the powder is weighed out and dissolved in 500 cm^3 of 0.1 M HCl solution by stirring for 1 h. The amount weighed is selected so that the final solution contains approximately 3×10^{-4} to 3×10^{-3} M in cimetidine; 15 cm^3 of cimetidine unknown solution is mixed with 3 cm^3 of 1 M acetate buffer (pH 5.2) and the resulting solution is analyzed by the standard-addition method, as previously described.

iii. *Ranitidine tablets:*
The procedure is the same as described for cimetidine tablets, but the powder is dissolved in 0.07 M phosphate buffer (pH 6.0). A 20.0-cm^3 aliquot of the resulting solution is analyzed by the standard-addition method, as previously described.

5.29   Clidinium Bromide and Similar Compounds

The drug substances listed in Table 5.19 have been analyzed by potentiometric titration with sodium tetraphenylborate solution by using a tetraphenylborate-selective membrane sensor.[82]

In all cases the end-point breaks exceeded 95 mV and the reported relative standard deviation varied from 0.2% (pyridostigmine) to 0.6% (prostigmine).

Table 5.19 Clidinium Bromide and Similar Compounds Assayed by Membrane Sensors

Compound	Formula (MM)	Therapeutic category	Optimum pH for determination
Clidinium bromide	$C_{22}H_{26}BrNO_3$ (432.4)	Anticholinergic	3–7
Prostigmine bromide (neostigmine)	$C_{12}H_{19}BrN_2O_2$ (303.2)	Cholinergic	5–10
Pyridostigmine bromide	$C_9H_{13}BrN_2O_2$ (261.1)	Cholinergic	3–10

Kina et al.[163] constructed two liquid-membranes sensors selective to prostigmine (neostigmine) cation by using prostigmine–tetraphenylborate and prostigmine–dipicrylaminate ion pairs, respectively, in nitrobenzene or 1,2-dichloroethane (10^{-4} M concentration). The liquids of the membrane sensors were prepared by using the ion-association extraction method.[171]

The selectivity and sensitivity of the prostigmine membrane sensors were estimated by measuring the EMF of the following electrochemical cell:

$$\text{SCE} \left| \begin{array}{c} \text{Reference} \\ \text{solution} \end{array} \right| \begin{array}{c} \text{Organic liquid} \\ \text{membrane} \end{array} \left| \begin{array}{c} \text{Sample} \\ \text{solution} \end{array} \right| \text{SCE} \qquad (5.I)$$

Table 5.20 Performance Characteristics of Prostigmine
Membrane Sensors[163]

Membrane solvent	Exchange site	Slope (mV decade^{-1})	Useful range (M)	Selectivity coefficients ($k_{\text{Pro, B}}^{\text{pot}}$)
Nitrobenzene	Tetraphenylborate	60	10^{-1}–10^{-5}	TEA: 0.61
	Dipicrylamine		(pH 4–10)	MCh: 0.16
				MNic: 0.005
1,2-Dichloroethane	Tetraphenylborate	60	10^{-1}–10^{-6}	TEA: 1.4
			(pH 4–10)	MCh: 0.16
				MNic: 0.005

Symbols: TEA, tetraethylammonium ion; MCh, methacholine; MNic, N-1-methyl
nicotinamide.

Electrode performances, including the selectivity coefficients are summarized in Table 5.20.

As can be seen from Table 5.20, selectivity or the sensitivity were the same for the two exchange sites with nitrobenzene as membrane solvent.

Analytical Procedures

i. *Direct measurement for prostigmine assay:*
   Standard solutions of 10^{-2} to 10^{-5} M prostigmine bromide are obtained by successive dilutions from a 10^{-1} M stock solution. A constant ionic strength ($I = 0.1$ M, adjusted with sodium nitrate) and pH between 4 and 10 must be used. The EMF measurements are made at room temperature in stirred solutions with an electrochemical cell of type (5.I). Millivolt readings (linear axis) are plotted vs. concentration (logarithmic axis) and the unknown concentration of the sample is determined from the calibration curve.

ii. *Potentiometric titration with 10^{-2} M sodium tetraphenylborate solution (tetraphenylborate-membrane sensor as indicator for all three compounds listed in Table 5.19; prostigmine-membrane sensor can be used as indicator for prostigmine assay):*
   In a 50-cm^3 beaker an aliquot of 25 cm^3 of the sample in the range 2.0×10^{-4} to 1.0×10^{-3} M for clidinium bromide and prostigmine bromide and 4.0×10^{-4} to 1.0×10^{-3} M for pyridostigmine bromide is pipetted. A volume of 5.0 cm^3 of the appropriate acetate buffer solution is added; after the potential is stabilized under stirring (about 1 min), the potentiometric titration is started. The amount of the pharmaceutical compound present in the sample is calculated in the usual way.

iii. *Pharmaceutical preparations assay:*

For tablet preparations, 20 tablets are weighed and powdered. An appropriate weighed amount of the powder (equivalent to about 0.5 mmol of active ingredient) is transferred to a 500 cm^3 beaker and stirred vigorously with about 400 cm^3 of water for 15 min. The solution is diluted to the mark in a 500 cm^3 volumetric flask and a 25.0-cm^3 aliquot is potentiometrically titrated as previously described.

5.30 Clobutinol

$$C_{14}H_{22}ClNO \ (MM = 255.7)$$

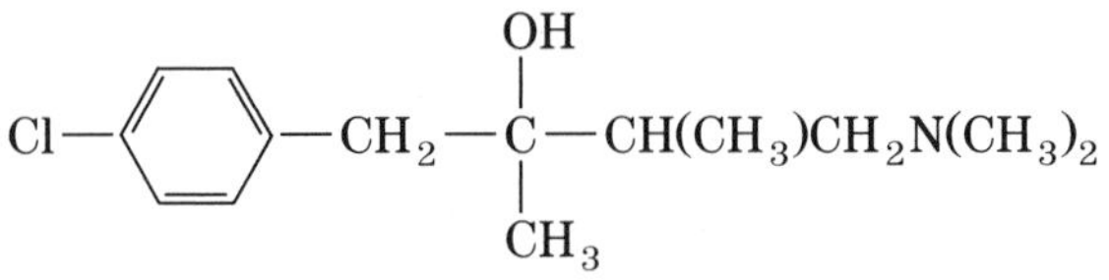

$$Cl-\hspace{-4pt}\langle\bigcirc\rangle\hspace{-4pt}-CH_2-\underset{\underset{CH_3}{|}}{\overset{\overset{OH}{|}}{C}}-CH(CH_3)CH_2N(CH_3)_2$$

Therapeutic category: cough suppressant

Discussion and Comments

A clobutinol-selective membrane sensor has been obtained from a PVC-matrix membrane containing clobutinol–tetraphenylborate ion-pair complex and dioctylphthalate as plasticizer in the mass ratio 14 : 1 : 2; the PVC content in PVC–tetrahydrofuran solution was 14.25%.[172] The sensor exhibits a Nernstian response to the clobutinol cation from $10^{-1.5}$ to $10^{-4.5}$ M in the pH range from 3.0 to 8.0. The selectivity coefficients, calculated by the separate solution method, show that only tripelenamine, diphenhydramine, and chlorpheniramine interfere in the electrode response ($k_{Clo, B}^{pot}$ = 1.41, 1.25, and 1.0, respectively.

Response times are about 2 min in $10^{-1.5}$ and $10^{-4.5}$ M solutions and instantaneous in the 10^{-2} to 10^{-4} M range. The membrane sensor has been used for clobutinol determination by direct potentiometry and in precipitation titration with sodium tetraphenylborate.

Analytical Procedures

i. *Direct potentiometry:*

Three standards of clobutional hydrochloride solutions (10^{-2}, 10^{-3}, and 10^{-4} M, respectively) are prepared by serial dilution from 0.1 M stock solution. The ionic strength is kept constant (e.g., to 0.1 M) with sodium nitrate solution. The clobutional–PVC membrane and

SCE are placed in the standards and the millivolt readings (linear axis) are plotted vs. concentration (logarithmic axis). The unknown concentration is determined from the calibration curve.

ii. *Potentiometric titration:*

The electrode pair (as before) is introduced into the sample solution (30 to 40 cm^3, approximately 5×10^{-3} M, and pH 5.0) and titrated with 5×10^{-2} *M* sodium tetraphenylborate solution. The end point corresponds to the maximum slope of the titration curve of EMF vs. titrant volume.

5.31   Clonidine

$$C_9H_9Cl_2N_3 \; (MM = 230.1)$$

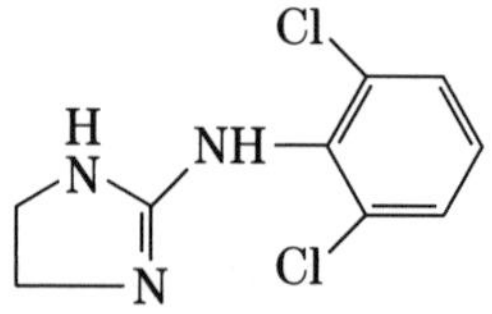

Therapeutic category: antihypertensive

Discussion and Comments

A Pt-coated PVC-membrane clonidine-selective sensor made with ion-pair complexes of clonidine–tetraphenylborate was described by Hu and Leng.[173] The electronanalytical properties of the membrane sensor (in particular, its stability) were studied and compared with those of conventional PVC-membrane clonidine-selective sensor. The sensor exhibits near-Nernstian response in the range 2×10^{-6} to 10^{-1} *M* clonidine and the limit of detection was 5×10^{-7} *M*. The sensor had satisfactory selectivity over many inorganic and organic cations tested.

Analytical Procedure

Three standards of clonidine hydrochloride solutions (10^{-2}, 10^{-3}, and 10^{-4} *M*, respectively) are prepared by serial dilutions from 0.1 *M* stock solution. Both the ionic strength and pH values are kept constant at 0.1 *M* ($NaNO_3$ or KNO_3) and approximately 5 (acetate buffer), respectively. Clonidine-membrane sensor and reference electrode (SCE) are immersed in the standards and the EMF readings (linear axis) are plotted vs. concentration (logarithmic axis). The unknown concentration is determined from the calibration curve.

5.32   Cloperastine

$$C_{20}H_{24}ClNO \ (MM = 329.9)$$

Therapeutic category: antitussive

Discussion and Comments

Cloperastine–dipicrylaminate ion-pair complex was used as electroactive material for construction of a PVC-coated copper-plate membrane sensor (dibutylphthalate, DBP, as plasticizer) sensitive to cloperastine.[174] The membrane sensor was activated in 5×10^{-4} M cloperastine solution for just 30 min before use. It exhibits a Nernstian response within the range 10^{-1} to 10^{-5} M with a slope of 58 mV decade^{-1}. The membrane sensor is negligibly affected by pH changes in the range 3 to 6, the plot E (in millivolts) vs. pH being used for basicity constant determination. It was reported that the reproducibility of the potential measurements was better than ± 1 mV in the range 10^{-2} to 10^{-4} M and the response times of the sensor varied from 5 s for 10^{-2} to 10^{-4} M to 30 s for very dilute solutions (10^{-5} to 10^{-6} M).

Cloperastine solutions containing between 0.04 and 0.9 mg cm^{-3} were determined by the direct potentiometric method with a good recovery (100.8 %) and a standard deviation of 2.9%. Cloperastine tablets were also assayed by this method with a recovery of 101.0% and a relative standard deviation of 4.7% ($n = 6$).

Analytical Procedure

Three standards of cloperastine hydrochloride solutions (10^{-2}, 10^{-3}, and 10^{-4} M, respectively) are prepared by serial dilutions from 0.1 M stock solution. Both the ionic strength and pH values are kept constant at 0.1 M (NaNO$_3$) and about 5 (acetate buffer), respectively. Cloperastine–PVC membrane sensor and reference electrode (SCE) are placed in the standards and the millivolt readings (linear axis) are plotted vs. concentration (logarithmic axis). The unknown concentration is determined from the calibration curve. For tablets assay, at least 20 tablets are weighed and finely powdered. An appropriate aliquot powder is then assayed following the same procedure.

5.33   Cyclizine

$$C_{18}H_{22}N_2 \ (MM = 266.4)$$

$$H_3C-N \overbrace{} N-CH \overset{\displaystyle C_6H_5}{\underset{\displaystyle C_6H_5}{\Big\langle}}$$

Therapeutic category: a powerful and long-acting
antagonist of histamine

Discussion and Comments

Campbell et al.[175] have described plastic-membrane sensors responding
to organic ions that were used to monitor the titrations of cyclizine
hydrochloride and other drug substances with sodium tetraphenylborate
solution. Sensors were constructed by coating a PVC film on a graphite
rod. The incorporation of nitrobenzene as plasticizer gave an electrode
responding to anions. Sensors containing both nitrobenzene and bis(2-
ethylhexyl)phthalate responded to both anions and cations, while giving a
negligible response to pH change.

The membrane sensors, treated as described in Campbell et al.[175] have
been used for at least six months with no evidence of deterioration.
Response times for freshly coated membrane sensors were less than 1
min and detection limits of 10^{-5} M were reported.

Lactose, starch, sodium starch glycolate, and magnesium stearate
present in the composition of cyclizine tablets had no effect on the assay.
Polyvinylpyrrolidone and gelatin gave high results and the potentiometric method is unsuitable for tablets containing these two excipients.
Cyclizine hydrochloride was 99.0% recovered from 50-mg tablets with a
standard deviation of 1.2%.

Analytical Procedure

The electrode pair (plastic-membrane sensor, containing both bis(2-eth-
ylhexyl)phthalate and nitrobenzene, as indicator and double-junction
silver–silver chloride reference electrode whose outer compartment is
filled with a saturated solution of sodium sulfate) is introduced into the
sample solution (30 to 40 cm^3, approximately 10^{-3} M) and titrated with
10^{-2} M sodium tetraphenylborate solution. The end point corresponds
to the maximum slope on the plot of EMF vs. tetraphenylborate volume.

5.34   Cyclophosphamide

$$C_7H_{15}Cl_2N_2O_2P \cdot H_2O \ (MM = 279.1)$$

Therapeutic category: antineoplastic; immunosuppressive

Discussion, Comments, and Procedure

See Section 5.13 and Dessouky et al.[135]

5.35   Cyproheptadine

$$C_{21}H_{21}N \ (MM = 287.4)$$

Therapeutic category: antihistaminic; antipruritic; appetite stimulant

Discussion and Comments

The preparation and characterization of a cyproheptadine membrane sensor was recently described by Bunaciu et al.[176] The sensor uses dinonylnaphthalenesulfonic acid as the counter-ion in the electroactive material. The PVC-membrane composition was 4.0% DNNS, 64.0%. o-nitrophenyloctyl ether, and 32.0% PVC (m/m). The internal filling solution was 10^{-3} *M* cyproheptadine hydrochloride of pH 5.0 (acetate buffer solution). The dinonylnaphthalenesulfonic acid in the PVC membrane was converted to ion-pair complex by soaking the sensor in cyproheptadine hydrochloride solution (10^{-2} *M*) for 24 h.

The EMF measurements were made with the following electrochemical cell:

$$\left.\begin{array}{c|c|c}
\begin{array}{c} \text{Cyroheptadine} \\ \text{membrane sensor} \end{array} &
\begin{array}{c} \text{Cyproheptadine} \cdot \text{HCl} \\ (C) \\ I = \text{const.; pH} = 5.0 \end{array} &
\text{SCE}
\end{array}\right. \qquad (5.\text{II})$$

where C is the cyproheptadine concentration, ranging from 10^{-6} to 10^{-2} M. The EMF is given by

$$E_1 = E_0 + 0.055 \log[\text{Cyp}^+] \qquad (5.53)$$

where the E_0 value is the conditional standard potential for the membrane sensor under the conditions of use of cell (5.II). The response is linear over the range 10^{-2} to 10^{-4} M with a near-Nernstian slope. The slope of curved part of the graph of E (in millivolts) vs. $p[\text{Cyp}^+]$ in the range 10^{-4} to 10^{-5} M is only about 40 mV decade^{-1}.

At pH values between 2.0 and 6.0 no significant changes in the membrane potential were observed (for different concentrations of cyproheptadine hydrochloride solutions). At pH values higher than 6.0 to 6.5, the cyproheptadine base in the aqueous test solution precipitates, so the EMF values are shifted toward more negative values.

The interference of various substances on the electrode response was studied by the mixed solution method and the respective coefficients, $k_{\text{Cyp}, j}^{\text{pot}}$, were calculated from the equation

$$k_{\text{Cyp}, j}^{\text{pot}} = (10^{\Delta E/S} - 1)[\text{Cyp}^+][j^{z+}]^{1/z} \qquad (5.54)$$

where ΔE is the change in potential in the presence of interfering ion j^{z+}, S is the slope of the calibration graph for the cyproheptadine ion Cyp^+, and $[\text{Cyp}^+]$ and $[j^{z+}]$ are the concentrations of the primary and interfering ions, respectively, at the same pH and ionic strength. Amino acids, scopolamine, lidocaine, vitamin B_1, and vitamin B_6 as well as common inorganic cations and most of the common excipients in pharmaceutical tablets do not interfere.

The membrane sensor proved useful in the potentiometric titration of cyproheptadine hypochloride in drug substances and in pharmaceutical preparations (tablets and syrup). The results of the assay of cyproheptadine in five samples of pharmaceutical preparations show that good recovery was achieved, with a high precision in the case of tablets (standard deviation 0.56%).

The sensor was also used to determine the content uniformity of cyproheptadine tablets (a relative standard deviation of 2.6% was reported).[176]

Analytical Procedures

i. *Potentiometric titration:*

The electrode pair (PVC-membrane cyproheptadine sensor and SCE) is placed into the partially aqueous (2.0% MeOH) sample solution of pH 5.0 (acetate buffer) (30 to 40 cm^3, approximately 5×10^{-3} M) and titrated with 5×10^{-2} M sodium tetraphenylborate solution. The end point corresponds to the maximum slope on the plot of EMF vs. tetraphenylborate volume.

For tablets assay, at least 20 tablets are weighed and finely powdered. An appropriate aliquot powder is then assayed following the same procedure.

ii. *Content uniformity assay of cyproheptadine tablets:*

Ten individual tablets are transferred to separate 25-cm^3 volumetric flasks and dissolved in 10 cm^3 partially aqueous solution (2.0% MeOH); then the suspension is diluted to volume with pH 5.0 acetate buffer solution. The contents of the volumetric flasks are transferred to 10 separate beakers and potentiometrically titrated as previously described. The cyproheptadine content of each tablet is calculated and the relative standard deviation is evaluated in the usual way.

5.36   Cystaphos and Cysteamine

$C_2H_7NNaO_3PS$ (MM = 179.1) C_2H_7SH (MM = 74.1)

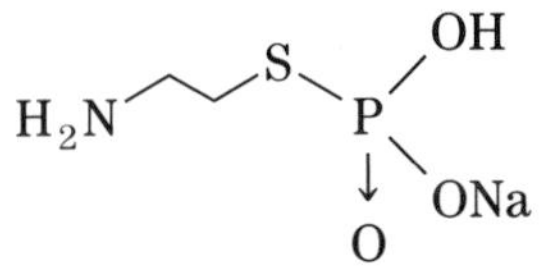

$$H_2N-CH_2-CH_2-SH$$

Therapeutic category: for prevention and treatment
of radiation sickness

Discussion and Comments

The quantitative determination of cystaphos is based on the reaction

$$H_2NCH_2CH_2SPO_3HNa \xrightarrow[H_2O]{H^+} H_2NCH_2CH_2SH + NaH_2PO_4 \quad (5.55)$$

Because the S — P bond is very labile in an acidic medium, the reaction takes place quantitatively to cysteamine and orthophosphate in less than 1 min at 100°C and pH 2 to 3. Ionescu et al.[177] studied conditions for the potentiometric titration of cysteamine produced by Reaction 5.55 using

10^{-2} M mercury(II) nitrate solutions as titrant and a Ag^+/S^{2-} membrane sensor as indicator electrode.

Figure 5.9 shows potentiometric titration curves of cysteamine under different conditions. The curves are well defined in all cases and the location of titrant volume may be easily made. Cysteamine determinations in 0.1 M sodium hydroxide are preferred because of improved stability in this medium. The titration curves correspond to a $1:2$ stoichiometry, according to

$$Hg^{2+} + 2\,H_2NCH_2CH_2SH \rightarrow (H_2NCH_2CH_2S-)_2Hg + 2\,H^+ \quad (5.56)$$

The possibility of using either a Ag^+/S^{2-}- or copper(II)-selective membrane sensor has been studied for direct determination of the cysteamine produced from S-(2-aminoethyl)thiophosphate hydrolysis.[177] The use of these sensors for direct potentiometric determinations of sulfur-containing organic compounds[178, 179] is based on the assumption that the electrode function arises from interactions between sulfur-containing functional groups and silver and copper ions, respectively, from the electrode membrane.

To establish the electrode function toward cysteamine, electrochemical cells of type (5.III) were used:

Ag^+/S^{2-}- or Cu^{2+}-selective membrane sensor	Cysteamine $I = 0.1\ M;\ (C)$	NaOH and PO$_4^{3-}$	Saturated KNO$_3$	SCE (5.III)

where C represents the cysteamine concentration varied in the 10^{-2} to 10^{-7} M range. The ionic strength of the solutions was kept constant with 0.1 M NaOH and 10^{-2} M orthophosphate solutions. The response of the Ag^+/S^{2-}-membrane sensor is linear within 10^{-2} to 10^{-4} M range (Figure 5.10). The slope of the electrode function is 87 mV decade^{-1}, which is consistent with 90 mV decade^{-1} for a $2:3$ Ag–cysteamine complex, formed according to

$$2\,Ag^+ + 3\,H_2NCH_2CH_2S^- \rightarrow Ag_2(H_2NCH_2CH_2S-)_3^- \quad (5.57)$$

The response time of this sensor is slow, especially in dilute cysteamine solutions (about 10 min in 10^{-4} M).

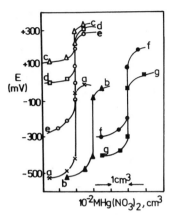

Figure 5.9 Potentiometric titration of 3.08 mg cysteamine under different conditions and using a silver sulfide crystal membrane sensor as indicator: (a) 0.1 M NaOH; (b) 0.01 M NaOH; (c) pH 2.5; (d) pH 4.0; (e) pH 7.5; (f) pH 6.0 (adjusted with hexamine); (g) pH 8.0 (adjusted with 2 cm^3 pyridine). Curves b, f, and g are displaced horizontally for clarity. (Reprinted from Ref. 177, p. 723, by courtesy of Marcel Dekker, Inc.)

The copper(II)-ion-selective membrane sensor presents a good response to cysteamine solution. The electrode function is stable and reproducible over a wide concentration range (10^{-2} to 10^{-6} M; curve b in Figure 5.10). In this case the slope is 58 mV decade^{-1}, in good agreement with 59.1 mV decade^{-1} for a redox interaction at the membrane–solution interface according to

$$2\,Cu^{2+} + 4\,H_2N(CH_2)_2SH \xrightarrow[-4\,H^+]{}$$

$$2\,H_2N(CH_2)_2SCu + H_2N(CH_2)_2SS(CH_2)_2NH_2 \quad (5.58)$$

Copper (II) oxidizes cysteamine to cystamine with the formation of $Cu^+H_2N(CH_2)_2S^-$.

(Details on determination of cystaphos crude samples may be found in Ionescu et al.[177] and Coşofreţ[98] [Section 18.3, pp. 334–338].)

Analytical Procedures

For the determination of cystaphos, 0.2 to 0.4 mmol of the compound is dissolved in 50 cm^3 distilled water; 1.0 cm^3 of 1 M perchloric acid solution is added and the sample boiled for 5 min. The cysteamine

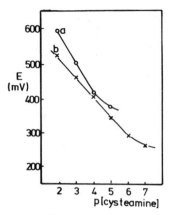

Figure 5.10 Response of the Ag^+/S^{2-} crystal membrane sensor (curve a) and copper(II) membrane sensor (curve b) to cysteamine. (Reprinted from Ref. 177, p. 727, by courtesy of Marcel Dekker, Inc.)

produced in the hydrolysis reaction is determined as follows:

i. *Potentiometric titration with 0.1 M mercury(II) nitrate solution:*
   5.0 cm³ of 1 *M* sodium hydroxide solution is added to the hydrolysis sample and is potentiometrically titrated with 0.1 *M* mercury(II) nitrate using a Ag^+/S^{2-}-crystal membrane as indicator sensor and SCE as reference electrode. The end point corresponds to the maximum slope on the titration curve.

ii. *Direct potentiometry:*
   The hydrolyzed sample is quantitatively transferred into a 100-cm³ volumetric flask; 10 cm³ of 1 *M* sodium hydroxide is added and the solution is diluted to volume with distilled water. Cysteamine standard solutions of 10^{-2}, 10^{-3}, and 10^{-4} *M* concentration are also prepared in 0.1 *M* sodium hydroxide. A calibration of *E* (in millivolts) vs. log[cysteamine] is plotted, using a copper(II)-ion-selective membrane sensor as indicator and SCE as reference. The sample concentration is determined from this graph.

5.37 DACCP

$$C_{15}H_{18}N_2O_6Pt \ (MM = 517.4)$$

Therapeutic category: anti-cancer agent

Discussion and Comments

A potentiometric method adequate for monitoring the degradation of 4-carboxylphthalato(1,2-diaminocyclohexane)-platinum(II) (DACCP) was recently developed by Nashed and Lindenbaum[180]; it is based on the measurement of benzene-1,2,4-tricarboxylic acid (trimellitic acid (TMA)), a degradation product of DACCP, with trimellitic-acid-selective membrane sensors. Liquid- and PVC-membrane sensors containing TMA–HDTMA (trimellitate–hexadecyltrimethylammonium) ion-pair complex as electroactive material were developed. For liquid-membrane sensor construction an Orion Model 92 liquid-membrane barrel and an Orion 92-81-04 porous membrane were used together with 0.1 M ion exchanger in octan-1-ol. For the PVC-type membrane sensor, which was constructed according to the general method of Moody et al.,[181] dioctylphthalate was found to be an adequate plasticizer.

The membrane sensors were conditioned by soaking in 0.1 M sodium trimellitate solution for 24 h before use and were also stored in this solution when not in use. The internal reference solution (0.1 M NaCl + 0.1 M NaTMA) were exchanged for freshly prepared solution every few days.

Both membrane sensors exhibited similar responses to trimellitate anion: detection limit of about 10^{-5} M and slopes of about 28 mV decade^{-1} (pH 7 to 9.5).

The results obtained for the direct potentiometric determination of TMA in pure solutions using a calibration graph, obtained with 12 solutions, each analyzed in replicate, in the concentration range 35 to 500 mg cm^{-3} yielding recoveries of 98.7% (standard deviation 1.8%).

The degradation of DACCP was monitored by the appearance of trimellitic acid (10^{-3} M aqueous solution of the drug at pH 7.5) as measured with TMA-selective sensor. The first-order rate constant for the appearance of TMA (or loss of drug) was found to be 6.6 × 10^{-4} min^{-1}, in good agreement ($\pm 5\%$) with values found by the other method. This method was not applicable at pH > 8.0 due to severe interference from DACCP and other degradation products.[180]

Analytical Procedure

To follow the DACCP degradation, 30 cm^3 of a 10^{-3} M aqueous solution of the drug at pH 7.5 (adjusted with 0.01 M phosphate buffer) is prepared. The degradation of DACCP is monitored by use of the appropriate TMA-membrane sensor and the Orion double-junction reference electrode, which measure the appearance of trimellitic acid. The concentration of TMA is calculated from the calibration graph obtain with 10^{-1}

to 10^{-5} M standard solutions of TMA at pH and ionic strength kept constant at 7 to 8 and 0.05 M, respectively (adjusted with solutions of sulfuric acid or sodium hydroxide and ammonium sulfate).[180]

5.38   Dextromethorphan Hydrobromide

$$C_{18}H_{26}BrNO \cdot H_2O \; (MM = 370.3)$$

Therapeutic category: antidepressant

Discussion and Comments

Higuchi et al.[182] found that plasticized-membrane sensors have a higher specificity for relatively hydrophobic organic cations and anions. They consider that any organic plasticizer matrix having limited hydrophilic character may be used as the gelling component of the membrane, its choice depending first of all on its compatibility with the desired liquid "plasticizer" components. The liquid components are chosen for their ability to solvate the ions of interest. The sensor response time is short, equilibrium being reached in less than 1 min in solutions having concentrations higher than 10^{-5} M. An electrode having a poly(vinylchloride) membrane plasticized with N,N-dimethyloleamide (Hallcomid 18-OL) gives a Nernstian response to tetrabutylammonium cation. The sensor proved useful in titrimetric analysis of organic cations such as dextromethorphan or diphenhydramine (see also Section 5.23) with sodium tetraphenylborate solution. The potential jump at the equivalence point is large and the end point is easily located.

Analytical Procedure

The electrode pair (PVC–amide membrane as indicator and SCE as reference) is introduced into the sample solution (30 to 40 cm^3, approximately 5×10^{-3} M) and titrated with 5×10^{-2} M sodium tetraphenylborate solution. The end point corresponds to the maximum slope on the titration curve.

5.39   Dibazol

$$C_{14}H_{12}N_2 \ (MM = 208.3)$$

Therapeutic category: vasodilator; antispasmodic; hypotensive

Discussion and Comments

Liquid-membrane and PVC-membrane dibazol-ion-selective sensors made with ion-pair complexes of dibazol with dicyclohexylnaphthalene sulfonate (DDCHNS), diisopentylnaphthalene sulfonate (DDPNS), diisobutylnaphthalene sulfonate (DDBNS), and tetraphenylborate (DTPB) as electroactive materials, were recently constructed by Yao.[183]

Calibration curves for the dibazol-ion-selective membrane sensors showed a near-Nernstian response in the approximately 10^{-2} to 10^{-5} M concentration range. The dibazol–dialkylnaphthalene sulfonate sensor with a high molecular weight (e.g., DDCHNS) showed a slightly wider linear range and greater response slope (the best plasticizer–solvent, DBP; solvent for liquid membranes, nitrobenzene).

Values of pH in the range 3 to 7 did not significantly affect the performance of the sensors for measuring dibazol concentrations. The response times of the PVC-membrane sensor were less than 20 s in the 10^{-2} to 10^{-4} M range and 20 to 50 s in 10^{-5} to 10^{-6} M solutions. The liquid-membrane sensor response times ranged from 30 s for solutions $> 10^{-4}$ M to 0.5 to 2 min for solutions $< 10^{-4}$ M.

The dibazol-selective membrane sensors were found to respond to a number of amines, alkaloids, and quaternary ammonium species, e.g., probanthine, tetrabutylammonium, chlorpheniramine, quinine, cinchonine, propranolol, and diphenhydramine. Consequently, these substances are likely to cause interference in the assay of dibazol, although they are rarely formulated in combination with dibazol.[183]

The membrane sensors can be used in the potentiometric determination of dibazol. The average recovery obtained, for the direct potentiometric assay of dibazol using the calibration curve, was 99.0% (relative standard deviation 1.4%). When sodium tetraphenylborate was used as titrant, an average recovery of 99.4% (relative standard deviation < 1%) was reported.[183] Dibazol tablets were also analyzed with good results and in agreement with the official method.

Analytical Procedures

i. *Direct potentiometry:*
The electrode pair (the dibazol–PVC membrane sensor is preferred because it eliminates the use of harmful organic solvents such as nitrobenzene) is introduced into the respective dibazol standard solutions (10^{-2}, 10^{-3}, and 10^{-4} M, respectively; pH about 6 to 6.5, 0.1 M $NaNO_3$ for keeping ionic strength at a constant value), and, under stirring, the respective EMF values are recorded. The graph of EMF vs. log[dibazol] is plotted and the unknown concentration of the sample solution is determined from this graph.

ii. *Potentiometric titration:*
The electrode pair (dibazol and SCE) is introduced into the sample solution of pH 6.0 to 6.5 (30 to 40 cm^3, approximately 5×10^{-3} M) and titrated with 5×10^{-2} M standard sodium tetraphenylborate solution. The end point corresponds to the maximum slope on the plot of EMF vs. titrant volume.

iii. *Assay of dibazol tablets:*
A sample of 25 to 30 tablets is finely powdered and a portion of the powder, equivalent to about 25 mg of dibazol, is transferred to a 50-cm^3 volumetric flask and diluted to volume with distilled water. Potentiometric measurements are made in replicate on the sample solution and on a standard solution containing approximately the same concentration of dibazol until reproducible (± 0.1 mV) values are obtained. (The dibazol dicyclohexylnaphthalene sulfonate poly(vinyl chloride) membrane sensor is recommended as indicator electrode, SCE as reference).

5.40   Diethylcarbamazine

$C_{10}H_{21}N_3O$ (MM = 199.3)

$$CH_3-N \overbrace{}^{} N-CON(C_2H_5)_2$$

Therapeutic category: anthelmintic

Discussion and Comments

The membrane sensors constructed by Campbell et al.[175] (see also Section 5.33), consisting of a graphite rod coated with a PVC film, may be used for the assay of diethylcarbamazine citrate in tablets. The sensor that contains both nitrobenzene and bis(2-ethylhexyl)phthalate as plasticizers gives a larger potential break in titrations than sensors that contain bis(2-ethylhexyl)phthalate alone.

Analytical Procedure

Twenty to thirty tablets are accurately weighed and finely powered. An amount of powder equivalent to 1 g of diethylcarbamazine citrate is accurately weighed out into a 100-cm^3 volumetric flask. The flask is half-filled with distilled water and shaken well; 20 cm^3 of 10% (v/v) acetic acid is added and the sample is heated for 5 min on a boiling-water bath, followed by 10 min in an ultrasonic bath. The sample is diluted to volume with distilled water. An aliquot of this solution is transferred to a 100-cm^3 volumetric flask to make approximately 10^{-3} M solution by dilution with distilled water; 30 to 40 cm^3 of this solution is titrated with 10^{-2} M sodium tetraphenylborate standard solution in the presence of PVC-membrane electrode sensor. The end point corresponds to the maximum slope on the plot of EMF vs. tetraphenylborate volume.

5.41   Diethyldithiocarbamate (Sodium Salt)

$$C_5H_{10}NNaS_2 \ (MM = 171.3)$$

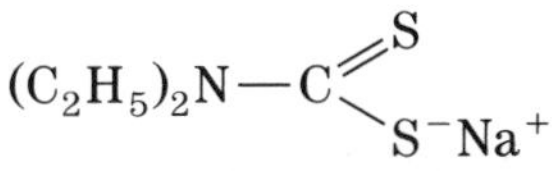

$$(C_2H_5)_2N-C\overset{\displaystyle S}{\underset{\displaystyle S^-Na^+}{\Big<}}$$

Therapeutic category: chelating agent; experimental
in Wilson's disease

Discussion and Comments

A membrane sensor sensitive to diethyldithiocarbamate (DDC) ions down to 10^{-5} M was prepared by precipitation of copper diethyldithiocarbamate and silver sulfide within a graphite rod.[184] The sensor was prepared by successive soaking of the graphite rod in saturated aqueous solutions of sodium sulfide and silver nitrate for 1 h. The rod was then washed thoroughly with deionized water and soaked in saturated 96% ethanolic solution of NaDDC for 1 h, removed, resoaked in aqueous copper sulfate solution for another 1 h, washed several times with deionized water and ethanol, and dried. The rod was inserted in a PVC sleeve and between measurements was stored in deionized water.

The sensor thus prepared showed a linear response to DDC ions with an average anionic slope of 55 mV decade^{-1} over the concentration range 10^{-1} to 10^{-5} M in 50 to 70% ethanol background. Fast and stable response (15 to 25 s for solutions $> 10^{-3}$ M and 30 to 40 s for solutions $< 10^{-3}$ M) was observed. It is worth mentioning that concentration of copper diethyldithiocarbamate in the graphite sensor had no significant influence on the electrochemical behavior of the sensor. No measurements of selectivity coefficients have been made because the membrane sensor was designated only for use as a sensor for multielement titrations.[73]

Analytical Procedure

Standard solutions of sodium diethyldithiocarbamate in the range 10^{-2} to 10^{-4} M are prepared by serial dilutions from 0.1 M stock solution. The solutions are prepared by keeping both pH and ionic strength at constant values. The standard solutions are transferred into 100-cm³ beakers containing Teflon-coated stirring bars. The coated DDC membrane sensor in conjunction with a double-junction reference electrode are immersed successively in the standards, and, under stirring, the EMF values are recorded. The graph of E (in millivolts) vs. log[DDC] is plotted and the unknown samples concentration is determined from this graph.

5.42   Digoxin

$$C_{41}H_{64}O_{14} \text{ (MM = 780.9)}$$

(secondary glycoside from *Digitalis lanata* Ehrh. or *D. orientalis* Lam., Scrophulariaceae)

Therapeutic category: cardiotonic

Discussion, Comments, and Procedure

A novel potentiometric enzyme immunoassay technique utilizing polystyrene beads in conjunction with a gas-sensing membrane electrode was described by Keating and Rechnitz.[185] The technique was illustrated with the measurement of digoxin via competitive inhibition of antidigoxin–horseradish peroxidase conjugate activity. The rate of CO_2 generation from the peroxidase–pyrogallol (the hydrogen donor) reaction in the presence of antibody-labeled horseradish peroxidase (HRP) enzyme with competition between the free digoxin to be determined and polystyrene-bead-immobilized digoxin was monitored. The entire assay principle is schematically outlined in Figure 5.11.

The ratio of digoxin to bovine serum albumin (BSA) in digoxin–BSA conjugate was 20 to 1 (the polystyrene beads were coated with digoxin–BSA by physical adsorption). The optimum conditions for the

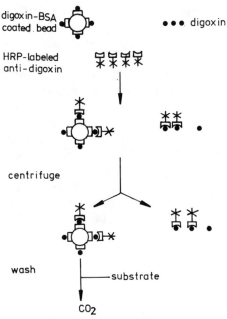

Figure 5.11 Schematic representation of the digoxin enzyme immunoassay. (Reprinted from Ref. 185, p. 3, by courtesy of Marcel Dekker, Inc.)

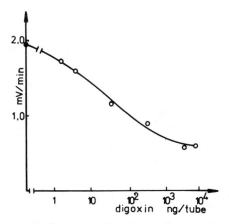

Figure 5.12 Typical standard curve for digoxin potentiometric immunoassay. (Reprinted from Ref. 185, p. 9, by courtesy of Marcel Dekker, Inc.)

enzymatic reaction in the presence of antibody–HRP conjugate were as follows: substrate (H_2O_2) concentration 3×10^{-3} M; pyrogallol concentration 3.2×10^{-3} M, pH 6.0 (phosphate buffer). The pH 6.0 is also convenient for the pCO_2 electrode.

The digoxin assay can be carried out by a competition between the digoxin–BSA coated beads and free digoxin, so that increasing amounts of free digoxin will slow the rate of CO_2 production, as monitored potentiometrically. Thus, the typical calibration curve for digoxin assay shown in Figure 5.12 was constructed by measuring CO_2 production rates of anti-digoxin–HRP bound to 50-μl aliquots of digoxin–BSA coated beads with varying amounts of free digoxin under constant conditions. It can be seen that the method gives a very sensitive assay for digoxin in the nanogram range and offers shortened analysis times when compared to previous methods using plastic beads.[186]

5.43   Dyestuffs

Cationic dyestuffs like those of pharmaceutical interest listed in Table 5.21 can be precipitated from aqueous solutions with an anionic reagent such as sodium tetraphenylborate. The titration of crystal violet with sodium tetraphenylborate solution was followed with commercially available ion-selective membrane sensors such as the Crytur 19-15 potassium-selective electrode based on a PVC membrane containing valinomycin and dipentylphthalate[187] or Orion tetrafluoroborate- or cyanide-membrane sensors.[9, 188]

Gurev and co-workers[189, 190] used as titrant picric acid or 3,5-dinitro-salicylic acid, and the titrations were monitored with a liquid-membrane sensor based on the Crystal Violet–picrate ion pair in nitrobenzene. Crystal Violet can also be determined by titration with sodium dodecyl-sulfate by applying a current between -40 and -100 μA. The applied current produces larger potential breaks. However, equilibration with an applied current for at least 3 min is recommended prior to starting a titration.[188]

Vytras and Dajkova[191] conducted a systematic study for the determination of cationic triarylmethane dyestuffs with sodium tetraphenylborate, using as indicator electrode an aluminum wire coated with a PVC membrane plasticized with 2-nitrophenyl ether, 2-ethylhexyl ether, or tricresylphosphate. As the triarylmethane dyestuffs tested are ionic compounds of high molecular mass and almost symmetrical charge distribution, most of the dyes were potentiometrically titrated without difficulties either in unbuffered or buffered solutions. For Methylene Blue, an end-point break of 120 mV was found[183] when a fluoroborate-selective membrane sensor was used as indicator electrode in a potentiometric

Table 5.21 Pharmaceutical Dyestuffs Assayed by Membrane Sensors

Dyestuff	Formula (MM)	Therapeutic category
Brilliant Green	$C_{27}H_{34}N_2O_4S$ (482.6)	Antiseptic

Crystal Violet (Gentian Violet)	$C_{25}H_{30}ClN_3$ (408.0)	Topical anti-infective; anthelmintic

Malachite Green	$C_{23}H_{25}ClN_2$ (364.9)	Topical antiseptic

Methylene Blue	$C_{16}H_{18}ClN_3S$ (373.9)	Antimethemoglosinemic antidote (cyanide)

titration at pH 1.3 to 2.0. A feasible pH medium for this dyestuff is 1.3 to 10; a cyanide solid-state membrane sensor can also be used.

Liquid-membrane ion-selective sensors for the determination of acidic and basic dyes have been previously developed by Fogg and co-workers.[192–195] They used as electroactive materials for the membrane sensors the ion-association complexes of the respective dyestuff with 12-tungstosilicate and tetraphenylborate, respectively (for details see also Coşofreţ,[98] pp. 212–214).

Analytical Procedure

The electrode pair (dyestuff-, potassium-, BF_4^--, or CN^--selective as indicator, with SCE reference) are introduced into the sample solution (30 to 40 cm^3, approximately 5×10^{-3} M) and the solution titrated under stirring with 5×10^{-2} M sodium tetraphenylborate. The end point corresponds to the maximum slope of the titration curve of EMF vs. tetraphenylborate volume.

5.44   Ephedrine and Related Compounds

The first ephedrine-selective sensors with either a liquid membrane or poly(vinyl chloride)-matrix membrane were constructed and characterized by Fukamaki et al.[196] Methylephedrine-membrane sensors have also been developed by the same authors. In both cases, tetraphenylborate was used as exchange site for ephedrine and methylephedrine, respectively (see Table 5.22).

An organic solvent solution of the tetraphenylborate salt of the appropriate cation was used as a liquid membrane. This was held in the bottom of a U-type glass tube or Orion electrode body (Model 92). The PVC-matrix membrane was prepared by mixing a 20% (m/m) solution of PVC in tetrahydrofuran with dioctylphthalate and the appropriate ion-pair complex (in the mass ratio 25 : 10 : 2). Methylephedrine-selective membrane sensors of both types exhibited Nernstian responses down to 10^{-4} M, whereas ephedrine-selective membrane sensors gave a Nernstian response down to 10^{-3} M. The membrane potentials of both sensors were independent of pH from 1.5 to 8.0.

Interferences by sodium, potassium, ammonium, and calcium ions are extremely low for both sensors. The presence of caffeine, antipyrine, aspirin, aminopyrin, vitamin C, etc. usually contained in drugs do not interfere.

The influence of co-ion (an ionic species with the same charge sign as the ion-exchange site in the liquid membrane) on the response of the membrane sensor has been given some theoretical clarification by the three-region concept of membrane potential.[197] In the presence of sodium salts of nitrate, iodide, or thiocyanate, the response of the methyl-ephedrine-selective sensor (liquid-membrane type) was no longer Nernstian because of a negative error. The order of increasing interference was in agreement with increasing extractibility of the ions in the nitrobenzene–water system, i.e., NO_3^-, I^-, SCN^-.[198]

Besides its use for the direct potentiometry of ephedrine or methyl-ephedrine, the PVC-membrane sensor (methylephedrine- and/or ephedrine-selective) has been found useful as an indicator electrode for the determination of the respective drug by precipitation titration with sodium tetraphenylborate standard solution.

Table 5.22 Ephedrine and Related Compounds Assayed
by Membrane Sensors

Compound	Formula (MM)	Therapeutic category
Ephedrine	$C_{10}H_{15}NO$ (165.2)	Sympathomimetic (L-form as adrenergic bronchodilator)
Methylephedrine	$C_{11}H_{17}NO$ (179.2)	Antitussive; antihistaminic
Epinephrine	$C_9H_{13}NO_3$ (183.2)	L-Form as adrenergic
Norepinephrine	$C_8H_{11}NO_3$ (169.2)	L-Form as adrenergic; vasopressor

New methods based on the use of selective membrane sensors for the determination of ephedrine and related compounds (Table 5.22) were developed by Ni[199] and Hassan and co-workers.[200, 201]

One of the new liquid-membrane sensors sensitive to ephedrine is based on the use of ephedrine–5-nitrobarbiturate ion-pair complex in nitrobenzene (10^{-2} M) as an electroactive material.[200] The body of an Orion Model 92 electrode equipped with an Orion 92-05-04 microporous membrane was used for ephedrine-sensor preparation (internal solution 2×10^{-2} M ephedrine hydrochloride–potassium chloride). The sensor gave a linear response within the range 10^{-2} to 10^{-5} M (detection limit 4.5×10^{-6} M, pH range 4 to 7, slope 55 mV decade^{-1}), which seems to be much better than that described by Fukamaki et al.[196]

Soluble drug excipients and diluents such as maltose, glucose, lactose, starch, and gelatin binder that are present in some tablets do not interfere with sensor response. Further, the sensor exhibits negligible interference from many nitrogen compounds such as amines, amides, and amino acids. The membrane sensor is, however, not really selective to ephedrine over some other alkaloids, such as strychnine, quinine, and caffeine ($k_{\text{Eph}, j}^{\text{pot}}$ = 2.8, 1.8, and 1.7, respectively).

The determination of 0.1 to 2000 μg cm^{-3} of ephedrine in aqueous solutions showed an average recovery of 99.2% (relative standard deviation 1.5%). The results obtained for the determination of ephedrine in some pharmaceutical preparations (injection and tablets) have shown an average recovery of 99.1% of the nominal value (relative standard deviation 1.7%).

Ephedrine–flavianate, as a novel electroactive material, was used for construction of a ephedrine liquid-membrane sensor. This yielded also good response to the neurotransmitters epinephrine and norepinephrine.[201] A 10^{-2} M ion exchanger in 1-octanol was used as liquid-membrane material. The same electrode body, microporous membrane, and internal solution were used. The sensor exhibited a linear response to ephedrine solutions of concentrations down to approximately 10^{-5} M with a near-Nernstian slope (55.2 mV decade^{-1}). The sensor also responds to epinephrine and norepinephrine cations because these compounds are similar to ephedrine, being aryl-β-ethanolamine derivatives (linear ranges 10^{-2} to 10^{-5} M for epinephrine and 10^{-2} to 10^{-4} M for norepinephrine; slopes 45.1 and 35.2 mV decade^{-1}, respectively).

Results for the determination of 0.1 to 2000 μg cm^{-3} ephedrine, epinephrine, and norepinephrine hydrochlorides in aqueous solutions using the sensor with the calibration graph method showed an average recovery of 98.7% (relative standard deviation 2.3%). The sensor was also used for the determination of ephedrine in some pharmaceutical injections, tablets, syrups, and eye drops containing 2 to 50 mg ephedrine per cubic centimeter or per tablet (pH of the measuring solution was adjusted to 4 to 7). The results obtained showed an average recovery of 97% of the nominal values and the standard deviation was 2.5%.

Analytical Procedures

i. *Ephedrine and methylephedrine assay by potentiometric titration:*

The electrode pair (ephedrine- and/or methylephedrine-selective membrane, with SCE as reference) is introduced into the sample solution (30 to 40 cm^3, approximately 10^{-3} M) and titrated with 10^{-2} M sodium tetraphenylborate solution. The end point corresponds to the maximum slope on the titration curve.

ii. *Ephedrine, epinephrine, and norepinephrine assay by direct potentiometry:*

Three standard solutions for each drug substance are prepared by serial dilutions from the respective 10^{-1} *M* stock solution. The standard solutions (10^{-2}, 10^{-3}, and 10^{-4} *M*) of each drug substance, with constant pH and ionic strength are transferred into 100-cm^3 beakers containing Teflon-coated stirring bars. The ephedrine–flavianate liquid-membrane sensor in conjunction with a reference electrode is immersed successively in the appropriate standards. The graphs of *E* (in millivolts) vs. log[drug substance] are plotted, and the unknown sample concentration of ephedrine, epinephrine, and norepinephrine is determined from the respective graph.

iii. *Ephedrine assay from pharmaceuticals (injections, tablets, syrups, eye drops):*

A portion equivalent to 25 to 75 mg of ephedrine is treated with 15 cm^3 of 0.05 *N* HCl and heated to 60°C for 3 min, and the pH is adjusted to 4 to 7. The solution is then diluted to 25 cm^3 with deionized distilled water and shaken. The EMF of the solution is measured with the electrode-pair system and compared with the calibration graph plotted as previously described.

5.45   Ethacrynic Acid

$$C_{13}H_{12}Cl_2O_4 \ (MM = 303.2)$$

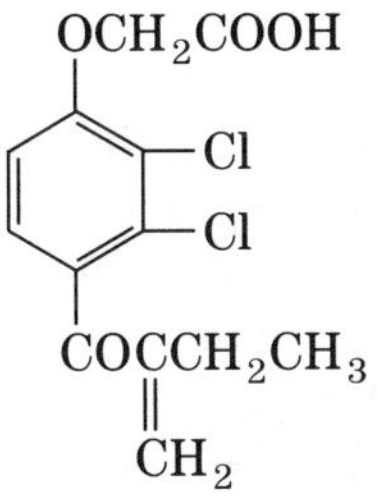

Therapeutic category: diuretic

Discussion and Comments

An ethacrynate-selective membrane sensor based on a PVC membrane and with inner graphite contact has been reported.[91] The PVC membrane was prepared from hexadecyltrioctylammonium iodide with dibutyl phthalate as the plasticizer. The sensor was activated for 1 h in 10^{-2} *M* ethacrynate solution. When the concentration of the site carrier in the membrane was 1%, a linear response was obtained within the range 10^{-3} to 1.6×10^{-5} M (detection limit 1.5×10^{-5} M, slope 55.6 mV decade^{-1}). Phenobarbital, amobarbital, phenytoin, and thiopental interfere in the sensor response ($k_{Eth,j}^{pot}$ = 0.37, 0.54, 2.3, and 7.6, respectively).

When the sensor was used to determine ethacrynic acid in the concentration range 0.06 to 0.27 mg cm^{-3}, an average recovery of 99.3% (coefficient of variation = 1.4%, n = 10) was reported.

Analytical Procedure

A stock solution of 10^{-2} M sodium ethacrynate is prepared by dissolving the respective amount of the compound in distilled water. The pH of the solution is adjusted to 9.0 with NaOH/HCl solutions; 10^{-3} and 10^{-4} M ethacrynate solutions are obtained from the stock solution by successive dilutions and by keeping pH at a constant value. The ionic strength of all three standards must also be kept constant (e.g., with sodium sulfate solution). The membrane sensor selective to ethacrynate in conjunction with SCE as reference is introduced into the standard solutions, and the EMF readings (linear axis) are plotted against concentration (logarithmic axis). The ethacrynate concentration from the unknown sample is determined from this graph.

5.46   Ethenzamide and Similar Compounds

A simple potentiometric method for the determination of drugs having a carboxyamide group (see Table 5.23) was described by Tagami and Fujita.[202]

The method is based on refluxing the sample in 20% HCl for a fixed time, when the carboxyamide is hydrolyzed and an equivalent amount of ammonium ion is liberated. After alkalinization (pH > 11), the converted ammonia is determined with an ammonia-gas-sensing electrode (Horiba, Model 5002-05 T) from the respective calibration curves (linearity was observed over the range of 2×10^{-5} to 10^{-2} M).

A mixture of ethenzamide, nicotinamide, pyrazinamide, or salicylamide and 20% HCl was boiled gently and the resultant ammonia in the decomposition solution was determined at various boiling times. In the case of ethenzamide and salicylamide, the electrode potentials reached a maximum at boiling times of 90 and 120 min, respectively. However, the electrode potentials of pyrazinamide and nicotinamide reached a maximum at the same boiling time of 10 min. In the case of decomposition with 10 and 5% HCl, the electrode potentials of nicotinamide reached a maximum at boiling times of 60 and 90 min, respectively.[202]

Table 5.23 Ethenzamide and Similar Compounds Assayed
by Membrane Sensors

Compound	Formula (MM)	Therapeutic category
Ethenzamide	$C_9H_{11}NO_2$ (165.2)	Analgesic

Nicotinamide	$C_6H_6N_2O$ (122.1)	Enzyme–co-factor vitamin

Pyrazinamide	$C_5H_5N_3O$ (123.1)	Antibacterial (tuberculostatic)

Salicylamide	$C_7H_7NO_2$ (137.1)	Analgesic

The determination of drugs listed in Table 5.23 in pure substances
were performed with average errors of 0.15, 0.08, 0.22, and 0.27%,
respectively, when amounts from 1 to 40 mg were analyzed.

The determination of commercially available nicotinamide injection
was carried out with a mean recovery of 107.2 ± 1.09% (according to
the *United States Pharmacopeia*[203] and the *Japanese Pharmacopoeia*,[204] nicotinamide injections contain not less than 95% and not
more than 110% of the stated amount of $C_6H_6N_2O$). In determining
other commercially available 10% nicotinamide powders, a suitable extraction was sought. (Extraction of nicotinamide is necessary to avoid
interference from auxiliary compounds; acetone was preferred as solvent

for the extraction.) The assay was performed on three samples of approximately 10% powder and the recoveries were 115.7, 118.5, and 115.3%, respectively, with a mean value of 116.5% (according to the *Japanese Pharmacopoeia*, description as a percentage of nicotinamide powder is not noted).

Campanella and co-workers[205–208] used heterogeneous membrane sensors with copper nicotinate as electroactive material in polyethene, silicone rubber, or paraffin, and liquid-membrane sensors based on an ester of nicotinic acid (cetyl nicotinate) for the determination of nicotinic acid derivatives.

The heterogeneous solid-membrane sensors have a calibration graph with a slope corresponding to a univalent cation. The liquid-membrane cetyl nicotinate sensor gave a calibration slope of 25 ± 1 mV decade^{-1} and was more selective than the solid-membrane sensor.

With an anion-exchanger liquid-membrane of 10^{-2} to 10^{-1} M solution of trimethylhexadecylammonium nicotinate (TMHAN) in decanol, liquid-membrane sensors sufficiently selective and very convenient for the titration of nicotinate with sulfuric acid and hydrochloric acid solutions were obtained: TMHAN may be prepared in solution in $1:1$ (v/v) water : acetone by the reaction

$$(CH_3)_3N^+(CH_2)_{15}CH_3Br^- + AgC_6H_4O_2N \rightarrow TMHAN + AgBr \quad (5.59)$$

The determination of nicotinamide with the TMHAN liquid-membrane sensor is based on the transformation of nicotinamide into nicotinate by chemical and enzymatic methods. The chemical method consists of alkaline hydrolysis or reaction with nitrous acid, whereas the enzymatic method utilizes nicotinamide hydrolase, which catalyzes

$$\text{nicotinamide} \xrightarrow{\text{H}_2\text{O}} \text{nicotinate} + NH_4^+ \quad (5.60)$$

Among the chemical methods, the reaction with nitrous acid is preferable to alkaline hydrolysis because the operations may be carried out at room temperature; the reaction is specific for the amide group and is quantitative. In addition, the reaction products may be readily eliminated from the reaction medium. Thus, the reaction

$$RCONH_2 + HNO_2 \rightarrow RCOOH + N_2 + H_2O \quad (5.61)$$

liberates nitrogen, whereas excess nitrous acid may be eliminated by using urea when nitrogen and carbon dioxide are expelled.

After addition of urea, it is possible to titrate potentiometrically the nicotinate ion formed, using the liquid-membrane sensor.[206]

Hadjiioannou and co-workers[209] described a simple potentiometric method for the determination of nicotinamide in multivitamin preparations using an ammonia-gas-sensing electrode (Orion, Model 95-10) for detecting the ammonia produced after alkaline hydrolysis. The rate of hydrolysis increases with temperature and sodium hydroxide concentration. A temperature of 50°C and 2 M sodium hydroxide solution were chosen as a compromise between speed of hydrolysis and experimental convenience. Amounts of nicotinamide in the 0.5 to 15 mg range per 5 cm^3 of sample solution can be determined with an average error of about ±1.7%. All B-complex vitamins (except vitamin B$_{12}$), when present with nicotinamide in the 1 : 1 ratio, have almost no effect on nicotinamide assay.[209] Cyanocobalamin has six CONH$_2$ groups per molecule and causes a positive error, but this has almost no effect on the nicotinamide assay because cyanocobalamin is present in much smaller amounts than nicotinamide.

The same authors[210] have conducted another interesting study on the alkaline hydrolysis of nicotinamide. The same ammonia-gas-sensing electrode was used to follow the formation of ammonia. A technique making use of simulated reactions has been developed to calibrate the electrode under dynamic conditions that overcome problems arising because of the relatively slow response of the sensor. A general expression has been derived for the pseudo-first-order rate constant ratio, over the concentration ranges 5×10^{-3} to 10^{-1} M nicotinamide and 0.1 to 0.5 M NaOH and the temperature range 22 to 31°C under constant ionic strength.

Two bacterial membrane sensors have been studied for the determination of nicotinamide with a linear response range of 2.8×10^{-4} to 2×10^{-2} M.[211] The strains used, although taxonomically different and differently improved (*Escherichia coli* mutated and *Bacillus pumilus* induced), present the same nicotinamide deaminase activity and may be used for analytical assays.

The enzymatic deamination of nicotinamide occurs according to the following equation:

$$\text{(nicotinamide)} + H_2O \xrightarrow[\text{nicotinamide deaminase}]{\text{bacterial}} \text{(nicotinic acid)} + NH_3 \quad (5.62)$$

The enzyme instability and the low enzyme level in microorganisms required an optimization of the microbiological and physicochemical factors involved in the construction and manipulation of the bacterial membrane sensor. The influence of physicochemical properties (type of membrane, temperature, pH, ionic strength of the buffers, co-factors and biochemical and microbiological factors [choices of strains, mutation or induction modes, growth media, incubation times, other enzymic inter-ferences, and storage]) on the sensor response were described in detail.[211]

An ammonia-gas-sensor (Tacussel pNH_3-1) was used for construction of the nicotinamide bacterial membrane sensors. Before each measure-ment, the bacterial sensor was conditioned in order to reach its baseline at pH 7.8 in a fresh buffered solution of 0.01 M TRIS–HCl containing 4×10^{-3} M magnesium chloride. Quantitative determinations of nicotin-amide were carried out in the same buffer at $30.0 \pm 0.2°C$ by direct potentiometry. Between measurements the sensor was stored at 30°C in brain heart infusion medium (BHI) under agitation (60 rpm). This storage mode insured the regeneration of the bacterial enzyme activity.[211]

The long-term stability (more than 100-fold higher than the purified enzyme, i.e., 5 and 3 days for *E. coli*–based sensor and *B. pumilus* [after induction]–based sensor, respectively) of both sensors was realized by the regeneration of living cells on the sensor itself.

The determination of nicotinamide in aqueous solutions (0.01 M TRIS–HCl, pH 7.8 buffer + 4×10^{-4} M $MgCl_2$), containing between 34 and 255 mg dm^{-3} nicotinamide, was performed by direct potentiometry with both sensors. The results demonstrated a perfect correlation be-tween the two membrane sensors (standard deviation less than 2%).

The quantitative determination of nicotinamide in pharmaceutical preparations have been possible on pulverized tablets without filtration and in multivitamin ampoules at lower concentrations of 1×10^{-3} M. It was recorded that the two strains have the same specificity for nicotin-amide and are not influenced by the presence of other vitamins of the B group.

Analytical Procedures

i. *Assay of drug substances:*

A mixture of 123 mg (approximately 10^{-3} mol) of pyrazinamide and 20 cm^3 of 20% HCl is boiled for 20 min. The resultant solution is adjusted to pH 6.5, diluted to 100 cm^3 in a volumetric flask and then 20 cm^3 of this solution is diluted to 100 cm^3 with water. A 50-cm^3 aliquot of the sample is transferred to an ~ 80-cm^3 cell and 5 cm^3 of 5 M NaOH is added and the mixture is incubated for 30 min at 20°C. Finally the ammonia electrode is immersed in the solution and the potential measurements are carried out. The ammonia concentration in the sample solution is determined from the calibration graph (constructed with standard solutions of pure drug, treated in the same way as before).

Assay procedures of ethenzamide, salicylamide, and nicotinamide were carried out in a similar manner. The ammonia concentration was determined from the calibration curve prepared from the correspond-ing standard drug solutions. (All the measurements are performed at

$20 \pm 0.1°C$ under stirring; the internal filling solution of the gas-sensing electrode is replaced with fresh solution before subsequent use.)

ii. *Assay of nicotinamide injection:*
A mixture of 3 cm^3 (equivalent to 150 mg of nicotinamide) of the injection and 25 cm^3 of 20% HCl is boiled for 30 min. The resultant solution is transferred into a 250-cm^3 volumetric flask and diluted to volume with distilled water. The ammonia concentration in a 50-cm^3 aliquot of the sample is determined using the ammonia-gas-sensing electrode. The nicotinamide content of the injection solution is determined from the calibration graph, prepared as previously described.

5.47   Ethionamide and Prothionamide

$C_8H_{10}N_2S$ (MM = 166.2) $C_9H_{12}N_2S$ (MM = 180.3)

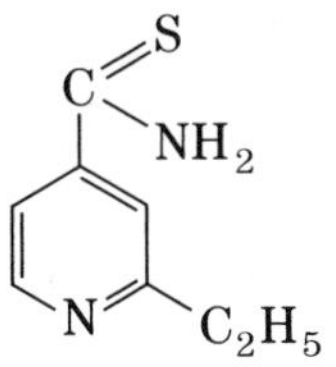

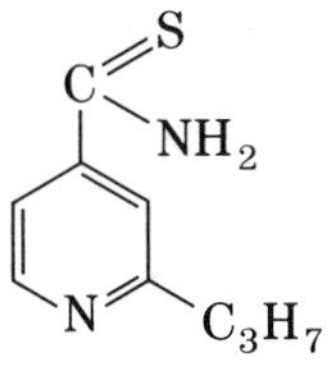

Therapeutic category: antibacterial (tuberculostatic) agents

Discussion and Comments

Both ethionamide and prothionamide contain a carbothionamido group that is decomposed into ammonium chloride, hydrogen sulfide, and the respective carboxylic acid, on heating with hydrochloric acid:

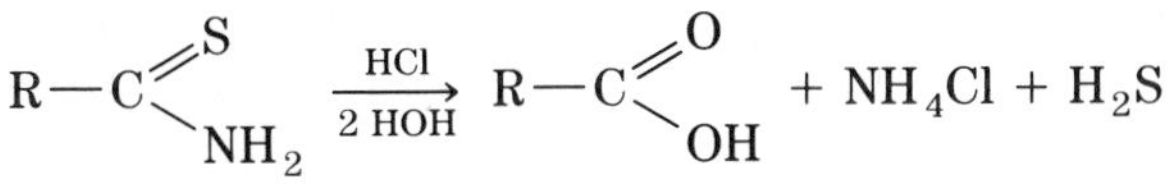

It was possible to use an ammonia-gas-sensing electrode (Horiba, Model 5002-05 T, or Orion, Model 95-10) to determine converted ammonia from ammonium chloride.[212] The decomposition of ethionamide and prothionamide (10^{-3} mol for each) was performed by heating on reflux with 20% HCl solution for 1 h. In the case of decomposition with 10% HCl, the electrode potential reached a maximum at a heating time of 3 h. After 1 and 2 h, the recoveries were 87 and 91%, respectively. After cooling and alkalinization with NaOH solution (pH > 11) the sample concentration was determined from the calibration graph, which was linear within the drug concentration range 2×10^{-5} to 10^{-2} M. The amounts of ethionamide and prothionamide in pure drug powder were estimated with the same average errors of 0.03% and the standard deviations were 0.15 and 0.20%, respectively. Determinations on both tablets were also carried out: the mean recoveries were 100.4% in both cases, and the respective standard deviations were 0.36 and 0.25%, respectively.

Analytical Procedures

i. *Decomposition of drugs:*
   A mixture of 166.24 mg (1×10^{-3} mol) of ethionamide or 180.26 mg (1×10^{-3} mol) of prothionamide and 20 cm^3 of 20% HCl is heated at reflux in an oil bath for 1 h. The solution is cooled, poured into a 100-cm^3 beaker, and diluted with approximately 50 cm^3 of distilled water. A drop of methyl orange indicator solution is added, with the beaker continuously cooled and the acid is cautiously neutralized with 10 N NaOH solution until the indicator begins to change color. The solution is then adjusted to pH 6.5 with dilute sodium hydroxide, using a pH-meter. The solution is poured into a 100-cm^3 volumetric flask and diluted to volume with distilled water. The concentrations of the final drug solutions are 1×10^{-2} M, corresponding to 1×10^{-2} M ammonia. Sample solutions for measurements are obtained by dilution of these solutions with distilled water.

ii. *Assay procedure:*
   A 50-cm^3 portion of the sample and standard solutions used is incubated for 30 min at 20°C and 1 cm^3 of 5 N NaOH solution is added before the electrode is immersed in the solutions. (After alkalinization, the solutions are stable for several hours in the cell with a rubber stopper.) The standard procedure is as follows: The electrode with 0.05 M ammonium chloride internal filling solution is washed with water and immersed for about 5 min in fresh 0.05 M NaCl solution acidified with dilute HCl at pH 4.0; then the old internal filling solution is replaced with fresh 0.05 M ammonium chloride solution. The electrode is placed in the first standard, and the potential is measured. After another washing, the electrode is placed in the second standard, and the potential is measured again. The sample concentration is determined from the calibration curve.

iii. *Assay of tablets:*
   Twenty tablets are weighed and finely powdered. A portion of the powder (equivalent to 166.24 mg of ethionamide and 180.27 mg of prothionamide) is accurately weighed, and 20 cm^3 of acetone is added. The solution is stirred and then centrifuged. The supernatant acetone solution is removed, and 20 cm^3 of acetone is added to the residue. The extraction procedure is repeated until the supernatant acetone solution is colorless. The acetone fractions are evaporated to dryness, and the resulting residue is refluxed for 1 h with 20 cm^3 of

20% HCl. As previously described, the acidic solution is adjusted to pH 6.5 and diluted to 250 cm^3 in a volumetric flask. The ammonia concentration in a 50-cm^3 aliquot of the sample is determined from the calibration curve.[211]

5.48   Fenol Derivatives

There are many analytical methods for derivative fenols (see Table 5.24) assay based on the use of membrane sensors.[187, 212-222]

8-Quinolinol and other organic compounds (e.g., aromatic hydroxy compounds, amines, and compounds containing active methylene groups) were potentiometrically titrated with 4-methyl-, 4-bromo-, or 4-nitro-benzenediazonium chloride solutions as titrant; an inorganic cation-selective indicator sensor, based on a PVC membrane plasticized with o-nitrophenyl octyl ether showed the best behavior during titration monitoring.[214] Two types of sensor were constructed: in one type, the membrane circle was attached to a plastic tube; in the second type, the membrane was formed by coating a porous graphite rod, sealed in a glass tube, with a PVC–plasticizer solution. No ion exchanger was added to the PVC solution during membrane fabrication. Before use in titrations with diazonium salts, the sensors were pre-conditioned by repeated titrations of the appropriate arenediazonium salt with sodium tetraphenylborate solution. Thus, the membrane plasticizer was gradually saturated with an ion pair (arenediazonium$^+$–TPB$^-$) formed on the basis of solvent extraction principles. In both cases a mixture of 10^{-2} M NaCl and 10^{-2} M sodium tetraphenylborate mixture was the solution used as internal electrolyte. The type of the sensor construction did not influence significantly the shape of the potentiometric titration curves. The solution of 4-bromobenzenediazonium salt was found to be most adequate as titrant: it is more stable than a solution of 4-methylbenzenediazonium salt and it can be used in a more alkaline medium than the 4-methyl-benzenediazonium chloride solution. Even in these circumstances the titrant must be prepared daily by a diazotization reaction with sodium nitrate solution (under cooling with ice at 0°C) and must be kept in an icebox. Additionally, the time necessary for one determination was longer than 1 h. From these reasons the method applicable to quinolinol does not seem to be more adequate than that based on copper complex formation[223] (see also analytical procedure i).

Potentiometric titrations of catechol with lead(II) nitrate using a Pb^{2+}-selective membrane sensor at an optimum pH range from 9.7 to 9.85 (borate buffer) proved successful from the analytical point of view.[216] The composition of isolated precipitate agreed with a 1 : 1 complex containing one molecule of water, as determined by elemental

Table 5.24 Fenol Derivatives Assayed by Membrane Sensors

Compound	Formula (MM)	Therapeutic category
Hydroxyquinoline (quinolinol)	C_9H_7NO (145.2)	Antibacterial; disinfectant
Clioquinol	C_9H_5ClINO (305.5)	Antiseptic; amebicide
Broxyquinoline	$C_9H_5Br_2NO$ (303.0)	Antibacterial; antiseptic
Brobenzoxaldine	$C_{17}H_{11}Br_2NO_2$ (442.0)	Antibacterial; antiseptic
Cathechol (pyrocathechol)	$C_6H_6O_2$ (110.1)	Topical antiseptic
Picric acid	$C_6H_3N_3O_7$ (229.1)	Disinfectant

analysis. Under identical conditions, resorcinol and hydroquinone did not yield either a precipitate or a titration curve. A fourfold excess of the isomers did not interfere in the titration 0.02 mmol of catechol. The mean recovery for catechol within the range 1 to 10 mg was 98.2%, with a standard deviation of 0.27%.

A picrate-ion-selective membrane sensor with an active membrane of tetrapentylammonium picrate in 2-nitromethylbenzene has been reported by Hadjiioannou and Diamandis.[221] The sensor was constructed by using the body of an Orion 92 electrode equipped with an Orion 92-81-04 membrane and 0.01 M electroactive material. The internal reference solution was 0.01 M sodium picrate with 0.1 M NaCl. The linear response of the sensor covered the range 10^{-2} to 10^{-5} M picrate (slope 58 mV decade^{-1}); between pH 3 and 10 the potential is practically independent of pH. At higher pH the potential increases slowly, probably because of complex formation between picrate and OH$^-$ ions. The sensor has a very high selectivity for picrate over chloride, fluoride, nitrate, bicarbonate, acetate, and iodate.

For picric acid and other 1,3-dinitro compounds and symmetrical trinitro compounds, a simple, selective, and accurate method based on the reaction with 0.2 M potassium cyanide followed by potentiometric titration of the excess of cyanide with silver nitrate has been described by Hassan.[222] Compounds containing three nitro groups in meta positions relative to each other react with two moles of the cyanide per mole of trinitro compound, but 2,4,6-trinitrophenol (picric acid) consumes three moles of cyanide. This is because acidic groups of p$K_a < 4$ quantitatively release one mole of hydrocyanic acid per trinitrophenol group. However, on adjusting the pH of trinitrophenol to 6 to 4 before applying the cyanide reaction, just two moles of cyanide are quantitatively consumed per mole of picric acid. Picric acid has been determined by this procedure with a relative standard deviation of $\pm 0.28\%$.

A new method for the separate micro-determination of chlorine, bromine, and iodine present together in an organic compound was described by Campiglio.[215] A 3- to 5-mg sample is burnt in an oxygen-filled flask and the products are absorbed in an alkaline solution of hydrazine. In the clioquinol case, iodide is oxidized with dichromate to free iodine, which is removed, absorbed in alkaline solution of hydrazine, and reduced again to iodide, while chloride remains in the original solution.

The apparatus for the halide separation is shown in Figure 5.13. It consists essentially of the absorber (L) and the connector (F) to be inserted into the combustion flask. The connector top is as small as possible, to facilitate the transfer of the iodine to the absorber. The tube G (8 mm O.D. and 2 mm I.D.) reaches down to 2 mm from the bottom of the flask, which allows the oxygen to bubble through the solution; combustion flasks of the same height are therefore required. The oxygen

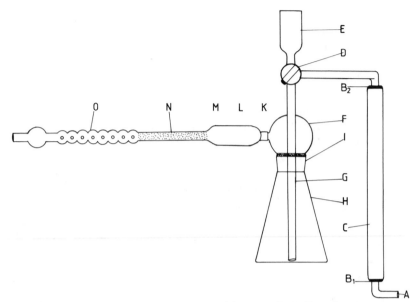

Figure 5.13 Apparatus for halide separation: (A) oxygen inlet; (B$_1$, B$_2$) 12/2 ball joints; (C) rotameter; (D) three-way stopcock (position I, ⊣ ; position II, ⊤ ; position III, ⊥); (E) funnel; (F) connector; (G) tube for reagent introduction and oxygen bubbling; (H) combustion flask; (I) NS26 ground joint; (K) NS7 ground joint; (L) absorber; (M) condensation chamber; (N) absorption tube packed with 2.5-mm-diameter glass beads (210 mm length); (O) shaped to Vigreux (110 mm length). (Reproduced from Campiglio, A., *Mikrochim. Acta 2*, 347, 1982. With permission.)

flow measured by means of the rotameter (C) is regulated through a needle valve at the exit from the cylinder. The reagents are introduced into the flask through the 9-cm^3-capacity funnel (E) and the stopcock (D). The absorber (L) consists of the condensation chamber (M) and the absorption tube (13 mm O.D. and 10 mm I.D.) packed with glass beads (N) and then shaped to Vigreux (O); it ends with a capillary of 2 mm I.D. The condensation chamber (70 mm length, 30 mm O.D., 27 mm I.D.; 43 cm^3 capacity, where a distillate volume up to 9 cm^3 can be collected), is cooled by a jet of tap water that then falls down inside a large drain funnel. For heating the flask, a 1-dm^3 beaker containing ethylene glycol is used as a bath heated by means of a Bunsen burner. The bath can be adjusted vertically to vary the temperature.[215] The separated halides are then potentiometrically titrated with 10^{-2} M AgNO$_3$ by using a sulfide-ion-selective membrane sensor. Accurate results within $\pm0.27\%$ were obtained, the corresponding standard deviations were 0.14 and 0.16% for iodine and chlorine, respectively.

The method represents a reliable alternative way for determination of halogens present together in an organic compound. It offers, in particular, two main advantages:

1. Only one sample of 3 to 5 mg is required for the determination of halogens.
2. The determination is direct for each halogen.

A 5 : 1 (m/m) mixture of broxyquinoline (I) and brobenzoxaldine (II) is the active principle of the drug Intestopan (Sandoz Ltd.), an intestinal antiseptic with potent antibacterial and amebicidal activity that does not upset the balance of the normal intestinal flora. Ionescu et al.[217] proposed for the determination of these two compounds a method that involves chelation of broxyquinoline with Cu^{2+}, followed by potentiometric titration of excess of Cu^{2+} with EDTA; brobenzoxaldine is hydrolyzed to 2-methyl-5,7-dibromo-8-hydroxyquinoline, which is then similarly determined. An Orion 94-29 copper(II)-selective membrane sensor was used to monitor the Cu^{2+} consumption.

For the determination of compound (I) an acetonic : water (2 : 1, v/v) mixture was chosen. Other organic solvents (e.g., methanol and dioxan) were tested but, in these solvents, the copper(II) chelates are produced as a gel that deposits on the Cu^{2+}-membrane sensor surface and affects the electrode response. The presence of water in the reaction medium is essential because the membrane sensor does not work properly in a completely non-aqueous medium. For choosing the pH for the potentiometric titrations, the effect of pH on the potential jump has been taken into account. When pH increases, the hydrolysis of the benzoate ester (compound II) in alkaline medium starts even at room temperature and causes interference in the determination of compound I in the presence of compound II. As a compromise, pH 5.6, adjusted with acetate buffer, was chosen; compound I is hydrolyzed in alkaline medium to 2-methyl-5,7-dibromo-5-hydroxyquinoline by heating with 0.1 M sodium hydroxide for 30 min.

Replicate analysis of Intestopan tablets gave a mean recovery of 99.6% for broxyquinoline (relative standard deviation 1.1%) and 99.9% for brobenzoxaldine (relative standard deviation 1.6%).

Analytical Procedures

i. *Determination of quinolinol:*
   The sample solution (about 30 to 40 cm^3 in a 150-cm^3 beaker; concentration approximately 10^{-3} M) is adjusted to pH 5.5 with 0.1 M acetate buffer and potentiometrically titrated with 10^{-2} M copper sulfate solution (a copper(II)-selective membrane sensor is used as indicator electrode and SCE as reference). The titration end point corresponds to the maximum slope on the plot of EMF vs. titrant volume.

ii. *Determination of picric acid by direct potentiometry:*
Standard solutions of 10^{-2} to 10^{-4} M concentration are prepared
by serial dilution of 10^{-2} M sodium picrate prepared by neutralizing a suitable picric acid solution with sodium hydroxide to a pH of
about 6. A constant ionic strength (I = 0.1 M, adjusted with potassium nitrate) must be used. The standard solutions are transferred
into 150-cm^3 beakers containing Teflon-coated stirring bars. The
picrate and the reference (Ag/AgCl or SCE) electrodes are immersed successively in the standards, and the EMFs measured. The
picrate concentration in the sample is determined from the calibration graph (E vs. [picrate]).

iii. *Determination of picric acid with potassium cyanide:*
The weighed sample (approximately 2 to 10 mg) neutralized with
0.05 M sodium hydroxide to pH 6 to 7 is transferred to a Pyrex test
tube (20 × 2 cm); 1.0 cm^3 of 0.2 M potassium cyanide solution is
pipetted to the sample solution and mixed thoroughly. The tube is
placed in a boiling-water bath for 10 min with occasional shaking.
The tube is cooled and 5 cm^3 of doubly distilled water added. The
contents is shaken and quantitatively transferred to a 250-cm^3
beaker. The tube is washed with aliquots of doubly distilled water
adding to a total of 30 to 40 cm^3. The pair of electrodes (Ag$^+$/S^{2-}-
or cyanide-selective membrane as indicator and double-junction
electrode as reference) is placed in the sample solution and the
excess of cyanide is potentiometrically titrated with 0.02 M silver
nitrate solution. A value of twice the volume of the titrant corresponding to the first inflection point is used to calculate the cyanide
concentration. A blank titration is performed (2 mol KCN ≡ 1 mol
picric acid).

iv. *Determination of clioquinol:*
A 3- to 5-mg sample is wrapped inside a piece of paper and placed
in the platinum basket. The flask is charged with 5 cm^3 of 0.2 N
NaOH and 0.5 cm^3 of 3% hydrazine sulfate; oxygen is blown into it
for 25 s and the sample is burnt in the usual way, according to
Schöniger's method. After standing with occasional shaking for 30
min, the flask is opened and the stopper and the platinum basket are
washed with 15 cm^3 of distilled water. The solution is treated with
three drops of 30% H_2O_2 and boiled gently under mixing for 3 min;
the flask is then cooled with tap water.

For the separation of iodide, the solution is neutralized by dropwise addition of 1 N HNO$_3$ in the presence of methyl red, diluted
with 15 cm^3 of distilled water and acidified with 3 drops of concentrated HNO$_3$; the flask is then fixed with a clamp to a stand and is
equipped with the connector (see Figure 5.13). A mixture of 4.5 cm^3
of 20% NaOH and 2 cm^3 of 3% hydrazine sulfate solution is now
entirely sucked into the absorber and 2 cm^3 of it is allowed to flow

into the condensation chamber. The absorber is then attached to the connector and inclined slightly down by adjustment of the flask position; finally, the rotameter is attached to the connector. 5 cm^3 of 0.5 N K$_2$Cr$_2$O$_7$ solution is added through the funnel into the flask, oxygen (20 to 25 cm^3 min^{-1}) is bubbled through the solution, and the condensation chamber of the absorber is cooled with tap water. The ethylene glycol bath is then raised (just to dip $\frac{2}{3}$ of the flask in it) and heated with a burner to 115 to 116°C; the solution is boiled at this temperature for 12 min under occasional control of the absorber cooling and oxygen flow. The bath is then lowered. After 2 min, the stopcock is turned to position III and the cooling and the oxygen flow are stopped. The absorber is removed, the absorbtion solution is poured into a 100-cm^3 beaker, and the absorber is rinsed five times each with 5 cm^3 distilled water. The solution is then neutralized under magnetic stirring by dropwise addition of concentrated HNO$_3$ in the presence of methyl red. After cooling, the solution is acidified under stirring with five drops of concentrated HNO$_3$ and finally titrated with 0.01 M AgNO$_3$ (sulfide-selective membrane sensor as indicator, double-junction electrode as reference).

For the determination of chloride, after disconnection of the rotameter, the flask is removed from the stand, the connector is removed, and its bubbling tube is washed internally (through the funnel) and externally with 1 cm^3 of distilled water, respectively; the volume of solution is about 28 cm^3. The flask is cooled with tap water to room temperature and the solution is treated with 4 cm^3 of 3% hydrazine sulfate solution and then allowed to stand for 15 min in order to obtain a complete dichromate reduction (indicated by the appearance of a blue color). If the reduction is incomplete for 15 min or if it is desirable to save time, the solution is heated slowly under mixing (and eventual addition of another five drops of hydrazine) up to 70°C until the blue color appears. After cooling, the solution is transferred to a 100-cm^3 beaker, the flask is rinsed with 5 cm^3 of isopropyl alcohol and then with 3 cm^3 of acetic acid, and the solution is finally potentiometrically titrated as previously described for iodide. A blank is needed. The titrant volume at the equivalence point is determined from the maximum slope on the titration curve.

v. *Determination of catechol:*

Samples containing up to 0.1 mmol of catechol are pipetted or weighed into a 50-cm^3 beaker containing a Teflon-coated stirring bar. They are diluted to approximately 25 cm^3 with distilled water and the pH is adjusted to 9.5 with the borate buffer (pH 9.75) using a pH-meter. Stirring is provided by a magnetic stirrer. Potentiometrically titrations are performed at room temperature, at a rate of 0.5

cm^3 min^{-1} titrant (0.01 M lead nitrate solution), using a lead-ion-selective membrane sensor.

vi. *Determination of broxyquinoline:*
The sample (0.15 to 0.20 mmol) is dissolved in 30 cm^3 of acetone and 15 cm^3 of distilled water, and 5.0 cm^3 of 0.05 M copper(II) sulfate solution is added. The pH is adjusted to 5.6 with 30 cm^3 of 0.05 M acetate buffer and the excess of Cu^{2+} is potentiometrically titrated with 0.05 M EDTA using a copper(II)-ion-selective membrane sensor.

vii. *Determination of brobenzoxaldine:*
The sample (0.20 to 0.25 mmol) is dissolved in 30 cm^3 of acetone; 10 cm^3 of 0.1 M sodium hydroxide is added and the mixture is hydrolyzed by refluxing on a hot-water bath for 30 min. After cooling to room temperature, the sample is quantitatively transferred to a 100-cm^3 beaker with 25 cm^3 of 2 : 1 v/v acetone–water mixture, the pH is adjusted to 6.0 with 0.1 M hydrochloric acid and then to 5.6 with 15 cm^3 of 0.05 M acetate buffer (pH 5.6). At least 7.0 cm^3 of 0.05 M copper(II) sulfate (accurately measured) is added, the mixture is mechanically stirred for a few minutes, and the excess of Cu^{2+} is potentiometrically titrated as described under vi.

viii. *Assay of Intestopan tablets:*
At least 10 tablets of the same batch are finely powdered and a portion of the mixed powder equivalent to about 100 mg of broxyquinoline and 20 mg of brobenzoxaldine is accurately weighed and is dissolved in 30 cm^3 of acetone. The mixture is hydrolyzed and titrated as described for brobenzoxaldine assay to yield the volume of copper(II) solution equivalent to the sum of broxyquinoline and brobenzoxaldine. Another portion of the powder, equivalent to about 100 mg of broxyquinoline is accurately weighed and analyzed as described for the determination of broxyquinoline, to yield the amount of broxyquinoline present. The content of brobenzoxaldine is then calculated by difference.

5.49   Ferulic Acid

$$C_{10}H_{10}O_4 \ (MM = 194.2)$$

Therapeutic category: cardiotonic

Discussion and Comments

Three quaternary ammonium salts have been tested as counter-ions for ferulate anion, in order to develop a PVC–ferulate membrane sensor with good analytical performance.[224] Among them, the quaternary ammonium salt 7402, $[CH_3(C_nH_{2n+1})_3N]^+Cl^-$ ($n = 9, \ldots, 11$), gave the best results with respect to the linear response range, limit of detection, and selectivity. When concentrations of 15 and 54% of quaternary ammonium compound and plasticizer (DOP), respectively, were used in the PVC membrane, a linear response range within 8.8×10^{-5} to 10^{-1} M (detection limit 5.5×10^{-5} M) with a slope of 59 mV decade^{-1} was achieved. The sensor was unaffected by pH changes in the range 3.0 to 5.3 and gave fast response times (less than 60 s). It was found that anions, such as NO_3^-, Br^-, and I^- interfere strongly with the sensor response.

Pharmaceuticals containing ferulate were satisfactorily determined by direct potentiometry.

Analytical Procedure

Standard solutions of 10^{-2} to 10^{-4} M concentration are prepared by serial dilution of 10^{-1} M sodium ferulate prepared by neutralizing a suitable ferulic acid solution with sodium hydroxide to pH of about 7.0. A constant ionic strength ($I = 0.1$ M, adjusted with sodium sulfate) must be used. The standard solutions are transferred into 150-cm^3 beakers containing Teflon-coated stirring bars. The ferulate and the reference (Ag/AgCl or SCE) electrodes are immersed successively in the standards, and the EMFs measured. The ferulate concentration in the sample is determined from the calibration graph (E vs. log[ferulate]).

5.50   Filcilin

$$C_{21}H_{28}BrNO_5 \ (MM = 454.3)$$

Therapeutic category: anti-ulcer agent

Discussion and Comments

The liquid-membrane and PVC-membrane selective sensors for filcilin were described by Zhu et al.[225] Both sensor types are based on the use of tetraphenylborate–filcilin ion-pair complex as the electroactive material. When a 10^{-2} M solution of tetraphenylborate–filcilin in nitrobenzene was used as liquid membrane, the sensor displayed a Nernstian-type response only in the range 10^{-2} to 10^{-4} M with a sub-Nernstian slope of 52 mV decade^{-1}. The PVC-membrane sensor (DBP as plasticizer) showed Nernstian response in 10^{-2} to 10^{-5} M filcilin solutions over a pH range 2.3 to 6.7 with a cationic slope of 58 mV decade^{-1}. The sensor gave rapid response, good reproducibility, stability and precision. Most other substances tested did not interfere, but apomorphine, N-butyl-scopolammonium, and quinine interfere slightly ($k_{\text{Fil, B}}^{\text{pot}} = 0.23, 0.27$, and 0.38, respectively) (separate solution method).

Using the direct potentiometric method for the filcilin assay in pure solutions, an average recovery of 99.7% (standard deviation 1.8%) was reported. A good recovery of filcilin from tablets (10 mg per tablet) was also found (standard deviation 0.43%).

Analytical Procedure

Standard solutions of 10^{-2} to 10^{-4} M concentration are prepared by serial dilution of 10^{-2} M filcilin hydrochloride prepared by dissolving a suitable amount of drug substance with distilled water. A constant ionic strength ($I = 0.1$ M, adjusted with sodium nitrate) must be used. The pH of all standard solutions is adjusted to approximately 5.0 with acetate buffer. The standard solutions are transferred into 150-cm^3 beakers containing Teflon-coated stirring bars. The filcilin-membrane sensor together with a reference electrode (Ag/AgCl or SCE) are immersed successively in the standards and the EMF values measured. The filcilin concentration in the sample is determined from the calibration graph (E vs. log[filcilin]).

5.51 Fluorouracil and Its Derivatives

Fluorouracil and some of its derivatives (Table 5.25), very well known antineoplastic agents, were quantitatively assayed after a fluorine release, with a fluoride-selective membrane sensor.[226, 227] Jones et al.[226] have described conditions for quantitative ionization of fluorine with sodium biphenyl reagent (dimethoxyethane–sodium biphenyl complex). The released fluorine is then determined by direct potentiometry. Although the reaction mechanism is not well understood, it is believed that decomposition of organofluorine compounds with sodium biphenyl reagent involves reductive cleavage of fluorine.[228] The reaction is almost instantaneous at room temperature and no significant differences in fluorine concentrations are found for reaction times ranging from 1 to 30 min. An alcoholic acetate buffer solution containing sodium chloride

Table 5.25 Fluorouracil and Its Derivatives Assayed
by Fluoride Membrane Sensor

Compound	R_1	R_2	Formula (MM)
Fluorouracil	H	F	$C_4H_3FN_2O_2$ (130.08)
Floxuridine	(sugar)	F	$C_9H_{11}FN_2O_5$ (246.2)
Trifluorothymidine	(sugar)	CF_3	$C_{10}H_{11}F_3N_2O_5$ (296.2)

(pH 5 to 5.5) is appropriate for background reaction mixture, and the sensor response to fluoride ion in the medium was linear throughout the working range of 10^{-4} to 10^{-5} M, or 0.02 to 0.20 mg F$^-$ per 100 cm^{-3} (slope 58.1 mV decade^{-1}). Equilibrium may be reached in less than 30 s in stirred solutions.

Analytical Procedure

An accurately weighed sample equivalent to 16 mg fluorine is dissolved in 70 to 80 cm^3 tetrahydrofuran or ethylene glycol dimethyl ether and diluted to 100 cm^3; 15 cm^3 of this solution are pipetted into a 200-cm^3 volumetric flask. The contents of a 15 cm^3 vial of sodium biphenyl reagent is added and the solution shaken. After 5 to 10 min, excess reagent is destroyed with 5 to 10 cm^3 of 2-propanol and the solution diluted to volume with 2-propanol. An aliquot of 10 cm^3 of this solution is further diluted to 100 cm^3 with alcoholic acetate buffer solution, pH 5 to 5.5.

A reagent blank is prepared by diluting 15.0 cm^3 of sample solvent and 15 cm^3 of sodium biphenyl reagent to 200 cm^3 with 2-propanol as in the sample preparation. This solution is used to prepare fluoride standards. These (0.02, 0.05, 0.10, 0.15, and 0.20 mg F$^-$ per 100 cm^3, respectively) are prepared by serial dilution from a stock solution containing 1 mg F^- cm^{-3}; 10 cm^3 of reagent blank, prepared as before, are added to each working standard solution before the final dilution. Freshly prepared solutions are used for each analysis.

The sample is transferred to a 150-cm^3 beaker containing a Teflon-coated stirring bar. The fluoride membrane sensor together with the reference electrode (SCE modified with a mixture of 70 cm^3 of standard KCl and 30 cm^3 of 2-propanol) is immersed in the solution and the EMF is measured after the stabilization of potential to ± 0.1 mV. The unknown concentration is determined from the calibration graph plotted with the EMF values recorded for standard solutions (E vs. $\log[F^-]$).

5.52   Flurazepam and Medazepam

$C_{21}H_{23}ClFN_3O$ (MM = 387.9) $C_{16}H_{15}ClN_2$ (MM = 270.8)

Therapeutic category: hypnotic, psychotropic; minor tranquilizer

Discussion and Comments

For flurazepam see the previous section and Jones et al.[226] Medazepam has been determined by potentiometric titration with sodium tetraphenylborate at pH 2.5 by using a tetraphenylborate-membrane electrode.[82] The end-point break is relatively large (190 mV) and the determinations were performed with a relative standard deviation of 0.4%.

Analytical Procedures

i. *Flurazepam determination:*
   See the procedure from Section 5.51.

ii. *Medazepam determination:*
   A 25-cm^3 aliquot of the sample in the concentration range 2.0×10^{-4} to 1.0×10^{-3} M is pipetted into a 50-cm^3 beaker; 5 cm^3 citrate buffer solution of pH 2.5 is added, and after the potential is stabilized the sample is potentiometrically titrated with 10^{-2} M sodium tetraphenylborate solution using a tetraphenylborate-membrane sensor.

The amount of medazepam present in the sample is calculated in the usual way. For tablet preparations, 20 tablets are weighed and powdered. An appropriate weighed amount of the powder (equivalent to about 5.0 mmol of active ingredient) is transferred to a 500-cm^3 beaker and stirred vigorously with about 400 cm^3 of water for 15 min. The solution is diluted to the mark in a 500-cm^3 volumetric flask and a 25-cm^3 aliquot is potentiometrically titrated as previously described.

5.53   Formaldehyde and Hexamine

$$CH_2O \ (MM = 30.03) \qquad C_6H_{12}N_4 \ (MM = 140.2)$$

Therapeutic category: antiseptics

Discussion and Comments

No other analytical methods have been developed for formaldehyde and hexamine since the papers of Ikeda[229] and Koupparis et al.[230] The method of Ikeda is based on an argentimetric potentiometric titration by using an iodide-selective membrane sensor (Orion, Model 94-53). Iodide ions are generated by the reaction with iodine solution in an alkaline medium:

$$HCHO + I_2 + 2\ KOH \rightarrow 2\ KI + HCOOH + H_2O \qquad (5.63)$$

Formic acid and methanol (which may be present in formaldehyde solution) in equimolar amount to that of formaldehyde did not interfere, whereas an excess of 10 times gave a positive error of less than 1%. However, acetone and acetaldehyde react with iodine and accordingly interfere strongly.

The kinetic method for the determination of formaldehyde and hexamine with a cyanide-ion-selective membrane sensor (Orion Model 94-06) proposed by Koupparis et al.[230] is based on the well-known addition reaction of hydrogen cyanide to carbonyl compounds. A very dilute solution ($1.6 \times 10^{-5} \ M$ cyanide) is treated with a large excess of unknown or standard formaldehyde solution, and the reaction is monitored with cyanide sensor. The time required for the potential to change by a preselected amount (e.g., 8 mV) is measured automatically and related directly to the formaldehyde concentration:

$$\frac{dE}{dt} = k\frac{RT}{F}[\text{HCHO}] \tag{5.64}$$

All terms, except [HCHO], in Equation 5.64 are constant. Also, [HCHO] remains practically constant during the reaction because formaldehyde is in large excess over cyanide, so that the rate of change of potential is essentially constant. Therefore, dE/dt may be replaced by $\Delta E/\Delta t$, which is the measurable parameter.

The kinetic method[230] for formaldehyde was also applied for the determination of hexamine. Hexamine was hydrolyzed on heating (60°C for 30 min) with acid to formaldehyde:

$$(\text{CH}_2)_6\text{N}_4 + 6\,\text{H}_2\text{O} + \text{H}^+ \rightarrow 4\,\text{NH}_4^+ + 6\,\text{HCHO} \tag{5.65}$$

In this case the following equation was deduced:

$$\ln\left[1 - \frac{(\Delta E/\Delta t)_t}{(\Delta E/\Delta t)_\infty}\right] = -k_{\text{obs}}t \tag{5.66}$$

Thus, if the left-hand side of Equation 5.66 is plotted vs. time, the observed rate constant k_{obs} of the hydrolytic reaction may be calculated from the slope of the curve. Values should be corrected for blank.

The average error for the kinetic determination of formaldehyde was about 1.3% and for hexamine about 1.6%. None of the substances used as diluents in hexamine manelate tablet preparations interfered. There was satisfactory agreement between the proposed method and the USP official method.

Analytical Procedures

i. *Argentimetric method:*
   10.0 cm^3 of about 0.05 M formaldehyde solution is pipetted into the titration cell and 10 cm^3 of about 0.1 M iodine–iodide solution and 5.0 cm^3 of 2 M potassium hydroxide are added. The mixture is stirred for 10 min. After adding 6 cm^3 of 2 M nitric acid in order to bring the pH < 2, distilled water is added to bring the volume to 100 cm^3. This solution is potentiometrically titrated with standard silver nitrate solution in the presence of iodide-selective membrane sensor. A blank run is made with an identical amount of iodine–iodide solution.

ii. *Kinetic determination:*

A solid-state double-switching network is used in conjunction with a recorder system for automatic time measurements (for details, see Efstathiou and Hadjiiouannou[231]).

a. *Formaldehyde*—15.0 cm^3 of composite KCN–EDTA solution and 4.0 cm^3 of buffer solution (pH 7.8) are added to the thermostated (30°C) reaction cell. The stirrer is started and after the potential has stabilized to a value of E_0 (after about 20 s) the recorder pen is adjusted to one side (lower potential) of the chart. The voltage reference sources of the control system are adjusted so that the time measurement starts at E_1 and stops at E_2 ($\Delta E = E_2 - E_1 = (E_0 + 12.0) - (E_0 + 4.0) = 8.0$ mV). The "start" button on the timer is reset and quickly a volume of 1.0 cm^3 of formaldehyde standard or sample solution is injected into the reaction cell. The analysis is computed automatically and the number on the timer is recorded. The cell is emptied by suction and rinsed with distilled water. The procedure is repeated for each analysis without changing the potential levels on the control system. For each series of unknowns, four standards are included.

b. *Hexamine*—10.0 cm^3 of standard or sample solution containing 20 to 100 mg hexamine and 10.0 cm^3 of 4.0 *M* sulfuric acid are added to 25-cm^3 vials with well-fitting stoppers. If solid samples are used, an accurately weighed amount of hexamine standard or sample is transferred to the vials; 20.0 cm^3 of 2.0 *M* sulfuric acid is pipetted and the vials are tightly stoppered, shaken, and immersed into a water bath thermostated at 60°C for 30 min. After cooling, 0.05 cm^3 of the hydrolyzed hexamine solution is assayed by the same procedure as for formaldehyde.

5.54   Gentamicin and Related Antibiotics

Many papers described accurate and sensitive potentiometric methods for antibiotics (Table 5.26) assay.[232-245]

A potentiometric method that uses a carbon dioxide gas sensitive electrode (HNU, Model 10-22-00, or Orion, Model 95-02) has been developed for microbiological assay of some antibiotics by Simpson and Kobos.[237, 238] The method is based upon the antibiotic inhibition of carbon dioxide production by a suspension of *Escherichia coli*, which is directly measured with the potentiometric gas sensor after an incubation period. The optimum incubation conditions chosen for bioassays of gentamicin, neomycin, streptomycin, and tetracycline were pH 7.8, $T = 37$°C, time of incubation 120 min, stock *E. coli* cell concentration of 4.5×10^8 cells per cubic centimeter and a stock nutrient solution of

Table 5.26 Gentamicin and Related Antibiotics Assayed by Membrane Sensors

Antibiotic	Formula (MM)	Therapeutic category
Gentamicin	Antibiotic complex produced by fermentation of *Micromonospora purpurea* or *M. echinospora*	Antibacterial
Amphotericin	$C_{47}H_{73}NO_{17}$ (924.1) (empirical formula)	Antifungal; antifungal used for systemic mycoses (vet.)
Doxycycline	$C_{22}H_{24}N_2O_2$ (444.4)	Antibiotic

Oxytetracycline	$C_{22}H_{24}N_2O_9$ (460.4)	Antibacterial; antimicrobial (vet.)

Tetracycline	$C_{22}H_{24}N_2O_8$ (444.4)	Antiamebic; antibacterial; antirickettsial

Neomycin	Antibiotic complex from which three components, neomycin A, B, and C have been separated; produced by *Streptomyces fradiae*	Antimicrobial; antifungal
Nystatin	$C_{47}H_{75}NO_{17}$ (926.1) (empirical formula)	Antifungal; antimycotic; growth promotant (vet.)

Table 5.26 Continued

Antibiotic	Formula (MM)	Therapeutic category
Kanamycin	Comprises three components: kanamycin A, the major component (usually designated as kanamycin) and kanamycin B and C, two minor congeners $C_{18}H_{36}N_4O_{11}$ (kanamycin A)	Antibacterial
Streptomycin	$C_{21}H_{39}N_7O_{16}$ (518.6) (Streptomycin A) $C_{27}H_{49}N_7O_{17}$ (743.8) (Streptomycin B)	Antibacterial; tuberculostatic

2.4 g/100 cm^3 nutrient broth. Under these optimum conditions the linear ranges of the log dose–response curves were 0.017 to 3.33 μg cm^{-3}, 0.17 to 3.33 μg cm^{-3}, 0.33 to 16.67 μg cm^{-3}, and 33 to 167 μg cm^{-3} for gentamicin, neomycin, streptomycin, and tetracycline hydrochloride, respectively. The results for the bioassay of tetracycline hydrochloride, gentamicin, and streptomycin in pharmaceutical preparations (capsules) were in good agreement with the manufacturer's claim and well within the uniformity of content for individual capsules as defined by USP.[246] The feasibility of applying the method to the determination of serum gentamicin has also been studied.[238]

The development of a novel microbiological method for the assay of some antibiotics in pharmaceutical preparations was described by the same authors.[239] The method is based on antibiotic inhibition of ammonia production by a suspension of *E. coli* in a nutrient solution with potentiometric detection of ammonia. An Orion Model 95-10 ammonia-gas sensor was used to monitor the ammonia produced by the bacterial cells at 25.0 ± 0.5°C. The response of the system was optimized for tetracycline hydrochloride and gentamicin. Prior to the potentiometric measurements, sodium hydroxide was added to the measurement cell to adjust the pH > 12. The ammonia concentrations were obtained from the measured E (in millivolts) and the average of two daily calibration curves.

A study of the effects of incubation parameters demonstrated that the optimum conditions for bioassay of pharmaceutical preparations of gentamicin are as follows: $T = 37°C$; time = 120 min; stock cell concentration = 1.6×10^8 cells cm^{-3}; stock peptone concentration = 6.0%; pH = 7.8. The following calibration ranges (in micrograms per cubic centimeter) were obtained for the ammonia system: tetracycline, 16.7 to

83.3; gentamicin, 0.083 to 3.33; streptomycin and neomycin, 0.33 to 33.3. It was found that the results of the bioassay of gentamicin in pharmaceutical preparations were in good agreement with the label claim and were within the uniformity content for individual capsules.

Polyene antibiotics, such as nystatin and amphotericin, cause leakage of cytoplasmic constituents, particularly of monovalent cations, from susceptible microorganisms.[235] A correlation between polyene concentration and K^+-ion efflux was found to exist but the measurement of K^+-ion leakage was thought unsuitable for an assay system, because of the distinct possibility of natural contamination by K^+ ions (which exist in solution in a hydrated form). The hydrated ionic radii of K^+ (0.232 mm) and Rb^+ (0.228 mm) are very similar, as are the ionic mobilities of these two ions. Rb^+ ions are known to be well tolerated by erythrocytes.[235] Rb^+ ion uptake was determined by a Rb^+-selective membrane sensor, obtained by modifying a K^+-selective membrane sensor, which is commercially available. The sensor exhibited Nernstian behavior for Rb^+ within the concentration range 10^{-1} to 10^{-5} M ($k_{Rb,K}^{pot} = 1.6 \times 10^{-1}$ and 2.0×10^{-1} by the separate-solution method and the mixed-solution method, respectively).

The efflux of Rb^+ ions from yeast cells subjected to a range of nystatin concentrations was monitored continuously over a 15-min period at 50°C. If the electrode potential after 10 min incubation was plotted vs. $pC_{nystatin}$, a straight line was obtained within a narrow concentration range.[235]

Antibiotics such as nystatin are believed to bind with the sterol present in biological membranes, leading to formation of pores from which leakage of cellular material kills the microbial cells. The application of this principle, which can provide a rapid and quantitative measurement of antifungal activity, was recently proposed by Mascini et al. in an interesting paper.[241] *Saccharomyces cerevisiae*, a very common strain, was used. This yeast strain was immobilized on an acetylcellulose membrane held on the surface of an oxygen or carbon dioxide sensor. This coupling gave a biosensor that was easily assembled and rapid in response. The addition of a certain amount of nystatin leads to the death of the cells and the slope of the electrochemical response is related to the concentration of nystatin (a linear response within the range 25 to 200 U cm^{-3} nystatin was obtained with a CO_2-based sensor).

The results obtained for a common pharmaceutical preparation containing nystatin, available as tablets and ointment, with the O_2-based and CO_2-based sensor with the help of the slope and the lag-time calibration curves, showed that the second method is more accurate. From the results obtained by Mascini et al.,[241] the method is considered to have adequate precision and accuracy for the control of bacteriological activity.

Table 5.27   Response Characteristics of Tetracycline Membrane
Sensors[240]

Parameter	Membrane sensor[a]		
	TC	DC	OTC
Usable range (M)	1.6×10^{-5}–10^{-2}	7.9×10^{-5}–1.9×10^{-3}	6.3×10^{-5}–6.3×10^{-3}
Slope (mV decade^{-1})	54	57	55
Intercept (mV)	244.0	209.5	160.0
Response time (s)	Spontaneous		10–20

^{a}Composition of polymer membranes, 7.0% ion pair, 46.5% DOP, and 46.5% PVC (m/m);
internal solution, 10^{-3} M antibiotic $+ 10^{-1}$ M NaCl.

The wide availability of *S. cerevisiae* makes the method of interest
because culture equipment and expertise are not required. The yeast
cells are destroyed by nystatin and a new yeast membrane is needed for
each analysis, but the preparation and the replacement of a membrane is
a simple matter provided that the bacteria can be obtained from bakeries
or grocery stores.[241]

A PVC-membrane sensor selective for hydrochlorides of tetracycline
(TC), doxycycline (DC), and oxytetracycline (OTC) were prepared by
using the respective ion-pair complexes with tetraphenylborate site carrier.[240] Table 5.27 summarizes the response characteristics of the tetracycline membrane sensors. The three investigated antibiotic membrane
sensors exhibited similar behavior in regard to effects of pH. The
electrode potential increases with the increase of pH up to approximately 2.5, after which the potential decreases markedly until pH values
of 7.0, 8.3, and 6.5 are reached for TC, DC, and OTC membrane sensors,
respectively. In alkaline media (up to pH 11.5) the potential readings
were nearly independent of the pH value. It was found that the three
antibiotics interfere with each other on using any of the three tested
sensors. All three membrane sensors were used successfully for the
determination of TC, DC, and OTC in their pure solutions and in the
pharmaceutical preparations of Tetracyn, Vibramycin, and oxytetracycline capsules using the standard-addition method. The standard deviation values varied from 0.8 to 1.3% for the analysis of pure solutions
($n = 10$) and 0.9 to 1.5% for the analysis of pharmaceutical preparations
($n = 5$).

Yao et al.[244] constructed membrane sensors sensitive to cationic or
anionic doxycycline by using the respective ion-pair complexes with
silicotungstate, tetraphenylborate, and cetyltrioctylammonium as electroactive materials. The DC-sensitive membrane sensors respond to the
monoprotonated DC over pH 1.5 to 3.5, whereas the doxycyclinate anion

membrane sensor responds to monovalent doxycyclinate anion with the
enolic group of the B cycle being dissociated over the pH of 8.0 to 10.5
with poor stability in potential. Microgram levels of DC in microvolume
samples can be determined using the inverted sensor of all-solid construction; $\log k_{ij}$ values for symmetric quaternary ammonium compounds were linearly related to the ionic potential and the induction
effect index of the ammonium ion according to the equation

$$\log k_{ij} = 12.465 - 38.24z/R + 365.21 \tag{5.67}$$

A general equation representing $\log k_{ij}$ vs. C-atom number of the alkyl
group of the quaternary ammonium compounds was given as

$$\log k_{ij} = 26.5 \times (2.7)^{n} - 38.24 \times (1.26n + 0.8)^{-1} + 7.48 \tag{5.68}$$

Membrane sensors selective to both cationic and anionic forms of
oxytetracycline (OTC) were also constructed[245]; ion-pair complexes of
OTC with silicotungstate, tetraphenylborate, dipicrylamine, and cetyltrioctylammonium were used as electroactive materials for PVC membranes. The membrane sensors respond to OTC cation (or anion) down
to 10 μmol dm^{-3} over the pH range 1.6 to 3.2 (8.2 to 11.0) with a slope
of 57.2 (58.0) mV decade^{-1}.

Ion-pair complexes of streptomycin with dicyclohexylnaphthalenesulfonate, silicotungstate, and phosphotungstate as well as dipicrylaminate
were used in order to construct PVC- and liquid-membrane
streptomycin-sensitive sensors.[242] For PVC-membrane-type sensors, a
copper plate was first electroplated with Pt, Au, or Ag and then coated
with the electroactive material (dibutylphthalate as plasticizer). The best
sensor was that containing streptomycin–dipicrylaminate as ion-
exchanger in a PVC membrane; its response is linear in the range
2×10^{-2} to 10^{-5} M with a detection limit of 2×10^{-6} M (pH range
5.4 to 7.2). In all cases cationic slopes of about 19 to 20 mV decade^{-1}
were found, which is in agreement with triprotonated streptomycin
species.

The responses of membrane sensors are highly affected by many
organic compounds, such as tetraalkylammonium derivatives or alkaloids, as well as by K^+ ions. Even so, the streptomycin sensors were used
in the direct potentiometric determination of streptomycin sulfate in the
range 0.434 to 71.51 mg dm^{-3}, with an average recovery of 98.6 to
99.1% and a mean standard deviation of 4.7 to 5.9%.

Streptomycin sulfate as well as kanamycin sulfate samples of pharmaceutical interest were determined with a chloranilate-sensitive liquid-
membrane sensor with a relative error of $\pm 1.0\%$ (relative standard

deviation 0.5%).[243] The method is based on the well-known reaction

$$BaCh + SO_4^{2-} \rightleftharpoons BaSO_4 + Ch^{2-} \qquad (5.69)$$

where Ch = chloranilate.

The equilibrium constant of this reaction is

$$K = K_{so\,BaCh}/K_{so\,BaSO_4} = 1.71 \times 10^{-4}/1.1 \times 10^{-10}$$

$$= 1.55 \times 10^3 \qquad (5.70)$$

The high value of K ensures a 99.9% release of Ch^{2-} ions from the insoluble barium chloranilate (BaCh) owing to the presence of SO_4^{2-} in the unknown sample. The amount of chloranilate released (pH 7.6, adjusted with 0.2 M urotropine) was measured with the chloranilate membrane sensor either by using a calibration graph or by the standard-addition method.

(The chloranilate-selective membrane sensor contains Aliquat 336S–chloranilate ion pair in decan-1-ol or 2-nitrotoluene as electroactive membrane and has the following main characteristics: slope 29.6 mV decade^{-1}; linear response 10^{-2} to 10^{-5} M; useful working pH range 5 to 12; lifetime 2 months; fast response time.)

Analytical Procedures

  i. *Direct potentiometry with antibiotic-selective membrane sensors (for tetracycline, doxycycline, oxytetracycline, and streptomycin):*
     Standard solutions of 10^{-2}, 5×10^{-3}, and 10^{-3} M concentration are prepared by serial dilution of 10^{-2} M of the respective antibiotic solution, prepared by dissolving suitable amounts of drug with distilled water. Both ionic strength and pH must be kept at constant values (for pH adjustment, see the preceding text). The standard solutions are transferred into 150-cm^3 beakers containing Teflon-coated stirring bars. The respective antibiotic membrane sensor together with a reference electrode (Ag/AgCl or SCE) is immersed successively in the standards, and the EMF values are measured. The antibiotic concentration in the sample is determined from the appropriate graph (E vs. log[antibiotic]).
 ii. *Determination of streptomycin and kanamycin (as sulfates) with chloranilate membrane sensor:*
     In the measuring cell are placed 100 cm^3 buffer solution of pH 7.6 (0.2 M urotropine) and about 100 to 300 mg of pure barium chloranilate. After stabilization of the electrode potential, the un-

known sample, which contains 20 to 50 mg of SO_4^{2-}, is added with stirring. The amount of chloranilate released is measured by using a calibration graph or by the standard-addition method.

iii. *Potentiometric microbiological assay of gentamicin, strepto-mycin, and neomycin with a CO_2-gas sensor:*
The dose levels are chosen to be within the linear range of the optimized log dose–response curves (see the preceding text). The standard and sample solutions are each run in duplicate at each dose level in a single bioassay. Three bioassays using freshly prepared solutions of standard and sample are performed for each antibiotic. A single dose of each pharmaceutical preparation is used for each assay.

iv. *Potentiometric microbiological assay of tetracycline with CO_2-gas sensor:*
Antibiotic solutions are prepared daily by dissolving the equivalent of 1 mg cm^{-3} of tetracycline hydrochloride in 10^{-3} M HCl. The ionic strength is adjusted to 0.1 M by the addition of NaCl. This stock solution is diluted 1 : 1 with phosphate buffer of pH 7.0 and the pH is adjusted with 0.1 M NaOH. The remaining solutions in the 1 to 250 μg cm^{-3} range are made by dilution of the 500 μg cm^{-3} stock solution with buffer. These antibiotic solutions are stored in the ice bath until needed.

The CO_2 sensor is calibrated at the beginning and end of each day by making additions of standard 0.1 M NaHCO$_3$ solution to a thermostated cell containing 1.0 cm^3 each of buffer, Difco nutrient broth, and a 100 μg cm^{-3} tetracycline hydrochloride solution. Prior to the potentiometric measurements, the pH of this solution is adjusted to < 4.5 with the addition of 1.0 cm^3 of 0.15 M HCl. One cubic centimeter each of the bacterial cell suspension, nutrient broth, and reference antibiotic solution are added to a 15-cm^3 sealed centrifuge tube, which is then placed in a thermostated shaker bath for varying lengths of time. Samples are prepared at 16-min intervals, because this time corresponds to the time required for the CO_2 sensor to return to baseline following a measurement. At the end of the incubation period, each sample is placed in an ice bath to cool for 1 min. The sample is then quickly transferred to the measuring cell with a 10-cm^3 hypodermic syringe. After 4 min to allow for tempera-ture equilibration, 1.0 cm^3 of 0.15 M HCl is added to adjust the pH to < 4.5, and the potentiometric measurement is made. The carbon dioxide concentrations are obtained using the measured potentials and the average of the two calibration curves. The carbon dioxide sensor is stored in buffer with stirring at room temperature between measurements.

The same procedure is used for the determination of pharmaceuti-cal tetracycline capsules. (A 2 × 2 parallel line bioassay is performed using two dose levels, 100 and 500 μg cm^{-3}, for reference and sample solutions, each being run in duplicate for a single bioassay. The bacterial cell suspension and the nutrient broth are prepared in pH 7.8 phosphate buffer).

v. *Nystatin assay:*

A probe (O_2 or CO_2 sensor), once tested, is immersed in 20 cm^3 of a 0.05 *M* phthalate buffer of pH 4.5 containing 2 g dm^{-3} glucose. After a few (usually 5 to 6) minutes, the membrane sensor reached a stable signal; in the case of CO_2 sensor, the potential obtained is reproducible at about ± 5 mV. Nystatin sample is dissolved in DMF and added to the phthalate buffer where the sensor is immersed. The concentration of DMF in the sample solution is kept below 0.66% (v/v). Electrode potential decrease is calculated from the linear portion of the response curve. The lag time is measured as the time between the introduction of DMF and the beginning of the signal increase or decrease (O_2 sensor).

For tablet determination, 10 tablets are weighed and finely powdered, and a portion is dissolved in a suitable amount of DMF to give a concentration of about 3 mg cm^{-3} nystatin by stirring for 3 to 4 h. The solution is then filtered and tested as previously described.

For ointments, a sample of about 200 mg is weighed and then extracted with several small aliquots of DMF by mixing intimately in a porcelain mortar. Then DMF is collected and diluted to a final standard volume to provide a concentration of about 3 mg cm^{-3} nystatin. The solution is tested as previously described.

5.55   Glucose

$$C_6H_{12}O_6 \text{ (MM} = 180.2)$$

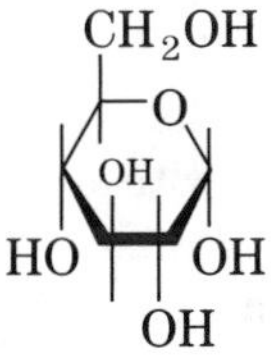

Therapeutic category: fluid and nutrient replenisher

Discussion and Comments

Many papers[247-262] on glucose determination, mainly in biological media, were published until now and most of them were discussed in Coşofreţ[98] (pp. 325–329).

Enzyme sensors obtained by binding enzymes on nylon net showed improved mechanical and analytical characteristics. The resulting membrane presents a high mechanical resistance and can be attached to, or removed from, the electrode surface several times without damage. The enzymes investigated by Mascini[263] were mostly oxidases and the enzyme membranes were coupled with the oxygen electrode. Some bioenzyme membranes were also evaluated for the determination of sucrose, maltose, starch, and lactose, by coupling a hydrolase enzyme and glucose oxidase to the oxygen electrode.

The chemical procedure for immobilizing single enzymes or two enzymes is described in detail in the work of Mascini.[263] The chemical binding is quite mild and leads to a final membrane with high enzyme activity, which gives results comparable to, or better than, common immobilization procedures. It is known that the glucose oxidase electrode is the best established enzyme electrode and the preceding immobilization procedure was well adapted to this enzyme. The calibration graph was linear in the range 4×10^{-3} to 2×10^{-1} g dm^{-3}. The lifetime of the enzyme-treated nylon membrane was over eight months even when the net was kept at room temperature in 0.1 M phosphate buffer of pH 7.0.

Galactose oxidase is not a specific enzyme for galactose but it oxidizes galactosides and some polysaccharides containing galactose end groups.[263, 264] However, in the absence of such compounds, the sensor is useful for rapid screening. The linear range for the galactose concentration was 5×10^{-2} to 5×10^{-1} g dm^{-3} in 0.1 M phosphate at pH 7.0.

Several disaccharidases (lactose, maltose, and sucrose) can be determined by using two immobilized enzymes: a disaccharidase specific for the substrate and the glucose oxidase to measure the liberated glucose. Glucoamylase hydrolyses α-1,4-glucan linkages in polysaccharides by removing successive glucose units. Maltose liberates two molecules of glucose, and soluble starch forms several glucose units. The measurements for the determination of maltose and starch have been made at pH 9 when linear calibration curves were obtained. The lifetime of both membranes was over three months.

The use of PVC membranes with entrapped glucose oxidase in conjunction with an iodide-selective membrane sensor (Orion, Model 94-53) for calibrating glucose-containing solutions according to the following reactions, was described[265]:

$$\text{glucose} + O_2 \xrightarrow[\text{oxidase}]{\text{glucose}} \text{gluconic acid} + H_2O_2 \qquad (5.71)$$

$$H_2O_2 + 2\,I^- + 2\,H^+ \xrightarrow[\text{or peroxidase}]{\text{Mo(VI) catalyst}} I_2 + 2\,H_2O \qquad (5.72)$$

The iodide membrane sensor used to monitor a decrease in background iodide levels permits the determination of glucose. Approaches with PVC-matrix membranes, containing glucose oxidase co-immobilized with peroxidase, have been described. These were compared with a membrane in which glucose oxidase has been chemically immobilized with glutaraldehyde on bovine albumin according to Equation 5.73,

$$
\begin{array}{ccc}
\underset{\substack{| \\ (CH_2)_3 \\ | \\ H-C=O}}{H-C=O} & \underset{\substack{(albumin) \\ H_2N-R'}}{H_2N-R} & \xrightarrow{-2\,H_2O} \quad \underset{\substack{\backslash \\ (CH_2)_3 \\ / \\ H-C=N-R'}}{H-C=N-R}
\end{array} \qquad (5.73)
$$

The optimum initial iodide concentration was found to be 10^{-4} M and a volume of 0.3 cm^3 of ammonium molybdate (10% solution in pH 5.0, phosphate buffer) to a total 20-cm^3 sample volume is recommended.

Table 5.28 summarizes the main features of the behavior of the three types of glucose oxidase membrane studied in association with the iodide-selective sensor for small increments of solutions of glucose (1 M) spiked into 10^{-4} M KI in pH 5.0, phosphate buffer solution.[265]

The data in Table 5.28 show that membranes of glucose oxidase chemically immobilized with glutaraldehyde to bovine albumin have

Table 5.28   Characteristics of Glucose Oxidase–Iodide Enzyme Sensors for Glucose Determination[265]

Property	Sensor type		
	Glucose oxidase in PVC	Glucose oxidase + peroxidase in PVC	Glutaraldehyde + glucose oxidase on bovine albumin
Range of glucose (M)	10^{-3}–10^{-2}	10^{-4}–10^{-2}	10^{-3}–10^{-2}
Mean slopes of sensors made from different membranes (mV decade^{-1})	72 (s.d. = 3.2, $n = 6$)	74 (s.d. = 6.0, $n = 5$)	43 (s.d. = 1.1, $n = 7$)
Response times (min)	2–12	2–10	8–10
Conditioning time for fresh enzyme membranes (h)	~ 12	~ 24	< 2
Wash times between samples (h)	1	0.5	0.2
Lifetime (days)	3	7	> 14

superior lifetimes and even more favorable conditioning and wash characteristics, but these suffer from lower slopes, hence lower sensitivity,
than the PVC-membrane systems.

Microbial sensors based on oxygen and CO_2 electrodes coupled with
immobilized *S. cerevisiae* were compared for measurements of glucose
and other carbohydrates.[266] With the oxygen sensor, the yeast works
under aerobic conditions but anaerobically with the CO_2 sensor. The two
metabolisms of the same strain make little difference to the lifetimes
(> 15 days), selectivities and response rates (5 to 10 min) of the sensor.
The effects of pH are very different owing to the pH sensitivity of the
CO_2 sensor. The oxygen-based sensor is more useful for low concentrations of glucose (0.01 to 1 mM) whereas the CO_2-based sensor is better
suited for 1 to 10 mM. With the oxygen-based sensor, the response time
is governed by the rate of metabolism, whereas with the CO_2-based
sensor the response time of the potentiometric CO_2 electrode is the
rate-determining step.[266]

Glucose (and other carbohydrates and some amino acids) is trans-
ported across the plasma membrane of the yeast cell of *S. cerevisiae*,
and it is metabolized through a glycolytic cycle. In both anaerobic and
aerobic processes, CO_2 is developed. Coupling this yeast with an oxygen
or CO_2 sensor is straightforward, and the assembled microbial electrode
responds to glucose by decreasing the oxygen concentration or by
increasing the production of carbon dioxide.

Methods for immobilizing glucose oxidase on cellulose acetate membranes were compared recently by Sternberg et al.[268] These membranes
were initially activated and were coupled to bovine serum albumin (BSA)
in order to increase the number of functional groups to which glucose
oxidase may couple and to impact a proteic environment to the enzyme
(Figure 5.14). Bovine serum albumin contains more than 57 active amino
groups per molecule and may be covalently grafted onto the free alde-
hyde groups obtained by cellulose acetate hydroxyl group oxidation
(Figure 5.15). The imine functions are subsequently reduced with boro-
hydride. The authors[268] compared three different procedures for cellulose membrane acetate activation and cellulose acetate–bovine serum
albumin membranes from each procedure that were reacted with identical activated glucose oxidase solutions. Comparisons of the properties of
glucose oxidase membranes prepared by these three procedures showed
that the best activities are obtained by periodate oxidation followed by
cyanoborohydride reduction (the other two procedures involved Ce(IV)
for oxidation [different concentrations] and BH_4^- for reduction). Glucose
oxidase was then coupled to cellulose acetate–bovine serum albumin
membranes by using an excess of p-benzoquinone as bifunctional
reagent. The excess of p-benzoquinone was separated by size exclusion
chromatography (Fig. 5.15). Most of the aforementioned coupling reactions take place at low temperature, at low ionic strength, and within the

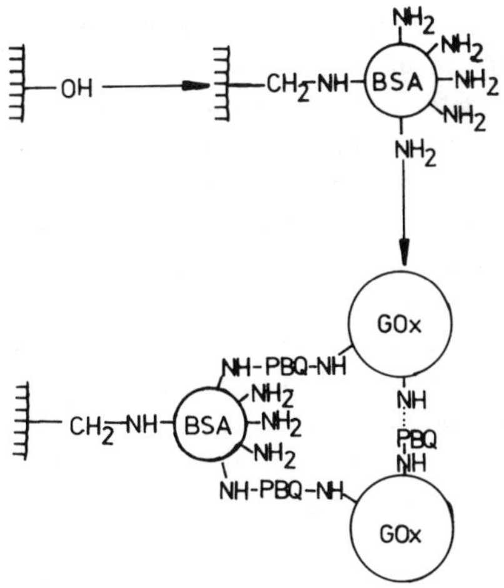

Figure 5.14 General scheme of glucose oxidase coupling on cellulose acetate membranes: enhancement of cellulose acetate surface groups by covalent bovine serum albumin. (Reprinted with permission from Sternberg, R., Bindra, D. S., Wilson, G. S., and Thévenot, D. R., *Anal. Chem.*, 60, 2781, 1988. Copyright 1988 American Chemical Society.)

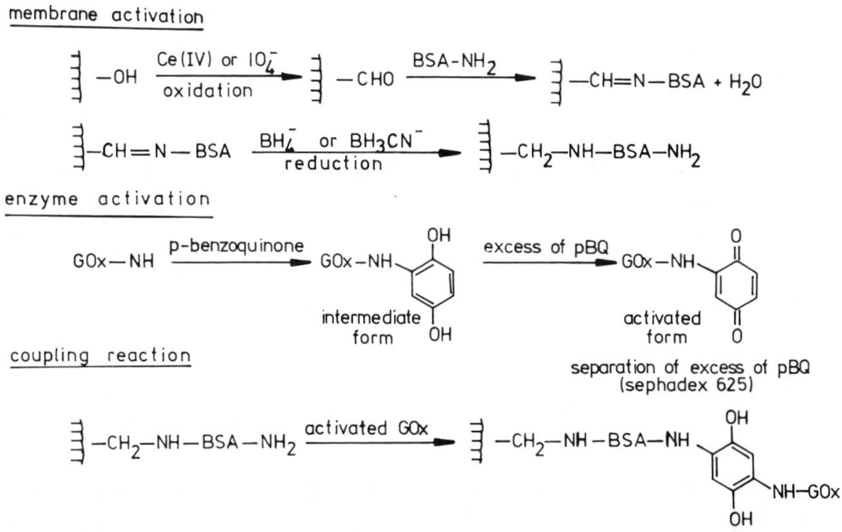

Figure 5.15 Chemical reactions involved in cellulose acetate–bovine serum albumin–*p*-benzoquinone–glucose oxidase membrane preparation. (Reprinted with permission from Sternberg, R., Bindra, D. S., Wilson, G. S., and Thévenot, D. R., *Anal. Chem.*, 60, 2781, 1988. Copyright 1988 American Chemical Society.)

physiological pH range. This coupling procedure is fairly reproducible and allows the preparation of thin membranes (5 to 20 μm) showing high surface activity (1 to 3 U cm^{-2}) that are stable over a period of 1 to 3 months. Electrochemical and radiolabeling experiments showed that enzyme inactivation as a result of immobilization is negligible. Linear ranges of calibration curves of sensors prepared with such membranes usually reach 2 to 3 mM glucose, indicating that for higher glucose concentrations the enzymatic reaction is the rate-limiting step. Higher linear ranges have been obtained either with less-active cellulose acetate membranes using a glucose oxidase entrapment procedure or with collagen–glucose oxidase membranes covered with nonenzymatic cellulose acetate membranes allowing external diffusion restriction.[261]

Vicinal dihydroxy compounds (α-diols and glucose and other carbohydrates) are known to react selectively and stoichiometrically with periodate at room temperature (Malaprade reaction) according to the equation

$$HOCH_2(CHOH)_n CH_2OH + (n + 1)HIO_4$$

$$\rightarrow (n + 1)HIO_3 + 2\,HCHO + n\,HCOOH + H_2O \quad (5.74)$$

These compounds were allowed to react with an excess of sodium periodate and the consumption of the latter was followed by monitoring the change in potential of a periodate liquid-membrane sensor (10^{-2} M nitron–periodate ion-pair in nitrobenzene as electroactive material; sensor displays a near-Nernstian response for 10^{-2} to 10^{-5} M periodate with an anionic slope of 55 mV decade^{-1}).[267]

Graphs of potential vs. time showed that a reaction time of 15 min at room temperature is adequate for the quantitative oxidation of glucose at pH 3 to 6. No interferences were caused by any of the reaction products such as IO_3^-, HCHO, and HCOOH. The reaction of disaccharides (e.g., sucrose, lactose, and maltose) with periodate did not reach completion under these conditions.

Analytical Procedures

i. *PVC glucose oxidase–iodide enzyme sensor:*
   Known volumes of solutions of glucose (1 M in pH 6, phosphate buffer solution) are spiked into solutions (20 cm^3) of 10^{-4} M potassium iodide in pH 5 phosphate buffer solution in which the iodide–enzyme as the indicator sensor and a reference electrode fitted with an outer 10% potassium nitrate solution junction are immersed. A 0.3-cm^3 volume of ammonium molybdate (10% solution in pH 5 phosphate buffer) is also added to the iodide solution when there is no peroxidase in the electrode membrane, and oxygen is bubbled through the cell in these situations (about 150 cm^3 min^{-1}). No oxygen

bubbling is required when the iodide–enzyme electrode membrane contained co-immobilized peroxidase, as Reaction 5.72 procedes well without it: such solutions are magnetically stirred.

The EMF values are recorded and used to construct the calibration curve, E vs. log[glucose]. The sample concentration is determined from this graph. (After each spiking run and direct measurement, the iodide–enzyme and reference electrodes are washed by immersing in pH 5 buffer solution until the EMF reaches the background value characteristic of the buffer.)

ii. *Periodate (IO_4^-)-liquid-membrane sensor:*
A 10-cm^3 portion of a standard 10^{-2} M $NaIO_4$ solution is transferred into a 50-cm^3 beaker and the periodate-membrane sensor in conjunction with the reference electrode (e.g., double-junction Ag/AgCl) is immersed in the solution. After stabilization of the potential within ± 2 mV, a 1.00-cm^3 aliquot containing 0.05 to 2.5 mg cm^{-3} of glucose (pH 3 to 6) is added, the solution stirred and the potential reading recorded after 20 min. A blank experiment is conducted under the same conditions using a 1.00-cm^3 aliquot of water instead of the sample test solution. The potential reading is recorded after the same reaction time. The concentration of IO_4^- in the presence and absence of glucose are calculated from calibration graph prepared with aliquots (25 cm^3) of 10^{-2} to 10^{-6} M standard $NaIO_4$ solutions. The concentration of glucose is calculated from the concentration of the periodate consumed (1 mol $IO_4^- \equiv 0.2$ mol of glucose).

5.56   Glycols

Ethylene glycol	Propylene glycol
$C_2H_6O_2$ (MM = 62.07)	$C_3H_8O_2$ (MM = 76.1)

$$H-\overset{\overset{\displaystyle H}{|}}{\underset{\underset{\displaystyle OH}{|}}{C}}-\overset{\overset{\displaystyle H}{|}}{\underset{\underset{\displaystyle OH}{|}}{C}}-H$$

$$CH_3-\overset{\overset{\displaystyle H}{|}}{\underset{\underset{\displaystyle OH}{|}}{C}}-\overset{\overset{\displaystyle H}{|}}{\underset{\underset{\displaystyle OH}{|}}{C}}-H$$

Therapeutic category: mainly used as a constituent of antifreeze solutions (EG); useful solvents of low toxicity for some vitamins, barbiturates, and other substances that are sufficiently soluble in water or are unstable in aqueous solutions

Discussion and Comments

A perchlorate-ion-selective membrane sensor (Orion, Model 92-81) responds also to periodate. Based on this property the sensor was successfully applied in potentiometric titration of vicinal glycols.[269, 270] These are oxidized with periodate and the reaction rate is followed by this membrane sensor; the time required for the reaction to consume a fixed amount of periodate and therefore for the potential to increase by a preselected amount (e.g., 25.0 mV) is measured automatically and related directly to the vicinal glycol concentration.

Two other simpler procedures for glycol determinations are based on the use of periodate membrane sensors.[267, 271] One sensor[271] has a liquid membrane of Capriquot (tri-n-octylmethylammonium)-periodate in nitrobenzene (0.1 mM) (linear response range 10^{-1} to 10^{-7} M IO_4^-; slope 60 mV decade^{-1}) and the other[267] used 10^{-2} M nitron–periodate in nitrobenzene as electroactive material (linear response range 10^{-2} to 10^{-5} M; slope 55 mV decade^{-1}). The first sensor was used in potentiometric titration of vicinal glycols with 0.1 M sodium periodate as titrant and the second in determination by direct potentiometry, according to Equation 5.74. Glycols were allowed to react with an excess of sodium periodate and the consumption of the latter was followed by monitoring the change in potential of the periodate liquid-membrane sensor.

Potential–time graphs showed that a reaction time of 15 min at ambient temperature is adequate to the quantitative oxidation of most α-diols. A recovery of 97.8% (standard deviation 1.6%) was reported for the determination of ethylene glycol in the range of 0.1 to 1.0 mg.

Analytical Procedures

i. *Potentiometric titration:*
To a weighed sample in a 50-cm^3 beaker, 5 cm^3 of 0.1 M sodium bicarbonate and distilled water are added to a final volume of 25 cm^3. The mixture is potentiometrically titrated with 0.1 M sodium periodate (periodate-ion-selective membrane sensor as indicator and SCE as reference). The end point corresponds to the maximum slope on the titration curve.

ii. *Direct potentiometry:*
A 10-cm^3 aliquot of a standard 10^{-2} M sodium periodate solution is transferred into a 50-cm^3 beaker and the nitron–periodate liquid-membrane sensor in conjunction with the reference electrode (e.g., double-junction Ag/AgCl) is immersed in the solution. After stabilization of the potential within ± 2 mV, a 1.0-cm^3 aliquot containing 0.05 to 2.5 mg cm^{-3} of the respective glycol (pH 3 to 6) is added, the solution is stirred, and the potential reading is recorded after 20 min. A blank experiment is conducted under the same conditions using a 1.0-cm^3 aliquot of water instead of the sample test solution. The potential reading is recorded after the same reaction time. The concentrations of IO_4^- in the presence and absence of the glycol are calculated from the calibration graph prepared with aliquots (25 cm^3)

of 10^{-2} to 10^{-6} M standard $NaIO_4$ solutions. The concentration of the respective glycol is calculated from the concentration of the periodate consumption (1 mol $IO_4^- \equiv$ 1 mol glycol).

5.57   Glycopyrrolate

$$C_{19}H_{28}BrNO_3 \text{ (MM = 398.4)}$$

Therapeutic category: anticholinergic

Discussion and Comments

The preparation and properties of ion-selective membrane sensors (PVC- and liquid-membrane types) based on ion-pair complex of glycopyrronium ion (Gly) with dicyclohexylnaphthalenesulfonate (GDCHNS), diisopentylnaphthalenesulfonate (GDPNS), diisobutylnaphthalenesulfonate (GDBNS), and tetraphenylborate (GTPB) were described by Yao and Liu.[272] The glycopyrronium dialkylnaphthalenesulfonate sensor with electroactive material of higher molecular weight has a larger linear range and a greater response slope (e.g., for GDCHNS sensor, Nernstian-type range 6.3×10^{-2} to 1.3×10^{-5}, slope 57.5 mV decade^{-1}, and detection limit 6.3×10^{-6} M).

However, the differences between the electroactive materials tested are small compared with the solvent effects. The response of the Gly-selective liquid-membrane sensor was measured for different liquid ion exchangers in several solvents, to test the effect of solvent dielectric constant ϵ on the electrode slope. It was observed that, for organic solvents from a homologous series, the slope S increases with increasing dielectric constant, according to the following regression equation:

$$S = C_0 - C_1/\epsilon \tag{5.75}$$

(values of C_0 and C_1 for phthalic esters and n-alcohols have been reported[272]).

Study of the optimum concentration of the electroactive material for sensor response showed that at concentrations of 5×10^{-4} to 5×10^{-3} M GDCHNS, in dibutylphthalate as PVC-membrane plasticizer, the sensor had a Nernstian response and gave constant and stable potential readings.

For the liquid ion exchanger in nitrobenzene, the slope reached a maximum at concentrations of 10^{-4} to 10^{-3} M.

At pH values between 4 and 8, no significant effect on membrane potentials was observed.[272] Above pH 8, formation of glycopyrronium base caused the potentials to become more negative and at pH values below 3.5 to 4, the sensors began to respond to hydrogen ions.

Compounds such as cinchonine, quinine, propranolol, dibazol, diphenhydramine, imidazole, berberine, tetrahydropalmitine, chlorpheniramine, and tetrabutylammonium ion interfere in the sensor response, but insignificant interferences were noted for other substances.

Determination of glycopyrrolate by direct potentiometry was performed with an average recovery of 98.7% and a standard deviation of 1.2%. The results obtained by potentiometric titration with sodium tetraphenylborate solution showed an average recovery of 98.9% and a mean standard deviation of 1.0%. These results were in good agreement with those obtained by the official USP method.

Analytical Procedures

i. *Direct potentiometry:*

Standard solutions of 10^{-2} to 10^{-4} M concentration are prepared by serial dilution of 10^{-2} M glycopyrronium bromide prepared by dissolving a suitable amount of drug substance with distilled water. A constant ionic strength ($I = 0.1$ M, adjusted with sodium nitrate) must be used. The pH of all standard solutions is adjusted to approximately 5.0 with acetate buffer. The standard solutions are transferred into 150-cm^3 beakers containing Teflon-coated stirring bars. The glycopyrrolate membrane sensor together with a reference electrode (Ag/AgCl or SCE) is immersed successively in the standards and the EMF values measured. The glycopyrrolate concentration in the sample is determined from the calibration graph (E vs. log[glycopyrrolate]).

ii. *Potentiometric titration:*

The pair of sensors (glycopyrrolate-selective as indicator and SCE as reference) is introduced into the sample solution (30 to 40 cm^3 of about 10^{-3} M, pH = 4 to 8) and titrated with sodium tetraphenylborate solution (10^{-2} M). The EMF is plotted as a function of titrant volume to obtain the equivalence point.

5.58   Guanidine and Its Derivatives

The importance of guanidine in the biological and medical fields has led to guanidinium membrane sensors[273, 274] with good performances. For

Table 5.29   Guanidine and its Derivatives Assayed by Membrane Sensors

Compound	Formula (MM)	Therapeutic category
Guanidine	CH_5N_3 (59.07)	Striated muscle stimulant

$$H_2N-\underset{\underset{NH}{\|}}{C}-NH_2$$

| Buformin | $C_6H_{15}N_5$ (157.2) | Hypoglycemic |

$$\underset{C_4H_9}{\overset{H}{>}}N-\underset{\underset{NH}{\|}}{C}-NH-\underset{\underset{NH}{\|}}{C}-NH_2$$

| Metformin | $C_4H_{11}N_5$ (129.0) | Hypoglycemic |

$$\underset{CH_3}{\overset{CH_3}{>}}N-\underset{\underset{NH}{\|}}{C}-NH-\underset{\underset{NH}{\|}}{C}-NH_2$$

| Phenformin | $C_{10}H_{15}N_5$ (205.2) | Hypoglycemic |

$$\underset{C_6H_5(CH_2)_2}{\overset{H}{>}}N-\underset{\underset{NH}{\|}}{C}-NH-\underset{\underset{NH}{\|}}{C}-NH_2$$

| Proguanil | $C_{11}H_{16}ClN_5$ (253.7) | Antimalarial |

$$Cl-\text{〈}\text{〉}-NH-\underset{\underset{NH}{\|}}{C}-NH-\underset{\underset{NH}{\|}}{C}-NH-CH(CH_3)_2$$

the determination of its derivatives (see Table 5.29) liquid-membrane copper(II)-ion-selective sensors[275, 276] have been used.[277]

Among various crown ethers tested as complexing agents for guanidinium cation $(H_2N)_3C^+$ (e.g., dibenzo-24-crown-8, dibenzo-27-crown-9, tribenzo-27-crown-9, or dibenzo-30-crown-10) the best was dibenzo-27-crown-9 (DB27C9), which led to most selective guanidinium membrane sensor[273] (the composition of PVC membrane was $1:2:0.1$ [m/m] PVC : di-n-butylphthalate : electroactive material). The linear response was over the range 10^{-1} to 10^{-4} M with selectivity coefficients of about 10^{-2} for most alkali and alkaline earth metal ions. The EMF readings were nearly constant in the pH range 4 to 10 and the sensors in use for over five months were still functional.

Table 5.30   Composition of PVC Membranes of Five Types of Guanidinium-Selective Sensor[274]

Membrane Sensor type	Membrane mass composition (mg)			
	Solvent mediator	PVC	DB27C9	GUTPB
1	300^a	150	5	—
2	300^a	150	5	5
3	300^a	150	5	10
4	300^a	150	—	10
5	300^b	170	—	40

^{a}DPB.
^{b}Dioctylphenyl phosphonate.

Study of the effect of solvent mediator for DB27C9-based membrane sensors showed that dibutylphthalate gave the best fast response with near-Nernstian characteristics, followed by dioctyladipate and dioctyl-sebacate; 2-nitrophenyloctyl ether, 2-nitrophenylphenyl ether, and dioctylphenyl phosphonate were unsatisfactory.[274]

The effect of guanidinium tetraphenylborate (GuTPB) in sensor membranes with dibutylphthalate as solvent mediator was also studied. Addition to DB27C9 membranes of ion pairs produced less-sensitive sensors compared with those based on the crown ether alone, but the selectivity (separation solution method) toward $(H_2N)_3C^+$ over some metal ions was improved. GuTPB-based membrane sensors were more sluggish in response.

Of the five types of guanidinium membrane sensor (see Table 5.30), type 1 gave the fastest response (< 15 s), whereas membrane sensors of types 2 through 5 were more sluggish, with response times of up to 1 min at low ion activity and more than 2 min at ionic activities of over 5×10^{-3} M.[274]

All membrane sensors store well in 0.1 M guanidine hydrochloride, but by four months they show a 12 to 16% decrease in slope (initial characteristics for type 1: $S = 58 \pm 0.5$ mV decade^{-1}; linear range 0.10 to 100 mM; detection limit 30 μM; membrane resistance 25 MΩ; operational pH range 3 to 11).

The method using a copper(II) membrane sensor for the determination of guanidine derivatives is based on the formation of $[Cu(Big)_2]X_2$ complexes (Big = biguanide compounds) by the reaction between copper(II) amine complexes and biguanides. There are several ways in which this method can be used for the determination of biguanides[277]:

1. Use of an excess of copper(II) amine in order to precipitate $[Cu(Big)_2]X_2$ and subsequent determination of the excess copper(II)

by potentiometric titration with EDTA using a copper(II)-ion-selective indicator sensor.

2. Transformation of the biguanide salt into the free base, which is then determined by direct potentiometric titration with an aqueous solution of copper sulfate using a copper(II) membrane sensor as indicator electrode.

3. Separation of the water-insoluble complex and dissolution of a precisely known amount in a suitable solvent and subsequent determination of copper(II) ions in the solution by a convenient method (e.g., based on a copper(II)-ion-selective membrane sensor).

Methods 2 and 3 are lengthier because an accurate conversion of the biguanide salt into free base is needed. Also, the filtration and drying stages are time-consuming and susceptible to errors. For method 1 the $[Cu(Big)_2]SO_4$ complex is formed by adding an excess ammoniacal copper sulfate solution (of known concentration) to aqueous biguanide hydrochloride. The excess of copper(II) is determined by potentiometric titration with 5×10^{-2} M EDTA using a copper(II) membrane sensor. Two potential breaks were observed on the titration curves. They correspond to excess copper(II) sulfate and complexed copper(II), respectively.

Analytical Procedures

i. *Guanidine assay:*
Standard solutions of 10^{-2} to 10^{-4} M concentration are prepared by serial dilution of 10^{-1} M guanidine hydrochloride, prepared by dissolving a suitable amount of drug substance with distilled water. A constant ionic strength ($I = 0.1$ M, adjusted with sodium nitrate) must be used. The pH of all standard solutions is adjusted to approximately 5.0 with acetate buffer. The standard solutions are transferred into 150-cm³ beakers containing Teflon-coated stirring bars. The guanidinium membrane sensor together with a reference electrode (Ag/AgCl or SCE) is immersed successively in the standards and the EMF values measured. The guanidine concentration in the sample is determined from the calibration graph (E vs. [guanidine]).

ii. *Buformin, phenformin, and proguanil (as hydrochlorides) assay:*
The accurately weighed sample (50 to 100 mg) is transferred into a 100-cm³ beaker and dissolved in a minimum quantity of distilled water. A volume of 6.0 cm³ of 5×10^{-2} M copper(II) sulfate (in 1 M ammonia) is added and the mixture is potentiometrically titrated (under stirring) with 5×10^{-2} M EDTA solution (Cu^{2+}-selective

membrane sensor as indicator; SCE as reference). The electrode potential is recorded as a function of titrant volume, and the titrant volume corresponding to the first potential break, evaluated in the usual way, is used to calculate the amount of alkyl-1-biguanide in the sample.

iii. *Metformin (as hydrochloride) assay:*
The accurately weighed sample (50 to 100 mg) is transferred into a 100-cm^3 beaker and dissolved in a minimum quantity of distilled water and 6.0 cm^3 of 5×10^{-2} M copper(II) sulfate (in 1 M ammonia) is added. The precipitate is filtered on a sintered glass filter and washed with distilled water. The excess of copper(II) sulfate is potentiometrically titrated with 5×10^{-2} M EDTA in the presence of a Cu^{2+}-selective membrane sensor (SCE as reference). The titrant volume, evaluated in the usual way, is used to calculate the amount of metformin in the sample.

5.59   Haloperidol and Related Fluorine Tranquilizers

The butyrophenone neuroleptic drugs are widely administered because they exert a relatively potent antipsychotic effect and show a greater selectivity when compared with phenothiazine derivatives. The butyrophenone molecule contains fluorine, which allows an essay by the determination of this element. The fluoride-ion-selective membrane sensor was proposed by Hopkala and Przyborowski.[278] The membrane sensor was used for the determination of fluorine butyrophenones (haloperidol and trifluperidol) as well as for the determination of bis(4-fluorophenyl)butyl-piperidines (pimozide and penfluridol) (see Table 5.31).

The mineralization of the compounds listed in Table 5.31 was carried out by oxidation, after Schöniger, in a polyethylene flask. The combustion products were absorbed in water. For the direct potential measurements of released fluoride ions, a phthalate buffer at pH 5.2 with cyclohexane-1,2-diamino-N,N,N',N-tetraacetic acid as metal-ion complexing agent was used; the ionic strength ($I = 1.5$ M) was adjusted with potassium nitrate. The procedure proved useful for the determination of haloperidol in tablets.

Potentiometric titration of released fluoride ions with standard lanthanum nitrate was carried out in ethanolic medium, thus ensuring a slight solubility of formed lanthanum fluoride. The equivalence point that corresponds to a 1 : 3 molar ratio was established by Gran's method. Good recovery and precision were obtained in both potentiometric methods.

Table 5.31 Haloperidol and Related Fluorine Tranquilizers Assayed by Fluoride Membrane Sensors

Compound	Formula	MM
Haloperidol	$C_{21}H_{23}ClFNO_2$	375.9
	F—◯—$COCH_2CH_2CH_2N$◯$\overset{OH}{}$—◯—Cl	
Trifluperidol	$C_{22}H_{23}F_4NO_2$	409.4
	F—◯—$COCH_2CH_2CH_2N$◯$\overset{OH}{}$—◯—CF_3	
Pimozide	$C_{28}H_{29}F_2N_3O$	461.6
	F—◯—CH—$CH_2CH_2CH_2N$◯N—◯ (with $\overset{O}{}$=C—NH); F	
Penfluridol	$C_{28}H_{27}ClF_5NO$	524.0
	F—◯—$CHCH_2CH_2CH_2$—N◯$\overset{OH}{}$—◯—CF_3, Cl; F	

Analytical Procedures

An accurately weighed sample (2 to 14 mg) is placed on a filter paper (Whatman No. 1) covered with a polyethylene film (1 × 1.5 cm) and rolled. The sample is burned up in a 1-dm^3 polyethylene flask in O_2, after Schöniger, and the combustion products are absorbed for 30 to 40 min in 15 cm^3 of distilled water or in a mixture of 5 cm^3 of distilled water and 10 cm^3 of phthalate buffer of pH 5.2 in the case of haloperidol tablets.

i. *Direct potentiometry:*

The following are pipetted to each 100-cm^3 volumetric flask: phthalate buffer solution of pH 5.2, constant ionic strength ($I = 1.5$ M, KNO$_3$), and containing CyDTA as complexing agent (10 cm^3); working solutions of NaF (10 to 500 μg F$^-$); and H$_2$O up to the mark. The mixtures are transferred into polyethylene beakers containing Teflon-coated stirring bars and the EMF values are recorded (F$^-$-selective membrane as indicator sensor, SCE as reference electrode). The results are used for the preparation of calibration graph (E vs. log[F$^-$]). Fluoride ions released after the combustion of the drugs are measured similarly.

ii. *Potentiometric titration:*

After the mineralization of the drug sample, the absorption solution is transferred into a 100-cm^3 polyethylene beaker, diluted with ethanol to 80% and titrated with an ethanolic-aqueous 10^{-2} M lanthanum nitrate solution, using the same electrode pair as for direct potentiometry. The titrant volume is evaluated by Gran's method.

5.60   Hydralazine

$$C_8H_8N_4 \ (MM = 160.2)$$

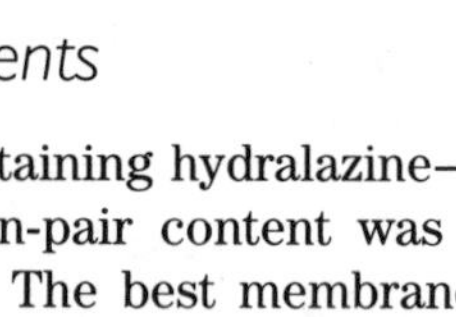

NH—NH$_2$

Therapeutic category: antihypertensive

Discussion and Comments

Four PVC membranes containing hydralazine–tetraphenylborate ion-pair complex, in which the ion-pair content was varied between 2.44 and 9.09%, were prepared.[279] The best membrane sensor contained 4.76% ion-pair complex and dioctylphthalate as plasticizer (the electrode body was filled with 0.1 M NaCl–10^{-3} M hydralazine hydrochloride as internal solution). The sensor showed a Nernstian-type response with a slope of 57.5 mV decade^{-1} at 20°C over the concentration range 4×10^{-4} to 10^{-1} M hydralazine. The membrane sensor can be used at pH 2.1 to 6.0; at pH < 2.1 the potential decreases, presumably because of diprotonated species. At pH > 6.0 the decrease in potential can be attributed to the conversion of hydralazinium cation HL$^+$ into free hydralazine.

The inorganic cations did not interfere because of large differences in ionic size, mobility, and permeability compared with HL^+.[279]

The membrane sensor was successfully applied for the determination of hydralazine in pure solutions and in Ser-Ap-Es antihypertensive tablets, which contain hydralazine hydrochloride, by the standard-addition method and potentiometric titration (mean recovery 101.1 and 98.0%, respectively; standard deviation 2.4 and 1.0%, respectively).

Analytical Procedure

The pair of sensors (hydralazine-selective as indicator and SCE as reference) is introduced into the sample solution (30 to 40 cm^3 of about 5×10^{-3} M, pH = 4 to 5) and potentiometrically titrated with 5×10^{-2} M sodium tetraphenylborate solution. The EMF changes are recorded and plotted as a function of titrant volume to obtain the equivalence point. The titrant volume at this point is evaluated from the maximum slope on the titration curve and it is used to calculate hydralazine concentration.

5.61 Isofluorphate

$$C_6H_{14}FO_3P \ (MM = 184.2)$$

$$\begin{array}{c} (CH_3)_2CHO \\ \searrow \overset{\displaystyle O}{\overset{\displaystyle \|}{P}}-F \\ (CH_3)_2CHO \nearrow \end{array}$$

Therapeutic category: cholinergic (ophthalmic)

Discussion and Comments

Diisopropyl fluorophosphate (isofluorphate) is contained in the giant axon of squid nerve tissue. This tissue can be coupled with a fluoride membrane sensor to produce a sensor selective to isofluorphate.[280] Diisopropyl fluorophosphatase (DIFPase, E.C. 3.8.2.1), present in squid nerve tissue and micro-organisms, can catalyze the hydrolysis of diisopropyl fluorophosphate (DIFP) to produce fluoride:

$$\begin{array}{c} (CH_3)_2CHO \\ \searrow \overset{O}{P} \\ (CH_3)_2CHO \nearrow \searrow F \end{array} + H_2O \rightarrow \begin{array}{c} (CH_3)_2CHO \\ \searrow \overset{O}{P} \\ (CH_3)_2CHO \nearrow \searrow OH \end{array} + H^+ + F^-$$

$$(5.76)$$

Therefore, it was expected that a convenient DIFP sensor could be constructed by immobilization of the squid nerve tissue on the surface of a fluoride-ion-selective sensor.

Squids (*Todarodes pacificus*) have stellate nerves containing giant axons. Squid nerve tissue was separated carefully from the squid body and was minced with a razor blade for about 30 min. This minced paste was sandwiched between two dialysis membranes and placed over the fluoride-ion-selective membrane by binding the dialysis membranes tightly to the electrode body with parafilm. The thickness of the immobilized paste was < 0.3 mm. After construction the nerve tissue sensor was stored in a TISAB I (Orion) citrate buffer solution adjusted to pH 7.1 for 3 h at 0°C to eliminate any fluoride contained in the squid nerve tissue.[280]

It was found that the hydrolysis of DIFP is negligible (< 0.5% in 1 h) at pH 7.1 and 0°C; thus the sensor was stored in a refrigerator at 0°C when not in use. To prevent putrefaction of the tissue and evaporation of the substrate from the sample solution, all experiments performed by Uchiyama et al.[280] were conducted at 0°C.

A sensor with 4.5 mg of squid nerve tissue has a relatively fast response time with an electrode slope of 44 mV decade^{-1} (pH 7.1) over the range 2×10^{-5} to 7×10^{-3} *M* DIFP (detection limit is 8×10^{-6} *M*). The sensor exhibited a surprisingly good lifetime. Its characteristics remained constant for about 16 days and then decreased gradually.

DIFP was determined at three concentration levels, with a relative standard deviation of less than 2.6%.

Analytical Procedure

Standard solutions in the range 7×10^{-3} to 2×10^{-5} *M* are prepared from pure diisopropyl fluorophosphate. Total ionic strength adjuster (TISAB, Orion) is used as the buffer solution to provide constant background ionic strength and to demask fluoride; the pH of all solutions is adjusted to 7.1 by using citrate buffer. Adequate aliquots of the standard solutions are transferred into 100-cm^3 double-jacket vessels containing magnetic stirring bars. The electrode pair is introduced successively into the standards and the EMF values are recorded vs. concentration (at 0°C). The unknown sample concentration of DIFP is determined from the calibration curve.

5.62   Isoniazid

$$C_6H_7N_3O \ (MM = 137.1)$$

Therapeutic category: antibacterial (tuberculostatic)

Discussions and Comments

Two potentiometric methods with membrane sensors have been described for isoniazid determination.[281, 282] In both, the reducing property of isoniazid is used. In the method of Koupparis and Hadjiioannou[281] isoniazid is quantitatively oxidized with an excess of chloramine-T (CAT), followed by measurement of unconsumed CAT with a CAT-selective membrane sensor (see also Section 5.20 and Coşofreţ,[98] pp. 243–244). The method was used for the determination of isoniazid in the 1 to 100 μmol range (10^{-4} to 10^{-2} M) with an error and precision of about 1 to 2% (between pH 5 and 6, two moles [four equivalents] of CAT are consumed per mole of isoniazid). The method has been tested for the determination of isoniazid in pharmaceutical preparations such as injection solutions and tablets.

A kinetic determination method for isoniazid and other reducing agents is based on oxidation with 2×10^{-4} M potassium iodate in 0.25 M sulfuric acid.[282] The time required for the recorded cell potential to increase from 5 to 45 mV (production of iodide ions) is measured by an iodide-selective membrane sensor (Orion, Model 94-53). Calibration graphs were constructed by plotting the reciprocal time ($1000/t$, s^{-1}) vs. concentration. The linear range for isoniazid determination was within 10^{-4} to 10^{-3} M with a relatively small slope of only 10.8 dm^3 mol^{-1} s^{-1} (correlation coefficient 0.998 for $n = 7$).

Analytical Procedures

i. *Chloramine-T membrane sensor:*
   10.0 cm^3 of 2.5×10^{-3} M chloramine-T (25 μmol), 50 cm^3 of phosphate buffer of pH 6.0, and 10.0 cm^3 of the sample or standard containing 1 to 10 μmol of isoniazid are pipetted into a 50-cm^3 amber narrow-necked reagent bottle. The solution is shaken continuously at room temperature for about 50 min and then transferred into the reaction cell at 25°C. The magnetic stirrer is started and the EMF is recorded when it has stabilized to within ± 0.1 mV (in about 20 s). Four isoniazid standards in the 10^{-4} to 10^{-3} M range are included. The excess of chloramine-T is found from a graph of E vs. log(micromoles of CAT in excess), that is, log($25 - 2 \times$ micromoles of isoniazid).

ii. *Iodide membrane sensor:*
   10.0 cm^3 of the potassium iodate working solution (2.0×10^{-4} M) and 10.0 cm^3 of 0.5 M H_2SO_4 are pipetted into the measuring cell and the potential is recorded under stirring; 0.1 cm^3 of the sample solution is injected and the change in potential is recorded with time. The cell is emptied and the procedure is repeated for each sample

solution. The time required for the recorded cell potential to increase from 5 to 45 mV is measured.

For calibration, aliquots of the appropriate standard solutions are introduced into the measuring cell and the change in potential is measured as before. The calibration graph is constructed by plotting the reciprocal time $(1000/t, \text{s}^{-1})$ vs. concentration.

5.63   Ketamine

$$C_{13}H_{16}ClNO \ (MM = 237.4)$$

Therapeutic category: general anesthetic

Discussions and Comments

Various ion exchangers containing ketamine (e.g., tetraphenylborate, silicotungstate, picrate, and reineckate) have been investigated for use as electroactive material in a ketamine liquid-membrane sensor.[283] Among them, tetraphenylborate- and silicotungstate-based sensors (1,2-dichloro-ethane as solvent) showed the best behavior with respect to range, response time, and reproducibility. Near-Nernstian ranges within 10^{-2} to $1.5 \times 10^{-5} \ M$ (detection limits $5 \times 10^{-6} \ M$) were reported.

The sensors were not affected by pH changes in the range 2.2 to 6.0. Ketamine at levels of milligrams per cubic centimeter or parts per million was determined with a relatively low error by potentiometric titration and direct potentiometry.

Analytical Procedure

The pair of sensors (ketamine-selective as indicator and SCE as reference) is introduced into the sample solution (30 to 40 cm³ of about $5 \times 10^{-3} \ M$, pH about 5.0) and potentiometrically titrated with $5 \times 10^{-2} \ M$ sodium tetraphenylborate solution. The EMF changes are recorded and plotted as a function of titrant volume to obtain the equivalence point. The titrant volume at this point is evaluated from the maximum slope on the titration curve and it is used to calculate the unknown concentration.

5.64   Levamisole

$$C_{11}H_{12}N_2S \text{ (MM} = 204.3)$$

Therapeutic category: L-(−)-form as an anthelmintic;
D-form as an antidepressant

Discussion and Comments

A PVC-membrane levamisole-selective sensor based on levamisole–tetraphenylborate, levamisole–reineckate, levamisole–picrate, and levamisole–$[HgI_4]^{2-}$ ion-pair complexes as active materials was proposed by Shen and Liao.[284] The effects of various site carriers, solvent mediators, and other factors have been discussed in detail. The best sensor was found to be that containing tetraphenylborate as anion site carrier (Nernstian range 10^{-1} to 2×10^{-5} M, detection limit 1×10^{-5} M, slope 60 mV decade^{-1}). A 0.5% ion-exchanger concentration in the PVC membrane was used. In the pH range 4 to 8 negligible potential changes were recorded. Among many inorganic and organic cations investigated as interferents, only ketamine, quinine, and novocaine showed high interference.

Pure levamisole solution samples were determined with a good accuracy by both direct potentiometric method and potentiometric titration (standard deviation 1.7 and 1.2%, respectively). When 5×10^{-2} M sodium tetraphenylborate solution was used as titrant for levamisole assay in tablets, an error of 0.4% was reported.

Analytical Procedures

The pair of sensors (levamisole-selective as indicator and SCE as reference) is introduced into the sample solution (30 to 40 cm^3 of about 5×10^{-3} M, pH about 5.0) and potentiometrically titrated with 5×10^{-2} M sodium tetraphenylborate solution. The EMF changes are recorded and plotted as a function of titrant volume. The titrant volume at the equivalence point is evaluated from the maximum slope on the titration curve and is used to calculate levamisole concentration.

5.65   Loxapine and Clothiapine

$C_{18}H_{18}ClN_3O$ (MM = 327.8) $C_{18}H_{18}ClN_3S$ (MM = 343.9)

X = O (loxapine);   X = S (clothiapine)

Therapeutic category: tranquilizers

Discussion and Comments

Sensitive membrane sensors for loxapine and clothiapine, two drugs from the dibenzodiazepine class, were developed by the use of a graphite spectroscopic rod coated with a film of PVC, previously dissolved in cyclohexanone and containing the ion pairs tetraphenylborate–loxapine and tetraphenylborate–clothiapine, respectively.[285] A mixture of 1:1 nitrobenzene–dioctyladipate was used as plasticizer. Both sensors display linear response within 10^{-2} to 10^{-5} M with a sub-Nernstian slope of 45 mV decade^{-1}.

Organic substances containing the pyrrolidine, piperazine, or piperidine group display some interference, whereas other types of neuroleptics of the benzodiazepine family do not interfere at all. Both sensors were used for the determination of the respective drug in pharmaceutical formulations (tablets, ampoules), by the direct potentiometric method (pH 4.6, acetate buffer). Recoveries between 98.4 and 102.0% were reported.

Analytical Procedure

Standard solutions of 10^{-2} to 10^{-4} M concentrations are prepared by serial dilution of 10^{-2} M loxapine and clothiapine (as hydrochlorides), respectively. The pH of all standard solutions is adjusted to 4.6 with acetate buffer, which also keeps the ionic strength at a constant value. The respective membrane sensor in conjunction with SCE is introduced into the standard solutions, and the EMF readings (linear axis) are plotted against concentration (logarithmic axis). The concentration of loxapine and clothiapine, respectively, from the unknown sample is determined from the appropriate graph.

For tablets of loxapine and clothiapine determination, at least 10 tablets are weighed and finely powdered and a portion of the powder, accurately weighed, is introduced into the electrochemical cell containing 50-cm^3 solution of acetate buffer of pH 4.6. After about 5 min of stirring for active principle dissolution the EMF is measured and compared with the calibration graph.

5.66   Mefloquine

$$C_{17}H_{16}F_6N_2O \ (MM = 378.3)$$

Therapeutic category: antibacterial, antipaludeen

Discussion and Comments

Solutions containing mefloquine, prepared in dilute aqueous (5×10^{-5} M sulfuric acid) could be quantified by direct measurements with the plastic-membrane selective sensor[286] prepared as described in Srianujata et al.[287] The potential response was linearly related to analyte concentration over three orders of magnitude (Nernstian slope) and measurements were made with $\pm 4\%$ accuracy and $\pm 2\%$ precision over the linear concentration range.

Analytical Procedure

A stock solution of 10^{-2} M mefloquine hydrochloride in 10^{-4} M sulfuric acid is prepared; 10^{-3} to 10^{-5} M mefloquine solutions (at the same pH value and ionic strength) are prepared by successive dilutions from the stock solution. Aliquots (25 cm^3) of standard solutions are transferred into 100-cm^3 beakers containing Teflon-coated stirring bars. The plastic membrane sensor together with a reference electrode (Ag/AgCl or SCE) is immersed successively in the standards and the EMF values recorded. The mefloquine concentration in the sample is determined from the calibration graph (E vs. log[mefloquine]).

5.67   Meperidine

$$C_{15}H_{21}NO_2 \text{ (MM = 247.4)}$$

$$H_3C-N \bigcirc\kern-1em\times \begin{array}{l} C_6H_5 \\ COOC_2H_5 \end{array}$$

Therapeutic category: narcotic analgesic

Discussion and Comments

A meperidine liquid-membrane sensor, containing meperidine-tetrakis(m-chlorophenyl)borate in p-nitrocumene (10^{-2} M), showed a Nernstian response in the range 10^{-2} to 7×10^{-6} M with a slope of 60 mV decade^{-1}.[83] An Orion liquid-membrane electrode body (Model 92) was used for electrode assembly with a Millipore LCWPO 1300 PTFE membrane. The internal aqueous solution was 0.01 M meperidine and 0.1 M sodium chloride (saturated with silver chloride).

The response of the membrane sensor is practically unaffected by changes in pH over the pH range 1 to 7. At higher pH, the potential changes markedly with pH because of progressive loss of the positive charge of the molecules with increasing pH. The graphs of potential (in millivolts) vs. pH were used to calculate the dissociation constant K_a of the cationic acid, because pK_a is equal to the pH where the initial concentration of the protonated species is halved, i.e., when the electrode potential decreases by $0.30S$ mV (S = sensor slope). A value of $pK_a = 8.57 \pm 0.6$ (at 20°C) was reported, in good agreement with the literature value for meperidine.

Meperidine was assayed with good results, by direct potentiometry in an injection preparation (0.100 g per 2 cm^3) (coefficient of variation 2.3%, $n = 4$).

Analytical Procedure

A 200-cm^3 volume of distilled water is pipetted into a 50-cm^3 beaker; the indicator sensor and the reference electrode are immersed in it and, after the potential is stabilized, various increments of 0.1 M meperidine hydrochloride are added. The EMF readings are recorded after stabilization following each addition, and the graph of E (in millivolts) vs. log[meperidine] is constructed. The slope of the membrane sensor is found by regression analysis of the linear part of the graph.

The commercial injection preparation is diluted with water so as to obtain a solution with a final concentration with respect to meperidine in the range 3.3×10^{-3} to 3.3×10^{-4} M. A 20.0-cm^3 volume of this solution is used for analysis. Subsequently, a second potential reading is obtained after the addition of a small volume of a concentrated standard meperidine hydrochloride solution. The initial concentration of the sample is calculated from the change in potential.

5.68   Meprobamate

$$C_9H_{18}N_2O_4 \ (MM = 218.3)$$

$$
\begin{array}{c}
CH_2OCONH_2 \\
| \\
H_3C-C-(CH_2)_2CH_3 \\
| \\
CH_2OCONH_2
\end{array}
$$

Therapeutic category: minor tranquilizer

Discussion and Comments

Three potentiometric analysis methods were described for the determination of meprobamate. All of them are based on its decomposition; when alkali medium was used for decomposition (Equation 5.77), the liberated carbon dioxide in the second reaction step, at pH 4.8, was determined by a carbon dioxide electrode[288]; when decomposition took place in acidic medium (Equation 5.78), the liberated ammonia in the second reaction step, at pH > 11, was determined by an ammonia-gas-sensing electrode[289, 290]:

$$
\begin{array}{c}
CH_2OH \\
| \\
H_3C-C-(CH_2)_2CH_3 + Na_2CO_3 + 2\,NH_3 \quad (5.77) \\
| \\
CH_2OH \quad \downarrow pH\ 4.8 \\
CO_2
\end{array}
$$

$$
\begin{array}{c}
CH_2OCONH_2 \\
| \\
H_3C-C-(CH_2)_2CH_3 \\
| \\
CH_2OCONH_2
\end{array}
\quad
\begin{array}{c}
\nearrow \ +4\,NaOH \\
\searrow \ +2\,HCl
\end{array}
$$

$$
\begin{array}{c}
CH_2OH \\
| \\
H_3C-C-(CH_2)_2CH_3 + 2\,NH_4Cl + 2\,CO_2 \quad (5.78) \\
| \\
CH_2OH \quad \downarrow pH > 11 \\
NH_3
\end{array}
$$

All methods have high specificity, CO_2 and NH_3 being easily monitored without sample separation from decomposition solutions. In the first case, a linear calibration curve was obtained within the concentration range 10^{-4} to 2.5×10^{-2} M meprobamate; in the second, the linearity of the calibration curve was obtained within 10^{-5} to 10^{-2} M meprobamate. For the decomposition of meprobamate in alkali medium, a boiling time of 90 min was required, when 1 M NaOH solution was used; in acidic medium (20% HCl solution) a boiling time of 2 h is required.

Analytical Procedures

i. *Alkaline decomposition:*

A mixture of 545.7 mg of meprobamate and 15 cm^3 of carbonate-free 1 N NaOH is placed in a 50-cm^3 round-bottomed flask. A reflux condenser equipped with a soda lime tube to prevent the entrance of CO_2 from the air is attached and the mixture is boiled gently in an oil bath for 90 min. The flask is then cooled, and the solution is poured into a 250-cm^3 beaker, then diluted with about 180 cm^3 distilled water. The solution is made slightly alkaline (pH 8.5) by adding 3 N H_2SO_4; pH measurement with a pH-meter is carried out in the beaker, which could be made gastight to prevent contamination from the air. The concentration of the final meprobamate solution is 10^{-2} M, corresponding to 2×10^{-2} M CO_2. Standard solutions for preparation of the calibration curve are obtained by dilutions of this stock solution with distilled water.

A mixture of about 200 mg of the sample (accurately weighed) and 10 cm^3 of 1 N NaOH solution is boiled for 90 min. As previously described, the resulting solution is made slightly alkaline and diluted to 250 cm^3 in a volumetric flask. An aliquot of 50 cm^3 of the sample solution is transferred into an approximately 85-cm^3 vial (3.5 × 9 cm); then 5 cm^3 of 0.1 M citrate buffer (pH 4.5) is added and the mixture is incubated for 30 min at 20°C. Finally, the potential measurement with the CO_2 electrode is carried out at 20°C. The CO_2 concentration in the sample solution is determined from the calibration curve, previously established using the standard $NaHCO_3$ solutions.

ii. *Acid decomposition:*

A mixture of 545.7 mg of meprobamate and 50 cm^3 of 20% HCl is placed in a 100-cm^3 round-bottomed flask and the mixture is boiled gently in an oil bath for 2 h. The flask was then cooled; then the solution is poured into a 250-cm^3 beaker and diluted with approximately 150 cm^3 of distilled water. A drop of methyl orange indicator solution is added and, while cooling the beaker continuously, the acid is continuously neutralized with 6 N NaOH solution until the indicator begins to change color. The solution is then adjusted to pH 6.5 with dilute NaOH solution using a pH-meter. The solution is poured into a 250-cm^3 volumetric flask and diluted to this volume with distilled water. The concentration of the final meprobamate solution is 10^{-2} M, corresponding to 2×10^{-2} M NH_3. Standard solutions for calibration are obtained by diluting this stock solution with distilled water.

A mixture of about 200 mg of the sample (accurately weighed) and 200 cm^3 of 20% HCl is boiled for 2 h. As previously described, the resultant solution is adjusted to pH 6.5 and diluted to 100 cm^3 in a volumetric flask; then 2 cm^3 of this solution is diluted to 100 cm^3 with distilled water. A 50-cm^3 portion of the sample is transferred to an approximately 85-cm^3 vial (3.5 × 9 cm), 1 cm^3 of 10 M NaOH or 2 cm^3 of 5 M NaOH is added, and the mixture is incubated for 30 min at 20°C. Finally, the NH$_3$-gas electrode is immersed in the solution and potential measurements are carried out. The vessel should be stoppered to inhibit loss of NH$_3$ and to prevent evaporation of water. The NH$_3$ concentration in the sample solution is determined from the calibration curve, previously prepared.

5.69   Mercury Compounds

Campiglio[291] used a microassay method based on combustion of 3 to 5 mg substance (see Table 5.32) in an oxygen flask and absorbing the products in 4 cm^3 concentrated nitric acid. The absorption solution was then boiled for 6 min, during which complete oxidation to mercury(II) occurs. Hg(II) ions were then potentiometrically titrated with 0.005 M potassium iodide solution in the presence of an iodide-selective membrane sensor as indicator. The results were within the usual ± 0.3% limit of error.

A simple and rapid procedure for measuring small amounts of phenylmercury(II) nitrate in aqueous solution was developed by Wood and Welles.[292] The method depends on the formation of insoluble phenylmercury(II) iodide (K_{so} = 9.7 × 10^{-16}) upon titration of phenylmercury(II) nitrate with potassium iodide:

$$C_6H_5\!-\!Hg^+ + I^- \rightleftharpoons C_6H_5\!-\!HgI \qquad (5.79)$$

The end point for Reaction 5.79 is detected potentiometrically using an iodide membrane sensor. The method is able to measure down to 0.000125% aqueous solution of phenylmercury(II) nitrate with 1% accuracy. Naphazoline hydrochloride, phenylephrine hydrochloride, fluorescein sodium, and antipyrine interfered with the method, whereas the common buffer systems, polyvinyl alcohol, sodium thiosulfate, Na$_2$EDTA, and chloramphenicol, had no effect.

Table 5.32 Organic Mercury Compounds Assayed
by Membrane Sensors

Compound	Formula (MM)	Therapeutic category
Merbromine	Consists chiefly of disodium 2,7-dibromo-4-hydroxy mercury fluorescein $C_{20}H_8Br_2HgNa_2O_6$ (750.7)	Weak disinfectant

Compound	Formula (MM)	Therapeutic category
Mersalyl acid	$C_{13}H_{17}HgNO_6$ (483.9)	Diuretic

Compound	Formula (MM)	Therapeutic category
Nitromersol	$C_7H_5HgNO_3$ (351.7)	Disinfectant

Compound	Formula (MM)	Therapeutic category
Phenylmercury(II) nitrate	$C_{12}H_{11}Hg_2NO_4$ (634.4) $C_6H_5-Hg-OH \cdot C_6H_5-Hg-NO_3$	Antibacterial; antifungal; preservative for ophthalmic solutions

Analytical Procedures

i. *For all compounds listed in Table 5.32:*
   The accurately weighed sample (3 to 5 mg) is mineralized by the
   Schöniger procedure with 4 cm^3 of concentrated nitric acid in the
   combustion flask. When combustion is complete, the flask is shaken

for 1 min and the contents left to rest for 15 min. A reflux condenser is fitted and the solution is heated to boiling for 6 min with occasional stirring. The flask contents are cooled and quantitatively transferred to a 150-cm^3 beaker, by which stage the solution volume is about 80 cm^3; 8.0 cm^3 of 30% potassium hydroxide solution is added with stirring so that the pH solution is between 0.8 and 1.0. This solution is potentiometrically titrated with 5×10^{-3} M potassium iodide. The end point corresponds to the maximum slope on the titration curve.

ii. *Phenylmercury(II) nitrate assay:*
   Five cubic centimeters of an aqueous phenylmercury(II) nitrate solution are transferred to a 50-cm^3 beaker equipped with magnetic stirring bar. The amount of phenylmercury(II) nitrate in solution ranges from approximately 0.05 to 2.5 mg; 20.0 cm^3 of distilled water are added to make the workable volume, and the solution is acidified with concentrated sulfuric acid. This solution is titrated with a solution of known concentration of potassium iodide (0.002%) saturated with iodine. The end point corresponds to the maximum slope on the titration curve.

5.70   Methadone

$$C_{21}H_{27}NO \text{ (MM = 309.4)}$$

$$C_2H_5C-\overset{\overset{\displaystyle C_6H_5}{|}}{\underset{\underset{\displaystyle C_6H_5}{|}}{\underset{\|}{C}}}-CH_2-\underset{\underset{\displaystyle CH_3}{|}}{CH}-N(CH_3)_2$$
$$O$$

Therapeutic category: narcotic analgesic

Discussion and Comments

The determination of methadone with a miniaturized hydrophobic cation plastic membrane sensor[287] or a coated-wire ion-selective sensor sensitive to methadone[30] is based on the hydrophobicity and ability of this compound (a γ-keto tertiary amine) to form cationic species at low pH values.

The membrane of cationic plastic selective sensor consists of a poly(vinylchloride)–dioctylphthalate mixture. Measurements of methadone (in aqueous samples or in urine) could be made either by direct potentiometry or by potentiometric titration with sodium tetraphenylborate solution. The response of the sensor toward methadone hydrochloride in water is linear from 10^{-6} to 10^{-2} M with a slope of 59 mV decade^{-1} (pH = 2 to 3).

The coated-wire ion-selective membrane sensor sensitive to methadone is based on dinonylnaphthalene sulfonic acid.[30] The membrane sensor showed a near-Nernstian response to methadone with an excellent linearity over the range 10^{-3} to 10^{-5} M (detection limit 10^{-6} M) and relatively short response time (30 to 60 s). The very high selectivity of this sensor over methylamphetamine, cocaine, and protriptyline was observed as predicted by inspection of the drug structures: methadone has the highest lipophilicity.

Analytical Procedure

The pair of sensors (hydrophobic cation plastic membrane or methadone coated wire as indicator and SCE as reference) is introduced into the sample solution (30 to 40 cm^3 of about 10^{-3} M, pH 2 to 3) and titrated with sodium tetraphenylborate solution (10^{-2} M). The EMF is plotted as a function of titrant volume to obtain the equivalence point. This is evaluated from the maximum slope of the titration curve and used to calculate the unknown methadone concentration in the sample solution.

5.71   Methotrexate

$$C_{20}H_{22}N_8O_5 \ (MM = 454.5)$$

Therapeutic category: antineoplastic; antimetabolite

Discussion and Comments

The use of an enzyme-cycling amplification procedure in conjunction with a pCO$_2$ membrane sensor for the determination of methotrexate was described by Seegopaul and Rechnitz.[293] The method is based on inhibition of dihydrofolate reductase enzyme, which couples with 6-phosphogluconic dehydrogenase to recycle the NADP$^+$/NADPH redox system (Figure 5.16). Inhibition of the reductase by methotrexate reduces the extent of cycling, which is then directly related to the drug concentration.[293]

Figure 5.16 illustrates the coupled enzyme mediation by the NADP$^+$/NADPH recycling system. In the presence of hydrofolic acid, dihydrofolate reductase converted β-NADPH to β-NADP$^+$, which was then regenerated to β-NADPH by 6-phosphogluconic dehydrogenase. In the process, 6-phosphogluconic acid was decarboxylated to yield ribu-

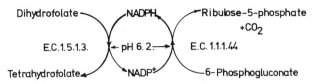

Figure 5.16 Schematic diagram of cycling system for the determination of methotrexate using dehydrofolate reductase (E.C. 1.5.1.3) and 6-phosphogluconic dehydrogenase (E.C. 1.1.1.44) with the pCO_2-gas-sensing electrode. (Reprinted with permission from Seegopaul, P. and Rechnitz, G. A., *Anal. Chem.*, 56, 852, 1984. Copyright 1984 American Chemical Society.)

lose-5-phosphate and CO_2, monitored by the CO_2-gas-sensing electrode. The overall reaction, therefore, provided a sensitive determination of the reductase enzyme through amplification and, consequently, the assay of methotrexate.

The following optimum parameters were found: dihydrofolate reductase activity 10 U dm^{-3}; β-NADPH concentration 2.5×10^{-4} M; dihydrofolic acid concentration 2.82×10^{-4} M; 6-phosphogluconic acid concentration 1.25×10^{-2} M; 6-phosphogluconic dehydrogenase activity 0.83 U cm^{-3}. These concentrations ensured rate dependence only on the folate reductase and, indirectly, on the methotrexate. A pH of 6.2 was the optimum value for the maximum enzyme activity.

All studies were conducted at 37°C (at 22°C, only one fourth of the maximum rate was obtained). Enzyme activity decreased linearly up to 15 μg dm^{-3} inhibitor (methotrexate) followed by a sharp deviation from linearity leading to a graduate leveling-off in available activity. The practical lower limit under the reported conditions is in the 1.5 μg dm^{-3} methotrexate region, e.g., well below the range of clinical interest.[293] In the concentration range 1.5 to 12 μg dm^{-3} methotrexate, the precision ranged from ± 1.9 to $\pm 5.2\%$. The recovery studies for the same concentration range showed values of 93 to 102%, with an average of 96%.

Analytical Procedure

One hundred microliters of 0.25 M 6-phosphogluconic acid, 50 μl of 0.01 M β-NADPH, 35 μl of 6-phosphogluconic dehydrogenase (1.65 units), and 2 μl of dihydrofolate reductase (0.02 units) are added to 1.5 cm^3 of 0.2 M citrate buffer (pH 6.2), containing 0.3 M KCl. Following solution mixing at 37 $\pm$ 0.1°C, various aliquots of 3000 μg dm^{-3} working methotrexate standard (1 to 30 μg dm^{-3}) are added and the total solution adjusted to 1.9 cm^3 with deionized water. After 5 min of preincubation, 100 μl of 5.64×10^{-4} M dihydrofolic acid is added to initiate the enzymatic reaction. Initial rates (in millivolts per minute) are then recorded. Blank determinations in the absence of methotrexate are carried out for correction of rate measurements. The initial rates are plotted against the corresponding methotrexate concentrations to pro-

vide a standard calibration curve. Unknown methotrexate samples are
similarly processed and their levels determined from the standard curve.

5.72   Metoclopramide

$$C_{14}H_{22}ClN_3O_2 \text{ (MM = 299.8)}$$

$$CONH(CH_2)_2N(C_2H_5)_2$$

OCH$_3$

Cl

NH$_2$

Therapeutic category: antiemetic

Discussion and Comments

A metoclopramide-selective PVC-membrane sensor based on the ion-pair
complex of metoclopramide with tetraphenylborate was prepared with
dioctylphthalate (DOP) as plasticizer (7.4% ion pair, 46.3% PVC, and
46.3% DOP).[294] When the electrode body was filled with a solution of
10^{-3} M metoclopramide hydrochloride and 0.1 M sodium chloride as
the internal solution, the Nernstian-type response of the sensor was
observed in the range $10^{-1.7}$ to $10^{-5.6}$ M (slope 50 mV decade^{-1}). The
response was rapid and reversible and the equilibrium response time was
5 to 10 s after the membrane sensor was placed in solution. There is a
negligible effect by pH within the range 2.5 to 7.3 where the sensor can
safely be used for metoclopramide assay; at pH values lower than 2.5,
the membrane sensor becomes progressively sensitive to the diprotonated species and the EMF readings decrease with decreasing pH. At
pH values higher than 7.3, the metoclopramide base precipitates and
consequently the concentration of protonated species decreases.

None of the investigated species (inorganic cations, sugars, amino
acids, and organic amines) was found to interfere in the sensor response.

The sensor proved to be useful in the potentiometric determination of
metoclopramide in pure solutions and in pharmaceutical preparations by
direct potentiometry using the standard-addition and potentiometric-
titration methods. Standard deviations of 0.81% for the standard-addition
method and 1.05% for potentiometric titrations prove that the membrane
sensor is very successful for the microdetermination of metoclopramide
in syrups and tablets (recovery 97.0 to 102.2% and relative standard
deviation 0.95 to 2.30%).

Analytical Procedures

i. *Direct potentiometry:*

The standard-additions method, in which small increments of a standard solution (10^{-1} M) of metoclopramide hydrochloride are added to 100-cm^3 samples of various concentrations, is used. The change in the millivolt readings is recorded after each addition and used to calculate the concentration of the metoclopramide sample solution.

For the analysis of metoclopramide formulations, aliquots of 2 to 5 cm^3 of syrup or 100 to 150 mg of tablet powder are quantitatively transferred into 150-cm^3 beakers, each containing 100 cm^3 of distilled water, and the standard-additions technique is applied as previously described.

ii. *Potentiometric titration:*

An aliquot of the metoclopramide containing 0.71 to 7.1 mg of metoclopramide is pipetted into a 150-cm^3 beaker. A 10-cm^3 aliquot of 0.1 M NaCl is added and the solution diluted to 100 cm^3 with distilled water. The resulting solution is titrated with 10^{-2} M standard sodium tetraphenylborate solution using the metoclopramide membrane sensor as indicator and SCE as reference.

For metoclopramide-containing preparations, 1 to 10 cm^3 aliquots of the syrup or 100 to 150 mg of the powdered tablets are transferred into 150-cm^3 beakers, each containing 100 cm^3 of water, and titrated as before. The end point corresponds to the maximum slope on the titration curve and it is used to calculate the metoclopramide concentration in the unknown sample.

5.73   Mitobronitol

$$C_6H_{12}Br_2O_4 \ (MM = 308.0)$$

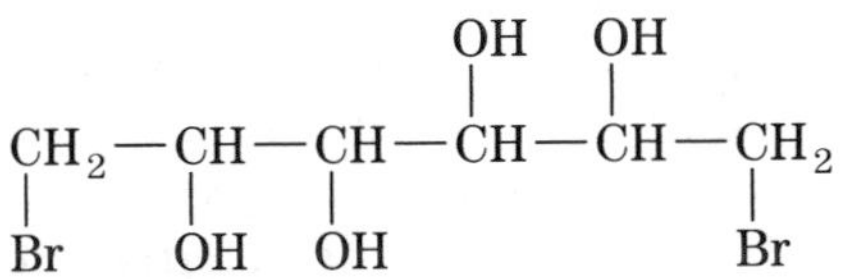

$$\underset{\underset{Br}{|}}{CH_2}-\underset{\underset{OH}{|}}{CH}-\underset{\underset{OH}{|}}{CH}-\overset{\overset{OH}{|}}{CH}-\overset{\overset{OH}{|}}{CH}-\underset{\underset{Br}{|}}{CH_2}$$

Therapeutic category: antineoplastic

Discussion and Comments

During the determination of the bromide contamination in aqueous solutions of mitobronitol (myelobromol) it was observed that the measured value by a bromide-selective membrane sensor as indicator electrode was highly affected by pH of solution in the range 9 to 11.5. Mitobronitol is not hydrolyzed up to pH 8.0, but bromide ion is completely released above pH 11.5 (Rakiás et al.[295]) (see Figure 5.17). This reaction is catalyzed in alkaline medium by Ag(I), which is dissolved from the surface of the silver bromide–based ion-selective sensor.

The hydrolysis constant (K_h) of mitobronitol was calculated from the electrode-potential–pH function by using the relationship

$$K_h = K_w/K \tag{5.80}$$

$$pH = 7.0 - 0.5pK + 0.5\log C \tag{5.81}$$

where K_w is the ionic product of water (10^{-14}), K is the dissociation constant of the acid formed during hydrolysis, and C is the bromide-ion concentration released from mitobronitol at a given pH. The K_h calculated was 1×10^{-11}.

Based on these experimental results of Pungor and co-workers[295], the ionic bromide contamination in mitobronitol can be determined by direct potentiometry with a bromide-selective sensor if the pH of the solution is between 1 and 8, whereas the bromine covalently bonded in the drug can be determined at pH > 11.5. In the latter case, the total bromide content of the solution is measured, and the mitobronitol can be determined from the results of the measurements at low and high pH. The analysis results obtained using this technique were in good agreement with those obtained by the potentiometric titration method using silver nitrate solution as titrant.

The method was successfully applied for the determination of some other compounds containing terminal halogen atoms (e.g., 2-amino-4-methyl-5-(2-iodoethyl)thiazole, 2-amino-4-bromomethyl-5-(2-bromo-ethyl)thiazole hydrobromide, and 2-amino-4-bromomethyl-5-(2-chloro-ethyl)thiazole hydrochloride).

Analytical Procedures

i. *Determination of free bromide:*
   An adequate amount (3.0802 g) of mitobronitol is dissolved in distilled water and the solution is diluted to a 1-dm^3 volumetric flask with distilled water. This solution has a concentration of 10^{-2} *M*. A 50-cm^3 aliquot of this solution is transferred into 100-cm^3 beaker containing a Teflon-coated stirring bar, and the EMF value is recorded with an ion-selective bromide membrane sensor as indicator in conjunction with a double-junction Ag/AgCl electrode. The EMF value is compared with a calibration graph prepared with potassium bromide solutions and it is used to calculate the content of free bromide ions in the test solution.

ii. *Determination of mitobronitol:*
   An aliquot of 10 cm^3 of approximately 10^{-2} *M* mitobronitol sample solution is pipetted into a 100-cm^3 volumetric flask; 10 cm^3 of 10^{-1} *M* sodium hydroxide solution is added and the volume is adjusted to

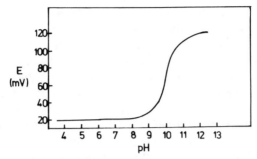

Figure 5.17 pH dependence of the potential of a bromide-selective membrane sensor (Radelkis, OP-Br-7112) in a 10^{-3} M solution of mitobronitol. (Reproduced from Rakiás, F., Tóth, K., and Pungor, E., *Anal. Chim. Acta*, 121, 93, 1980, Elsevier Science Publishers, Physical Sciences and Engineering Division. With permission.)

the mark with distilled water. A 50-cm^3 aliquot of this solution is transferred into a 100-cm^3 beaker containing a Teflon-coated stirring bar, and the EMF value is recorded as previously described. The EMF is compared with the calibration graph prepared with potassium bromide solutions (pH ~ 12.0) and is used to calculate the total content of bromide ions in the test solution. The concentration of mitobronitol in the sample is calculated by taking into account the content of free bromide ions as determined under procedure (i).

5.74 Moroxidine

$$C_6H_{13}N_5O \ (MM = 171.2)$$

$$HN = CNHCNH_2$$

Therapeutic category: antiviral

Discussion and Comments

A liquid-membrane moroxidine-selective sensor with moroxidine–silicotungstate in nitrobenzene (concentration 10^{-3} to 10^{-4} M) as active material was found to be one of the best sensors from the 17 moroxidine sensors studied by Yao and Gao.[296] Both liquid membranes and PVC membranes containing also dipicrylaminate, reineckate, and tetraphenylborate as site carriers were investigated.

The moroxidine liquid-membrane sensor based on silicotungstate displayed a linear response over the range 10^{-2} to 3×10^{-6} M (detection limit 5×10^{-7} M and slope 60 mV decade^{-1}) with a response time within 1 to 1.5 min. The PVC–moroxidine sensor containing the same electroactive material (dibutylphthalate as plasticizer) displayed a linear response over 10^{-1} to 10^{-5} M, but with a faster response (less than 20 s). Both sensors were useful in the pH range 6.0 to 9.5.

Many organic ions interfere in the moroxidine liquid-membrane sensor. Selectivity coefficients of 2.6×10^4, 9.6×10^3, 0.35, 1.0, 0.65, 18.3, 70.4, 85.6, 495, 7.3×10^3, and 1.2×10^5 were reported for berberine, promethazine, thiamin, vitamin B_6, $(CH_3)_4N^+$, atropine, procaine (cinchonine), ketamine, quinine, dibazol, and $(C_4H_9)_4N^+$, respectively. For the liquid-membrane sensors with moroxidine–tetraphenylborate as electroactive material, the electrode linearity slope can be related inversely to the dielectric constant of the membrane solvent (the following sequence was found: nitrobenzene, 61; m-nitrotoluene, 60; n-decanol, 50; chlorobenzene, 44).

Moroxidine was assayed with good results in aqueous solutions (recovery 99.9%, standard deviation 1.5%) or tablets (recovery 100.3%, standard deviation 2.7%).

Analytical Procedure

Standard solutions of 10^{-2} to 10^{-4} M concentration are prepared by serial dilution of 10^{-2} M moroxidine hydrochloride, prepared by dissolving a suitable amount of drug substance with distilled water. A constant ionic strength (e.g., 0.1 M, adjusted with sodium nitrate) must be used. The pH of all standard solutions is adjusted to about 7.0 with phosphate buffer. Aliquots of standard solutions are transferred into 100-cm³ beakers containing Teflon-coated stirring bars. The moroxidine membrane sensor together with a reference electrode (Ag/AgCl or SCE) is immersed successively in the standards, and the EMF values are measured. The moroxidine concentration in the sample is determined from the calibration graph (E vs. log[moroxidine]).

5.75 Nafronyl

$C_{24}H_{33}NO_3$ (MM = 383.5)

Therapeutic category: vasodilator

Discussion and Comments

A simple potentiometric method for the rapid determination of nafronyl drug in pharmaceutical preparations such as tablets was described.[297] Nafronyl-ion-selective membrane sensors with either nafronyl–dipicrylamine ion-pair complex in 1,2-dichloroethane or the nafronyl–dinonylnaphthalenesulfonic acid (DNNS) ion-pair complex in a PVC matrix as electroactive materials were used.

The sensor based on nafronyl–dipicrylamine was made by impregnating the support material (a graphite rod) with the electroactive material (5×10^{-3} M), whereas the sensor based on the nafronyl–DNNS ion-pair contained a membrane with 4.0% DNNS, 64.0% plasticizer (o-NPOE), and 32.0% PVC. The electrode body was filled with 10^{-3} M nafronyl oxalate solution at pH 4.5 (acetate buffer). Both sensors exhibited near-Nernstian responses to protonated nafronyl activity from 10^{-2} to about 10^{-5} M in pH ranges that depended on the nature of the electroactive material used in the membrane. In acidic medium the nafronyl–DNNS complex was stable, but the nafronyl–dipicrylamine complex was not, the dipicrylamine becoming protonated. The nafronyl liquid-membrane sensor is H^+-sensitive outside the pH range from about 5 to 7. At pH > 7 free base precipitates in the aqueous test solution, the concentration of unprotonated nafronyl species increases, and lower EMF readings are recorded for both membrane sensors.

Both sensors showed high selectivity with respect to most inorganic and organic cations tested. The selectivity coefficients $k_{\text{Nafr, B}}^{\text{pot}}$ with respect to glycine, L-histidine, methionine, nicotinamide, atropine, scopolamine, diethanolamine, and common inorganic cations, were all less than 10^{-3}.

The average recovery in analysis of six pure nafronyl oxalate samples, each in duplicate and weighing between 40 and 70 mg, was 100.6%, standard deviation 1.9%, when a nafronyl-selective membrane sensor was used as indicator in potentiometric titration with 5×10^{-2} M sodium tetraphenylborate solution. For nafronyl oxalate assay in three pharmaceutical products, a relative standard deviation of 1.0 to 1.5% was obtained.

Analytical Procedure

A 25-cm^3 aliquot of the nafronyl oxalate sample solution of about 5×10^{-3} M and adjusted to pH 4.5 with acetate buffer solution is pipetted into the reaction cell and titrated, under stirring, with 5×10^{-2} M sodium tetraphenylborate solution in presence of nafronyl membrane sensor as indicator. The end point corresponds to the maximum slope on the titration curve.

For tablets assay, at least five tablets are finely powdered and a portion of the powder equivalent to about 50 mg of nafronyl oxalate is weighed into a 50-cm³ beaker, dissolved in about 30 cm³ of pH 4.5 acetate buffer solution, and potentiometrically titrated as before.

5.76   Naphazoline

$$C_{14}H_{14}N_2 \; (MM = 210.3)$$

Therapeutic category: adrenergic (vasoconstrictor)

Discussion and Comments

A PVC-type naphazoline membrane sensor containing naphazoline–tetraphenylborate as ion exchanger and dibutylphthalate as plasticizer showed near-Nernstian response toward naphazoline within the 10^{-2} to 10^{-5} M range.[298] Its limit of detection is 2.0×10^{-6} M and the slope of the calibration curve is 57.5 mV decade^{-1}. The sensor was used in the pH range of 3 to 8. Its short response time (10 to 20 s for 10^{-2} M and 30 to 50 s for 10^{-3} to 10^{-5} M) and high selectivity over many inorganic as well as organic compounds (e.g., berberine, nicotinamide, atropine, chlorphenyramine, isoniazid, etc.) proved to be useful for naphazoline determination (recovery 98.9 to 100.3%, standard deviation 0.38%).

Analytical Procedure

Standard solutions of 10^{-2} to 10^{-4} M concentration are prepared by serial dilution of 10^{-2} M naphazoline hydrochloride prepared by dissolving a suitable amount of drug substance with distilled water. A constant ionic strength (e.g., 0.1 M, adjusted with sodium nitrate) must be used. The pH of all standard solutions is adjusted to approximately 5.0 with acetate buffer. Aliquots of standard solutions are transferred into 100-cm³ beakers containing Teflon-coated stirring bars. The naphazoline membrane sensor in conjunction with a reference electrode (Ag/AgCl or SCE) is immersed successively in the standards, and the EMF values are measured. The naphazoline concentration in the sample is determined from the calibration graph (E vs. log[naphazoline]).

5.77   Naproxen

$$C_{14}H_{14}O_3 \ (MM = 230.3)$$

Therapeutic category: anti-inflammatory; analgesic; antipyretic

Discussion and Comments

A naproxinate-ion-selective membrane sensor based on tetraheptylam-monium–naproxinate ion pair showed a linear response to naproxinate ion in the range 10^{-1} to 10^{-5} M at pH > 7.[299] The tetraheptylammo-nium–naproxinate was prepared by a phase transfer reaction (0.1 M aqueous solution of naproxinate was used) in methylene chloride for 10 h. After solvent removal, decanol was added to yield a 1 M solution of ion-pair complex. This solution was used to treat Norit charcoal to obtain a stationary support for the active ions.[299] The carbon paste, containing the electroactive material in decanol, together with a platinum wire provided the conduction path. A pre-conditioning storage of the sensor of 12 to 24 h in 10^{-2} M solution of naproxinate ion was required. The sensor was used to monitor the dissolution profiles for naproxen and gave similar results to other analytical methods. The technique involves monitoring of EMF values of the sensor vs. time, and the readings are correlated with the calibration curve.

Analytical Procedure

Standard solutions of 10^{-2} to 10^{-4} M concentration are prepared by serial dilution of 10^{-2} M sodium naproxinate obtained by dissolving a suitable amount of drug substance with distilled water. A constant ionic strength ($I = 0.1$ M, adjusted with sodium sulfate) must be used. The pH of all standard solutions is adjusted to approximately 8.0 with phosphate buffer. Aliquots (e.g., 25 cm^3) of these standards are transferred into 100-cm^3 beakers containing Teflon-coated stirring bars. The naproxinate membrane sensor in conjunction with a double-junction Ag/AgCl refer-ence electrode is immersed successively in the standards and the EMF values measured. The naproxinate concentration in the sample is deter-mined from the calibration graph (E vs. log[naproxen]).

5.78   Nicardipine and Related Coronary Vasodilators

Potentiometric coated-wire sensors[300] based on dinonylnaphthalene sulfonic acid were prepared for the drugs listed in Table 5.33 (except for nitrendipine; for this drug assay a cyanide-sensitive sensor was used, after a previous oxidation of the sample with an alcoholic iodine solution at 90 to 95°C[301]; the iodide produced was determined *in situ* by the cyanide membrane sensor, and its concentration was related to nitrendipine; the linear range was within 10^{-3} to 5×10^{-6} M concentration).

The existence of widely varying structures (e.g., the drugs listed in Table 5.33) among the calcium channel blockers has puzzled researchers who are trying to elucidate a well-defined mechanism of action. The coated-wire membrane sensors described by Cunningham and Freiser[300] are very simple and could prove useful in future investigations. The sensors were constructed as described in detail in Martin and Freiser,[302] except that PVC-insulated copper wires were substituted for coaxial cables. This facilitated the fabrication of large numbers of sensors (20 for each drug) with a negligible increase in noise from the absence of shielded wire. After manufacture, it was necessary to bathe the sensors in 10^{-3} to 10^{-4} M solutions of the protonated drug species for several days before stable responses could be obtained, during which time hydration of the polymer membrane occurred. Soaking in blank buffer for several minutes prior to calibration resulted in best reproducibility.[300] All the membrane sensors displayed nearly Nernstian response in the concentration of analyte above about 10^{-5} M. Detection limits of at least 10^{-5} M were observed for all of them (for nicardipine, 10^{-6} M) and response times ranged from a few seconds to several minutes for lower concentrations (all measurements have been performed at pH 4.0, acetate buffer).

The sensors can be applied for the respective drug assay in pharmaceuticals (no interferences from sample matrix observed) or for the elucidation of the mechanism concerning the way in which each of these drugs inhibit Ca^{2+} uptake by smooth muscle cells.[300]

Analytical Procedure

Standard solutions from 10^{-3} to 10^{-5} M of the respective drug are prepared by successive dilutions of 10^{-2} M adequate solution (all drugs as hydrochlorides). A constant ionic strength as well as pH is provided by using 0.01 M acetate buffer (pH 4.0). The EMFs of the standard solutions are measured, under stirring, using the appropriate drug coated-wire sensor in conjunction with a double-junction calomel electrode having 0.1 M ammonium nitrate in the external junction.

Table 5.33 Some Coronary Vasodilators Assayed by Membrane Sensors

Drug substance	Formula (MM)
Nicardipine	$C_{26}H_{28}N_3O_6$ (478.4)

Diltiazem	$C_{22}H_{26}N_2O_4S$ (414.5)

Lidoflazine	$C_{30}H_{35}F_2N_3O$ (491.6)

Nitrendipine	$C_{18}H_{20}N_2O_6$ (360.0)

Verapamil	$C_{17}H_{38}N_2O_4$ (454.6)

The unknown concentration of the sample is determined from the calibration graph, E vs. $\log C_{drug}$.

5.79 Penicillins

It is well known that antibiotic penicillins (see Table 5.34) are derivatives of thiazole and their structures were elucidated by hydrolysis reactions (see also Coşofreţ,[98] pp. 249–252). They hydrolyze easily, especially in acidic medium or under the influence of penicillinase. The hydrolysis results in opening of the β-lactam ring that is characteristic of these compounds:

(5.82)

Table 5.34   Some Penicillins Assayed by Membrane Sensors

Compound	R	Formula (MM)
Ampicillin (sodium)	C_6H_5—CH—NH$_2$ (with CH substituent)	$C_{16}H_{18}N_3NaO_4S$ (371.4)
Benzylpenicillin, or penicillin G (sodium or potassium)	C_6H_5—CH$_2$—	$C_{16}H_{17}N_2NaO_4S$ (356.4) $C_{16}H_{17}KN_2O_4S$ (372.5)
Nafcillin (sodium)	naphthalene—OC$_2$H$_5$	$C_{21}H_{21}N_2NaO_5S$ (436.5)
Penicillin V (potassium)	phenyl—O—CH$_2$—	$C_{16}H_{17}KN_2O_5S$ (388.5)

Some penicillin-selective membrane sensors have been constructed on the basis of Reaction 5.82, by immobilizing penicillinase on a pH glass electrode.[166, 303-313]

The sensors were prepared by immobilizing the penicillinase in a thin membrane of polyacrylamide gel molded around and in intimate contact with a H^+-sensitive glass membrane electrode. When this kind of sensor is placed into an aqueous solution of penicillin adjusted to pH 6.4, the immobilized enzyme hydrolyzes the penicillin to produce the corresponding penicilloic acid. This is a strong acid that releases protons and changes the pH at the surface of the glass electrode. The membrane sensor responds to penicillin over the range 5×10^{-2} to 10^{-4} M at an optimum pH 6.4.

A new method for the preparation of thin, uniform, self-mounted enzyme membranes, directly coating the surface of pH glass electrodes, was developed by Tor and Freeman[166] (for details, see also Section 5.27). The enzyme penicillinase is dissolved in a solution containing synthetic prepolymers (e.g., polyacrylamide–hydrazide with $\overline{MM}$ 100,000, copolymer acrylamide–methacrylamide–hydrazide). The electrode is dipped in the solution, dried and drained carefully. The backbone polymer is then cross-linked under controlled conditions to generate a thin (about 50 μm) enzyme membrane. Phosphate at a concentration of 10 mM (pH 8.0) was found to be the most adequate buffer, allowing for a linear calibration curve within the concentration range 4×10^{-5} to 1×10^{-3} M. The stability recorded for this enzyme sensor is significantly better than other sensors (mostly 1 to 3 weeks) of the same kind.

A poly(vinyl chloride) derivative that contained 1.8% carboxyl residue was used to prepare a potentiometric sensor sensitive to penicillin G.[314] A pH-sensitive-type electrode was constructed by covering a silver wire electrode with the membrane composed of the carboxyl-substituted poly(vinyl chloride) (31.0%), o-nitrophenyloctyl ether (63.0%), tridodecylamine (5.5%), and sodium tetraphenylborate (0.5%). The sensitive layer of the pH-sensitive electrode was modified with penicillinase to make a penicillin-sensitive sensor. Penicillinase was adsorbed irreversibly to the surface of the polymer membrane (PVC–COOH–TDA) by immersing the sensitive layer of the electrode in the penicillinase solution. Treatment of the pH-sensitive electrode with 0.3% penicillinase solution at pH 6.7 (1 mM phosphate buffer) resulted in favorable performance of the penicillin sensor. It was observed that when the pH-sensitive electrode, prepared by the use of parent PVC–TDA membrane, was treated with penicillinase solution, the PVC–TDA coated sensor responded to penicillin G to some extent, but the response disappeared after repeated measurements of the sample. This was presumably due to the desorption of penicillinase from the surface of the PVC–TDA membrane. These results suggested that the carboxyl groups in PVC–COOH serve as ionic sites for binding penicillinase by electrostatic attraction.[314]

The potentiometric response of the penicillin sensor was examined for 0.1 to 30 mM penicillin G in 5 mM phosphate buffer at pH 7.0. The potential reached steady-state values within 4 to 6 min and potential drift during the series of measurements was small. When not in use, the sensor was stored in a refrigerator at about 4°C and under such conditions the sensor was used for more than two months.

The possibility of enzymatically coupled ion-selective field effect transistors (ENFET) was postulated by Janata and Moss[315] in 1976. Later, Danielsson et al.[316] described a urea-sensitive device on the basis of their sensing FET and called their device an enzyme transistor. Caras and Janata[317] predicted that, as with conventional ion-selective electrodes (ISEs), ion-selective field effect transistors (ISFETs) can be made sensitive to different organic substrates by immobilizing a suitable enzyme layer over the surface of an ISFET gate. As an example, they constructed a device sensitive to penicillin, by depositing a co-cross-linked penicillinase–albumin layer over a pH-sensitive field effect transistor. The time response of the penicillin FET was between 30 and 50 s for τ_{95} and between 10 and 25 s for τ_{63}. The thickness of the albumin membrane affected the time response of the penicillin FET, but no systematic study of this effect was done. The degree of cross-linking and the concentration of albumin and enzyme loading showed no consistent effect on the response time.[317] The differential mode of measurement largely eliminates the temperature sensitivity and the effect of ambient pH variation. The linear range of the calibration curve and the sensitivity (at 0.02 M phosphate buffer, pH 7.2, and 37°C) were found to be comparable to the conventional penicillin-sensitive membrane macrosensors. The upper limit of the linear range of the response curve increases with high buffer capacity; however, the sensitivity of the sensor decreases. Long lifetime (two months), small size, and no need for a retaining membrane are other positive characteristics of this new device.

A diffusion-kinetic model for penicillin enzyme field effect transistor has been developed and solved numerically by Caras and Janata.[318] It was shown that the external buffer plays a critical role in the response of this device. It is also shown that the second weakly acidic group, $-N-H$, in the penicilloic acid (**II** in Reaction 5.82), with a pK_a value of 5.2, affects the response characteristics, namely, at the high substrate concentration.

The detection limit of the penicillin ENFET probe is well above the penicillin level found in biological fluids during the penicillin therapy. It is possible that this sensor may find its application in monitoring of the fermentation production of penicillin.[318]

Liquid- as well as PVC-membrane sensors sensitive to benzylpenicillin were also developed by Campanella et al.[319, 320] The membrane of the liquid sensor consisted of a solution in 1-decanol of the ion exchanger

Table 5.35 Characterization of PVC-Membrane Ion-Selective Benzylpenicillin Sensor in Standard Solutions of Potassium Benzylpenicillin and Comparison with the Data Obtained with the Corresponding Liquid-Membrane Sensor[320]

Parameter	PVC-membrane sensor	Liquid-membrane sensor
Response time (s)	15	15
Slope (mV decade^{-1})		
at 20°C and pH 6.0	58.2 ($\pm$0.1)	56.9 ($\pm$0.2)
Linear range (M)	5.80×10^{-4}–7.06×10^{-3}	1.96×10^{-4}–3.05×10^{-3}
Repeatability		
of measurements		
(as pooled s.d.%)		
in the linear range (%)	1.9	1.2

benzyldimethylcetylammonium–benzylpenicillin (BDMCABP), which was prepared from commercially available materials, and of benzyldimethyl-cetylammonium chloride (BDMCA). In the absence of BDMCA the life-time of the sensor is short (3 to 4 days) compared with 7 to 8 days when BDMCA is also present), but the slope was closer to the Nernstian values.

The liquid-membrane sensor was prepared by placing 0.1 cm^3 of a 0.01 M solution of BDMCABP in 1-decanol between two porous Teflon disks ($\phi = 13$ mm, thickness = 0.1 mm, pore size = 0.2 μm) and mounting these on the bottom of a sensor with Ag/AgCl as a reference, dipping in a 0.01 M solution of potassium benzylpenicillin and of potassium chloride in water. The electrical resistance of the liquid membrane was of the order of kilohms.

The PVC-membrane benzylpenicillin sensor is based on BDMCABP (5% m/m) in a matrix of PVC and bis(2-ethylhexyl)sebacate as plasti-cizer (0.01 M potassium benzylpenicillin and potassium chloride as internal reference solution). The electrical resistance of the PVC mem-brane was of the order of megohms.

The main analytical characteristics of the PVC-membrane sensor com-pared with those of liquid-membrane type are shown in Table 5.35. From the results obtained by Campanella et al.[320] it was concluded that the benzylpenicillin PVC membrane sensor has satisfactory analytical char-acteristics when evaluated in terms of its response time, precision, accuracy, linearity range, reproducibility, and selectivity. The selectivity coefficient values for several inorganic and organic anions (OH$^-$, H$_2$PO$_4^-$, HCO$_3^-$, Cl$^-$, SO$_4^{2-}$, NO$_3^-$, oxalate, citrate, acetate, and benzoate), obtained by the mixed-solutions method, were generally sufficiently low and no interference occurred.

The lifetime of this sensor was much longer (at least two months) than that of the liquid-membrane sensor. The sensor displayed narrow Nernst-type responses to other penicillins and cephalosporins (e.g., carbenicillin, 4.0×10^{-4} to 6.5×10^{-3} M; methicillin, 9.9×10^{-4} to 5.8×10^{-3} M); piperacillin, 4.0×10^{-4} to 8.7×10^{-3} M; cephalothine, 7.9×10^{-4} to 1.0×10^{-3} M; cephalexine, 3.6×10^{-4} to 1.2×10^{-3} M) with sub-Nernstian slopes.

Both benzylpenicillin-membrane sensors were successfully applied for the assay of antibiotics in synthetic matrices and in injectable and pellet preparations containing penicillin or cephalosporin antibiotics. Comparable results were obtained with direct potentiometric analysis and with the direct standard-addition method.

Potentiometric titration of penicillins with mercury(II) after alkaline or enzymatic hydrolysis has also been suggested, the iodide-ion-selective[321] or mercury(II)-ion-selective[322] membrane sensor being used. These methods are not of general applicability and are adversely affected by appreciable solubility of the metal–penicillin complexes.

Analytical Procedures

i. *General method with a penicillin enzyme sensor:*
Stock solutions (0.01 M) of various penicillins are prepared on the day of use. In order to obtain the solutions of the desired penicillin concentration, these stock solutions are diluted with appropriate buffer solution (see the preceding text). The penicillin enzyme sensor in conjunction with a reference electrode (Ag/AgCl or SCE) is placed successively in the standard solutions and, under stirring at a prescribed temperature, the EMF values are recorded. Graphs of E vs. $\log C_{\text{penicillin}}$ are plotted and the unknown concentration of the samples is determined from the respective calibration curve. (It is also possible to introduce the sensors—indicator and reference—into an appropriate buffer solution and under stirring to wait until a potential baseline is recorded. Then small increments of concentrated penicillin solutions are spiked into the cell and EMF values recorded as before.)

ii. *Benzylpenicillin membrane sensor (liquid or PVC type):*
The benzylpenicillin solutions and those of the other penicillins and their formulations are obtained by dissolving accurately weighed quantities of the potassium or sodium salt, or of the formulation, in distilled water and adjusting the pH to 6.0. If turbidity due to insoluble excipients is observed after dissolution of the formulations, filtration is carried out before the potentiometric measurement. After

calibration curve constructions, the assay procedure is the same as before (in all cases pH 6.0 is used).

5.80   Pentaerythritol Tetranitrate

$$C_5H_3N_4O_{12} \ (MM = 316.1)$$

$$O_2N-O-CH_2 \diagdown \qquad \diagup CH_2-O-NO_2$$
$$C$$
$$O_2N-O-CH_2 \diagup \qquad CH_2-O-NO_2$$

Therapeutic category: vasodilator

Discussion and Comments

A new potentiometric method based on the use of a membrane sensor for the pentaerythritol tetranitrate assay has not been reported since the work of Hassan.[323] His method is based on the fact that mercury in the presence of concentrated sulfuric acid quantitatively reduces nitrates to nitrogen(II) oxide and is itself converted to mercury(I) and/or mercury(II) ions. Finally, the mercury ions are potentiometrically titrated with 10^{-2} M potassium iodide solution in the presence of an iodide membrane sensor (e.g., Orion, Model 94-53) (for details see also Coşofreţ,[98] pp. 184–186). The overall stoichiometry of the reaction ($2\,NO_3^- \equiv 3\,Hg_2^{2+} \equiv 3\,Hg^{2+} \equiv 6\,I^-$) indicates that 1 cm^3 of 10^{-2} M potassium iodide solution corresponds to 0.413 mg nitrite group. A series of 10 replicate analyses of pentaerythritol tetranitrate in the 100 to 1000 μg range showed a mean relative standard deviation of 0.3% and an average recovery of 98.3%.

Analytical Procedure

Two to five milligrams of pentaerythritol tetranitrate is taken in a test tube (10 × 21 cm), 2 to 3 cm^3 of 96% sulfuric acid is added, and the air in the tube displaced with pure nitrogen. Three drops of mercury are added and the tube is shaken for 5 to 7 min at room temperature with continuous flow of nitrogen. The contents of the tube are transferred to a 250-cm^3 beaker, rinsed with 50 cm^3 of double-distilled water and under stirring, the solution is potentiometrically titrated with 10^{-2} M potassium iodide solution (electrode pair—iodide-selective, double-junction Ag/AgCl). The end point corresponds to the maximum slope on the E vs. V titration curve. (A blank is run in the same manner.)

5.81   Phencyclidine

$$C_{17}H_{25}N \text{ (MM} = 243.4)$$

Therapeutic category: anesthetic

Discussion and Comments

An ion-selective membrane sensor based on the high molecular weight ion-pairing agent, dinonylnaphthalene sulfonic acid (DNNS), which showed a great selectivity for various large organic cations,[324] has also been investigated for its response to phencyclidine cation.[325] The DNNS in polymer membranes (constructed as indicated in Martin and Freiser[324]) was converted to phencyclidine cation form by soaking the sensors in 10^{-3} M solution of phencyclidine hydrochloride. Nearly Nernstian response was obtained down to $10^{-4.2}$ M. This lower limit of linear response and the detection limit of $10^{-5.1}$ M are somewhat poorer than the 10^{-5} M and $10^{-5.9}$ M linear limits obtained with the DNNS-based dodecyltrimethylammonium sensor.[324]

The phencyclidine metabolites (4-phenyl-4-piperidinocyclohexanol and 1-(1-phenylcyclohexyl)-4-hydroxypiperidine) that result from monohydroxylation of one of the saturated rings in the parent compound do not interfere with the sensor response; the calculated selectivity coefficients ($k_{\text{Phen, B}}^{\text{pot}}$) were 7.7×10^{-3} and 2.9×10^{-2}, respectively. This was predicted because the added hydroxyl group greatly increases the hydrophilicity of the phencyclidine molecule and therefore decreases its ion-pair extractability.[325]

The membrane sensor was used in potentiometric titrations of phencyclidine with sodium tetraphenylborate solution: 10^{-3} and 10^{-4} M phencyclidine hydrochloride solutions were determined with a relative standard deviation of 0.6 and 0.21%, respectively. It also can be applied for the potentiometric determination of phencyclidine in biological materials (plasma, urine, brain tissue, etc.) after a previous simple preconcentration of respective samples.

Analytical Procedure

A 25-cm^3 aliquot of the phencyclidine hydrochloride sample solution of about 10^{-3} M and adjusted to approximately pH 5.0 with acetate buffer solution is pipetted into the reaction cell and titrated, under stirring with

10^{-2} M sodium tetraphenylborate solution in presence of phencyclidine membrane sensor and SCE as reference. The end point corresponds to the maximum slope on the titration curve.

5.82   Phenothiazines

Phenothiazine drugs are compounds with well-known neuroleptic activity. (Table 5.36 presents some for which membrane sensors were designed.) Chlorpromazine remains the most widely used antipsychotic drug throughout the world and continues to serve as a standard with which other neuroleptics are compared.[326] Other phenothiazine analogues possess an antidepressive rather than a neuroleptic action and are widely used for the treatment of depressive states. Of the various

Table 5.36   Phenothiazine Drugs Assayed by Membrane Sensors

Phenothiazine drug	R_1	R_2	Formula (MM)
Chlorpromazine	$-(CH_2)_3N(CH_3)_2$	$-Cl$	$C_{17}H_{19}ClN_2S$ (318.9)
Perphenazine	$-(CH_2)_3-N\underset{\smile}{\frown}N-CH_2OH$	$-Cl$	$C_{21}H_{26}ClN_3OS$ (404.0)
Promethazine	$-CH_2-CH-N(CH_3)_2$ $\quad\quad\quad\vert$ $\quad\quad\quad CH_3$	$-H$	$C_{17}H_{20}N_2S$ (284.4)
Thioridazine	$-CH_2CH_2-$ (H$_3$C-N piperidine ring)	$-H$	$C_{21}H_{26}N_2S$ (370.6)
Thioproperazine	$-(CH_2)_3-N\underset{\smile}{\frown}N-CH_3$	$-SO_2N(CH_3)_2$	$C_{22}H_{30}N_4O_2S_2$ (446.6)
Trifluoperazine	$-(CH_2)_3-N\underset{\smile}{\frown}N-CH_3$	$-CF_3$	$C_{21}H_{24}F_3N_3S$ (407.5)

phenothiazines, the piperazine derivatives (e.g., perphenazine and tri-fluoperazine) were found to be potent antiemetics.

The performance characteristics of phenothiazine membrane sensors have been described.[327-330] Those based on phenothiazine$^+$–tetraphenylborate$^-$ and phenothiazine$^+$–dinonylnaphthalene sulfonate$^-$ ion pairs, respectively, in a PVC matrix showed near-Nernstian responses over ranges depending on the nature of the phenothiazine drug. In all cases the usable range for quantitative determinations was 10^{-2} to 10^{-5} M.[327]

The selectivity of the phenothiazine-drug membrane sensors is related to the free energy of transfer of the phenothiazine drug cation between aqueous and organic phases. Responses of the membrane sensors are negligibly affected by the presence of a number of amino acids, neurotransmitters, alkaloids, and quaternary ammonium compounds containing fewer than four carbon atoms in each side chain. As expected by inspection of the phenothiazine drug structures, selectivity decreases in the order chlorpromazine, promethazine, and perphenazine. The greater selectivity of the chlorpromazine sensors over promethazine is due to the fact that chlorpromazine is 34.4 atomic mass units larger than promethazine. This correlation between atomic mass and hydrophobicity is not valid for perphenazine, which contains an —OH hydrophilic group in the aminic chain. This result supports the fact that selectivity is determined primarily by the partition coefficient of the protonated amine between organic and aqueous phases.

It was found that the linearity of the functions E (in millivolts) vs. pH depends also on the nature of the phenothiazine drug. For chlorpromazine and promethazine, the electrode responses are not affected by pH from acidic medium up to pH 6.8 and 7.6, respectively. For perphenazine, a linear graph of E (in millivolts) vs. pH was observed only in the range pH 4.5 to 6.5. At lower pH values the perphenazine membrane sensors become progressively sensitive to the diprotonated perphenazine species and the EMF readings decrease with pH decrease. At higher pH values the perphenazine, as well as the chlorpromazine and promethazine, free bases precipitate in the test aqueous solutions and consequently the concentration of unprotonated species gradually increases. As a result, lower EMF readings were recorded.

All phenothiazine-drug sensors proved useful in the potentiometric determination of the respective phenothiazine drugs, in pure forms or in pharmaceutical preparations, both by direct potentiometry (standard-addition method) and by potentiometric titrations. The better results for precision and accuracy were obtained by potentiometric titrations with 10^{-2} M NaTPB (standard deviation $\leq 0.9\%$). The standard-addition method is also recommended for its simplicity and rapidity (standard deviation $< 2.6\%$). In contrast to the approximately 2 h required for assay by the official method,[331] an electrode assay can be accomplished within 15 min.

A new liquid chlorpromazine-selective membrane sensor, containing as liquid ion exchanger the salt of chlorpromazine with eosin and tetraphenylborate in p-nitrocumene, displayed near-Nernstian response over the range 6×10^{-3} to 4×10^{-5} M in the working pH range of 1 to 6.[83] An Orion liquid-membrane electrode body (Model 92) was used as sensor assembly with a Millipore LCWPO 1300 PTFE membrane (0.01 M chlorpromazine cation + 0.1 M NaCl, as internal solution). The potentiometric selectivity coefficients for amitriptyline, promethazine, and thioproperazine, determined by the separate-solution method, were 0.2, 0.1, and 0.02, respectively.

Chlorpromazine, in some pharmaceutical preparations (e.g., injections, tablets), was assayed with good results by potentiometric titration with standard sodium picrate solution at pH 3.3.[9, 83]

New ion-pair complexes of trifluoperazine with dialkylnaphthalene sulfonates, tetraphenylborate, nitroarylsulfonates, heteropoly acids and halogeno–metal complex acids have been synthesized by Yao and Liu.[329] Their extraction coefficients in the water–membrane solvent systems were determined to be 10, and sensor slopes approach the theoretical values. Systematic studies on the sensor behavior showed that sensor performance depends greatly on the dielectric constants of both the electrode membrane solvent and the sample solution background rather than on the kind of ion pair incorporated. For homologous membrane solvents, the extraction coefficient and the liquid-membrane sensor slope increase with the increasing dielectric constant of the solvent; the slopes were found to be related linearly to the inverse dielectric constants of the membrane solvents according to the relations

$$S = 68.5 - 95.7\epsilon^{-1} \quad \text{(for } n\text{-alcohols)} \tag{5.83}$$

and

$$S = 69.5 - 62.0\epsilon^{-1} \quad \text{(for } o\text{-phthalates)} \tag{5.84}$$

A similar relationship exists between slopes of both the liquid-membrane sensors and the PVC-membrane sensors vs. dielectric constants of the mixed solvent backgrounds:

$$S = 101.5 - 3267\epsilon^{-1} \quad \text{(for methanol– and ethanol–water systems)} \tag{5.85}$$

$$S = 130.0 - 5500\epsilon^{-1} \quad \text{(for acetone–water system)} \tag{5.86}$$

Table 5.37 lists the site carriers used for preparation of electroactive materials with trifluoperazine as well as the main characteristics of the trifluoperazine membrane sensors obtained.

Table 5.37 Response Characteristics of Trifluoperazine
Membrane Sensors[329]

Site carrier	PVC-membrane sensor (DOP as plasticizer)		Liquid-membrane sensor (nitrobenzene as solvent)	
	Linearity range, pC (M)	Slope (mV decade^{-1}) (at °C)	Linearity range, pC (M)	Slope (mV decade^{-1}) (at °C)
$[HgI_4]^{2-}$	2.5–5.7	55.1 (14)	2.5–5.5	60.5 (20)
Picrate	2.5–5.7	56.4 (14)		
Picrolonate	2.5–6.0	57.4 (16)		
2,4-Dinitro-naphthol-7-sulfonate	2.5–5.7	58.9 (16)	2.5–5.5	60.1 (18)
Phosphotungstate	2.5–6.0	57.4 (15)		
Reineckate	2.5–5.7	56.7 (16)	2.5–5.4	59.9 (18)
$[BiI_4]^{2-}$	2.5–5.5	55.0 (16)	2.5–5.2	60.9 (21)
TPB	2.5–6.0	60.3 (25)	2.5–6.0	57.7 (21)
Silicotungstate	2.5–6.0	58.3 (18)	2.5–5.5	60.8 (18)
Phosphomolybdate	2.5–6.0	56.6 (22)	2.5–6.0	60.3 (22)
Dicyclohexylnaphthalene-sulfonate	2.5–5.7	60.4 (23)	2.5–6.0	61.9 (24)
Diisopentylnaphthalene-sulfonate	2.5–5.3	56.9 (25)		
Diisobutylnaphthalene-sulfonate	2.5–5.3	55.6 (25)	2.5–5.7	56.0 (24)^a

^{a}DBP as solvent.

It was found that some ion-pair complexes were sparingly soluble in
nitrobenzene, DBP, or other membrane solvents. Their respective liquid-
membrane sensors were, therefore, not prepared. The trifluoperazine
membrane sensors were negligibly affected by pH changes in the range
4.0 to 6.5, and their response times ranged from less than 30 s for
$\geq 10^{-4}$ M solutions to about 1 min for 10^{-5} to 10^{-7} M solutions (PVC
type) and from about 45 s to 1 to 1.5 min, respectively (liquid type).

Trifluoperazine aqueous solutions containing 14 to 530 ppm drug
substance were assayed by both direct potentiometric method and potentiometric titration with NaTPB standard solution (recovery 100.9%,
standard deviation 3.0% for direct method; recovery 100.8%, standard
deviation 1.7% for titration method). In both cases a PVC-membrane
trifluoperazine–dicyclohexylnaphthalene sulfonate sensor was used.

A PVC-membrane sensor of a coated-wire type was prepared that is
selective to thioridazine.[330] It showed a Nernstian response over a

thioridazine concentration range of 6.3×10^{-6} to 2.5×10^{-3} M at 25°C (pH range 2.1 to 7.0).

Analytical Procedures

i. *Direct potentiometry:*

The appropriate phenothiazine drug sensor and a reference electrode (e.g., SCE) are immersed in the respective phenothiazine-containing aqueous solutions (50 cm³) of ionic strength 0.1 M (adjusted with NaCl) and the appropriate pH (see the preceding text). After potential equilibration by stirring, the EMF value is recorded and compared with the calibration graph. As an alternative, the standard-addition method is used and for this purpose 0.5 cm³ of standard solution (10^{-2} M phenothiazine drug hydrochloride solution) is added. The change in millivolt reading is recorded and used to calculate the concentration of the respective phenothiazine drug.

For the direct potentiometric assay of tablets, typically five tablets containing phenothiazine hydrochloride as the active principle are finely powdered and transferred with 0.1 M NaCl solution into a 500-cm³ calibrated flask. A 25-cm³ aliquot of this solution is pipetted into a 100-cm³ beaker; then 25 cm³ of 0.1 M NaCl solution are added and the appropriate indicator sensor and the reference electrode are immersed in it, as previously described.

ii. *Potentiometric titration:*

A 10-cm³ aliquot of the respective phenothiazine drug hydrochloride solution (containing 1 to 10 mg) is pipetted into a 100-cm³ beaker. About 30 cm³ of 0.1 M NaCl solution is added and the resulting solution is titrated with a 10^{-2} M standard solution of NaTPB, using an appropriate phenothiazine drug membrane sensor as the indicator. The volume of titrant at the equivalence point is obtained in the usual way. For tablet analysis, 25- to 50-cm³ aliquots from stock tablet solutions are pipetted into a 100-cm³ beaker and potentiometric titrations are carried out as described.

5.83   Phenytoin

$$C_{15}H_{12}N_2O_2 \ (MM = 252.3)$$

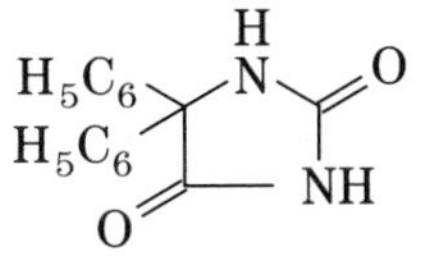

Therapeutic category: anticonvulsant; antiepileptic

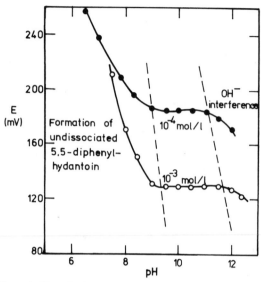

Figure 5.18   Effect of pH on PVC-membrane phenytoin-selective sensor at two different concentrations of sodium phenytoin solutions. (Reproduced from Coşofreţ, V. V. and Buck, R. P., *J. Pharm. Biomed. Anal.* 4, 45, 1986. With permission.)

Discussion and Comments

Phenytoin (5,5-diphenylhydantoin), after more than three decades of clinical application, remains one of the most effective antiepileptic drugs, with minimal sedative–hypnotic side effects.[332]

The construction and characterization of phenytoin membrane sensors of both PVC- and liquid-membrane types were recently described.[333–335] The phenytoin sensor based on its ion-pair complex with the quaternary ammonium cation tricaprylmethylammonium in a PVC matrix showed near-Nernstian response over 10^{-1} to 10^{-4} M range and a detection limit of 1.5×10^{-5} M (slope 56.3 mV decade^{-1}). The ion-pair complex was embedded in a PVC matrix containing o-nitrophenyloctyl ether as plasticizer.[334] The membrane composition was 7.7% electroactive material, 61.5% o-NPOE, and 30.8% PVC, and the electrode body was filled with 10^{-3} M sodium phenytoin solution of pH 10.0 (borax–NaOH buffer).

The plots presented in Figure 5.18 show that between pH 9.2 and 11.0 the potential is very little affected by pH changes. At high pH values, the potential decreased slowly because of hydroxide-ion interference. This interference was greater at 10^{-4} M sodium phenytoin solution than with 10^{-3} M. At lower pH, the potential increased sharply because the concentration of dissociated 5,5-diphenylhydantoinate was considerably diminished.

Among many inorganic and organic ions, only NO_3^- was found to produce slight interference with the sensor response ($k_{Ph^-, NO_3^-}^{pot} = 0.63$). The bulk of the excipients in pharmaceutical tablets or capsules, usually consisting of lactose or glucose diluent or cornstarch or gelatin binders, do not show any interference, nor do maltose, mannitol, or sugar.

The electrode proved useful for the assay of phenytoin content both in pure sodium phenytoin solutions and in pharmaceutical formulations using the potentiometric standard-addition method. The average recovery of six pure samples, each in triplicate and containing 27 to 77 μg cm^{-3}, was 100.2% and relative standard deviation was 1.8%. A high precision (relative standard deviation < 2.0%) was obtainable for determination of phenytoin in pharmaceutical formulations such as tablets and capsules. Usually the potentiometric assay could be accomplished within 15 min, in contrast to the 5 h required for assay by the official standard method.[336]

Linear response ranges within 10^{-2} to 10^{-4} M phenytoin were reported with hexadecyltrioctylammonium and hexadecyltriphenylphosphonium ion, respectively, as site carrier in the PVC membranes of coated graphite rod types (DOP as plasticizer). NO_3^-, phenobarbital, amobarbital, thiopental, and ethacrynic and salicylic acid showed more or less interference in the HTOA-based sensor response. The selectivity coefficients were 0.43, 0.32, 0.37, 31, 15, and 6.8, respectively.[335]

A phenytoin-selective liquid-membrane sensor with phenytoin cetyltrioctylammonium in n-heptanol as electroactive material was also proposed for phenytoin determination.[336] The calibration slope decreases with decreasing molecular weight of the active material studied, in the order phenytoin–cetyltrioctylammonium > phenytoin–dodecyltriheptylammonium > phenytoin–cetyltrimethylammonium > phenytoin-tetrabutylammonium. For homologues of alcohols and esters of phthalic acid, the calibration slope decreases with elongation of the carbon chain (e.g., for alcohol series n-butanol, n-pentanol, n-hexanol, n-heptanol, n-octanol, n-nonanol, and n-decanol, the following slopes were obtained: 62, 61, 58, 57, 53, 52.5, and 52 mV decade^{-1}, respectively).

The phenytoin membrane sensor proposed by Yao and Tang[335] was also applied to phenytoin assay in pharmaceuticals by the standard-addition method.

Analytical Procedures

Direct potentiometric assay of phenytoin drug substance and tablets: The phenytoin membrane sensor and a reference electrode (e.g., SCE) are immersed in the respective phenytoin aqueous solution (50 cm^3) of pH 10 (borax–NaOH buffer). After electrode equilibration by stirring, the EMF value is recorded and compared with the calibration graph. As an alternative, the standard-addition method is used and for this purpose

0.5 cm^3 of standard solution (10^{-2} M sodium phenytoin) is added. The change in millivolt reading is recorded and used to calculate the concentration of the phenytoin solution.

For the direct potentiometric assay of tablets (e.g., tablets containing 50 mg phenytoin), a portion of the powder (obtained by finely powdering five tablets from the same lot) equivalent to about 50 mg of phenytoin is transferred to a 500-cm^3 volumetric flask; 30 cm^3 of 0.1 M NaOH solution and 50.0 cm^3 of borax–NaOH buffer solution of pH 10.0 are added and the solution is made up to volume with distilled water (solution A). An aliquot of 20 cm^3 solution A is pipetted into a 100-cm^3 volumetric flask; 10.0 cm^3 of borax–NaOH buffer solution (pH 10.0) added and the solution made up with distilled water (solution B). An aliquot of 25.0 cm^3 solution B is pipetted into a 100-cm^3 beaker in which both indicator and reference electrodes are immersed. After electrode equilibration by stirring, and after recording the EMF, 1.0 cm^3 of 10^{-2} M sodium phenytoin standard solution of pH 10.0 is added; the change in EMF is recorded and used to calculate the phenytoin content of the tablets.

5.84   Procaine (Novocaine)

$$C_{13}H_{20}N_2O_2 \ (MM = 236.3)$$

$$H_2N\text{--}\langle\bigcirc\rangle\text{--}COO(CH_2)_2N(C_2H_5)_2$$

Therapeutic category: local anesthetic

Discussion and Comments

Simple, rapid, and accurate methods for the determination of procaine in pharmaceutical preparations have been described.[37, 84, 337–341] Some of them are based on the use of procaine-selective membrane sensors[337–340] and others on using mercury(II)- or silver(I)-ion-selective liquid-membrane sensors.[37, 341] Procaine liquid-membrane sensors[338] based on procaine–tetraphenylborate and procaine–dipicrylaminate ion-pair complexes, respectively, as electroactive membranes used nitrobenzene as solvent. This was found to be adequate because of the high partition coefficient, slow volatility, and high dielectric constant, which gives high membrane conductance. As support for the liquid membrane ($C = 10^{-3}$ M), a hydrophobic graphite bar made water-repellent was used. The response of both sensors was linear over the range 10^{-1} to 10^{-5} M with sub-Nernstian slopes of 47 and 48 mV decade^{-1}, respectively. The slopes of the curved parts of the calibration graphs were only about 25

Table 5.38 Main Characteristics of Procaine–PVC
Membrane Sensors[339]

Site carrier in membrane	Linear range (M)	Detection limit (M)	Slope (mV decade^{-1})
Tetraphenylborate	10^{-1}–1.4×10^{-5}	4.2×10^{-6}	55.5
Dipicrylamine	10^{-1}–1.0×10^{-5}	4.4×10^{-6}	55.4
$[HgI_4]^{2-}$	10^{-1}–8.9×10^{-5}	2.5×10^{-6}	54.9
Reineckate	10^{-1}–8.1×10^{-5}	9.6×10^{-7}	51.4

mV decade^{-1} for 10^{-5} to 10^{-6} M solutions. At pH values between 2 and
6, no significant change in the membrane potential was observed (for the
two sensors and for different concentrations of procaine hydrochloride
liquid-membrane sensors were found to be highly selective over amino
acids, benzoic and nicotinic acids, piperazine, urea, and triethanolamine,
whereas some alkaloids (codeine, caffeine, and atropine) interfere in
their response.

Mean absolute error for procaine hydrochloride determinations in
amounts varying between 4 and 12 mg in two pharmaceutical preparations (injectable solutions and tablets) did not exceed $\pm 0.5\%$, when the
potentiometric titration method using sodium tetraphenylborate as titrant
was used.

Electroactive materials, such as procaine ion-pair complexes with
tetraphenylborate,[339, 340] dipicrylamine, $[HgI_4]^{2-}$, or reineckate[339] have
been used to prepare procaine membrane sensors of PVC type. When
dinonylphthalate (DNP) was used as solvent mediator in the PVC membrane, the sensors displayed relatively large linear response, as can be
seen in Table 5.38.

PVC-membrane sensors are highly selective over theophyllin and
caffeine, but alkaloids such as cinchonine and atropine interfered in their
response.

Procaine was successfully determined in aqueous samples by both
potentiometric titration method (recovery 100.5%, standard deviation
1.8%) and standard-substraction method (recovery 99.8%, standard deviation 0.9%).

Analytical Procedures

i. *Direct potentiometry:*
   Standard solutions of 10^{-2} to 10^{-4} M procaine hydrochloride ($I =$
   0.1 M, NaNO$_3$; pH = 5.0) are prepared by serial dilution of 10^{-1} M
   stock solution of the drug. Aliquots of the standard solutions are

transferred into 100-cm^3 beakers containing Teflon-coated stirring bars. The procaine membrane sensor in conjunction with a reference electrode is placed successively in standard solutions and EMF values recorded. The graph E vs. log[procaine] is plotted and the unknown concentration of the sample is determined from this graph.

ii. *Potentiometric titration:*

The pair of electrodes is placed into the sample solution (approximately 30 to 40 cm^3 of about 10^{-3} M, pH 5.0) and this is potentiometrically titrated with 10^{-2} M sodium tetraphenylborate solution. The end point corresponds to the maximum slope on the E vs. volume titration curve.

Note: For injectable solutions as well as for tablet preparations, these procedures may be used, after adequate dilution step and powdering, respectively.

5.85   Procyclidine

$$C_{19}H_{29}NO \ (MM = 287.4)$$

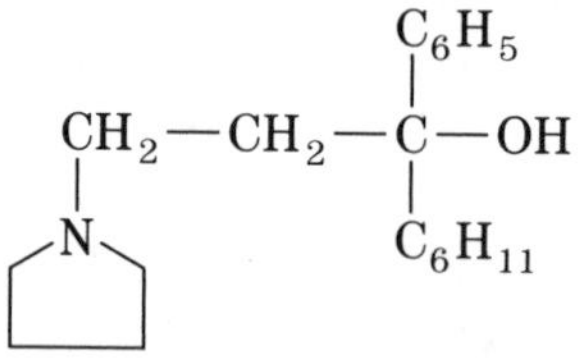

$$CH_2 - CH_2 - \underset{\underset{C_6H_{11}}{|}}{\overset{\overset{C_6H_5}{|}}{C}} - OH$$

Therapeutic category: antiparkinsonian agent

Discussion and Comments

The membrane sensors constructed by Campbell et al.[175] (see also Sections 5.33 and 5.40) and consisting of a graphite rod (150×6.5 mm) plasticized with PVC of low relative molecular mass (a mixture of bis(2-ethylhexyl)phthalate and nitrobenzene as plasticizer) may be used for the assay of procyclidine hydrochloride in tablets. The sensor that contains both plasticizers gives a larger potential break in potentiometric titrations than sensors that contain bis(2-ethylhexyl)phthalate alone.

Analytical Procedure

Twenty tablets are accurately weighed and finely powdered. An amount of powder equivalent to 0.08 g of procyclidine hydrochloride is accurately weighed and transferred into a 100-cm^3 calibrated flask. The flask is half-filled with distilled water and shaken well; 2 cm^3 of glacial acetic acid are added and the sample is heated for 5 min on a boiling water bath followed for 10 min in an ultrasonic bath. The sample is diluted to volume with distilled water. An aliquot of 25.0 cm^3 of this solution is pipetted into a titration cell and titrated with 10^{-2} M sodium tetraphenylborate standard solution in the presence of a PVC-membrane sensor as indicator and a double-junction Ag/AgCl reference electrode. The end point corresponds to the maximum slope on the plot of EMF vs. tetraphenylborate volume.

5.86   Propantheline Bromide

$$C_{23}H_{30}BrNO_3 \text{ (MM = 448.4)}$$

Therapeutic category: anticholinergic

Discussion and Comments

A liquid-membrane sensor containing propantheline–tetraphenyborate in 2-nitrotoluene (10^{-2} M) as electroactive material and a mixed solution of 0.01 M propantheline bromide $+0.1$ M sodium chloride as reference aqueous solution in an Orion liquid-membrane electrode body (Model 92) displayed a near-Nernstian response over the range 6×10^{-3} to 10^{-6} M with a slope of 58 mV decade^{-1} (pH range 1 to 9).[83] The same ion exchanger for a PVC membrane with o-nitrophenyloctyl ether as plasticizer was used, in continuous-flow systems, for potentiometric measurements.[343]

Chlordiazepoxide, which is frequently present in propantheline pharmaceuticals does not interfere with the membrane sensor at pH ≥ 6.

Propantheline is hydrolyzed in solutions at pH > 5, the decomposition products being xanthenecarboxylic acid and the quaternary 2-hydroxyethyldiisopropylmethylammonium cation. The response of the propantheline liquid-membrane sensor to the cationic product of hydrolysis has been studied by Hadjiioannou and co-workers.[83] Complete hydrolysis of 0.01 M propantheline solution was performed by heating in 0.01 M sodium hydroxide solution. After pH adjustment to 4.5, the propantheline concentration was determined with the sensor, but because no measurable propantheline was found it was concluded that the sensor only responds to non-hydrolyzed drug.

Both standard-addition and potentiometric titration with sodium picrate at pH 3.3 (acetate buffer) were used to determine propantheline in three pharmaceutical preparations (tablets, 15 or 20 mg per tablet), after a previous simple extraction procedure. (It was found that propantheline was not released quantitatively from the tablets by simple extraction with water or any buffer in the pH range 1 to 8.[83]) Propantheline bromide raw material was determined by the potentiometric method with a ClO_4^-–field effect transistor sensor, containing perchlorate–cetylpyridinium as an electroactive material, and was found to be of at least 99.4% purity (relative error 0.5%).[342]

Analytical Procedures

i. *Preparation of calibration graph:*
   A 20-cm^3 volume of distilled water is pipetted into a 50-cm^3 beaker; the propantheline membrane sensor and a Ag/AgCl reference electrode are immersed in it and, after potential equilibration, various increments of a 0.1 *M* propantheline bromide solution are added. The EMF readings are recorded after stabilization following each addition, and the plot of *E* vs. log[propantheline] is constructed. The slope of the sensor is found by regression analysis of the linear part of the graph.

ii. *Direct potentiometric (standard-addition method) assay of tablets:*
   At least five tablets are finely powdered and mixed with 25 cm^3 of dichloromethane and stirred for 30 min at room temperature. The mixture is filtered through a sintered-glass funnel and the solid residue is washed with 10 cm^3 of dichloromethane. The combined filtrate is evaporated to dryness by gentle heating under a stream of air. This residue is dissolved in 20 cm^3 of distilled water and the resulting solution is diluted with distilled water so as to obtain a final solution concentration in the range 3×10^{-3} to 3×10^{-4} *M*. An aliquot of 20 cm^3 of this solution is used for analysis. A potential reading is first recorded for this solution. Subsequently, a second potential reading is obtained after the addition of a small volume of a concentrated standard drug solution. The initial concentration of the sample is calculated from the change in potential.

iii. *Potentiometric titration:*
   An 18.0-cm^3 aliquot of the sample and 2.0 cm^3 of 1 *M* acetate buffer solution of pH 5.0 are pipetted into the titration cell. The potentiometric titration is performed under stirring with 10^{-2} *M* sodium tetraphenylborate or sodium picrate solution, at a flow rate of 0.36 cm^3 min^{-1} using propantheline liquid-membrane sensor as indicator. The end point corresponds to the maximum slope on the titration curve.

5.87 Propranolol and Related β-Blockers

Sensitive membrane sensors have been developed[300, 344-346] for the compounds listed in Table 5.39, which are well-known β-adrenergic blocker agents. All of these compounds are monovalent cations at physiological pH values, which makes them amenable to ion-selective membrane sensor potentiometry. Response characteristics were critically

Table 5.39 β-Blocker Drugs Assayed by Membrane Sensor

Compound	Formula (MM)	Ref.
Propranolol	$C_{16}H_{21}NO_2$ (259.3) 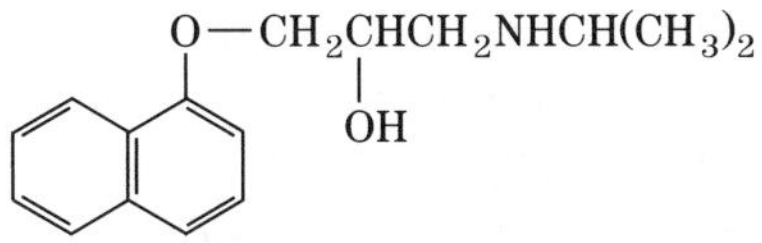	344–346

$$O-CH_2CHCH_2NHCH(CH_3)_2$$
$$|$$
$$OH$$

| Acebutolol | $C_{18}H_{28}N_2O_4$ (336.4) 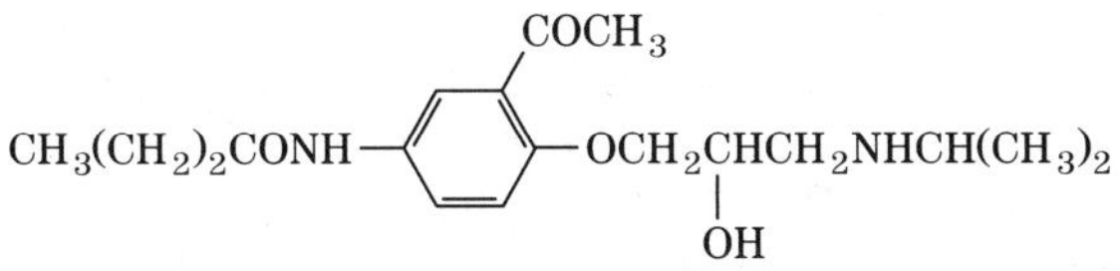 | 300 |

$$COCH_3$$
$$CH_3(CH_2)_2CONH- \;\; -OCH_2CHCH_2NHCH(CH_3)_2$$
$$|$$
$$OH$$

| Metoprolol | $C_{15}H_{25}NO_3$ (267.4) 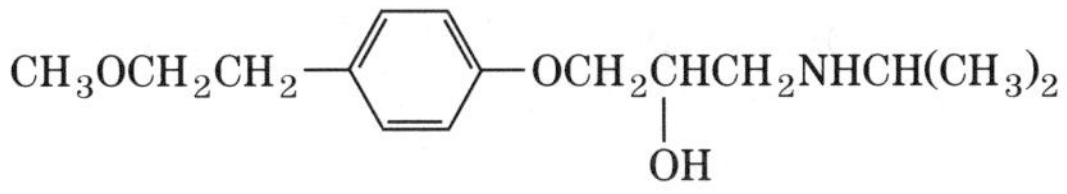 | 346 |

$$CH_3OCH_2CH_2- \;\; -OCH_2CHCH_2NHCH(CH_3)_2$$
$$|$$
$$OH$$

| Oxprenolol | $C_{15}H_{23}NO_3$ (265.3) 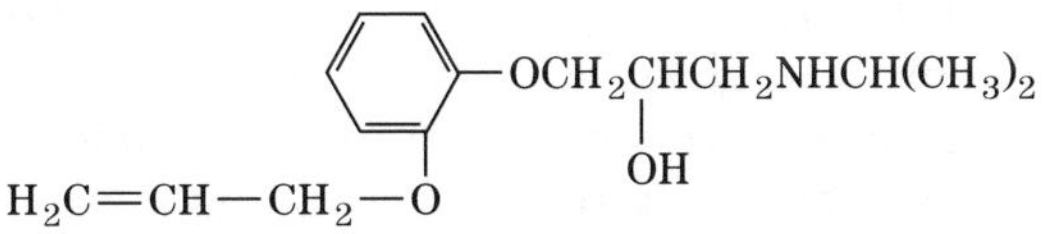 | 344 |

$$-OCH_2CHCH_2NHCH(CH_3)_2$$
$$H_2C=CH-CH_2-O \qquad OH$$

| Timolol | $C_{13}H_{24}N_4O_3S$ (316.4) 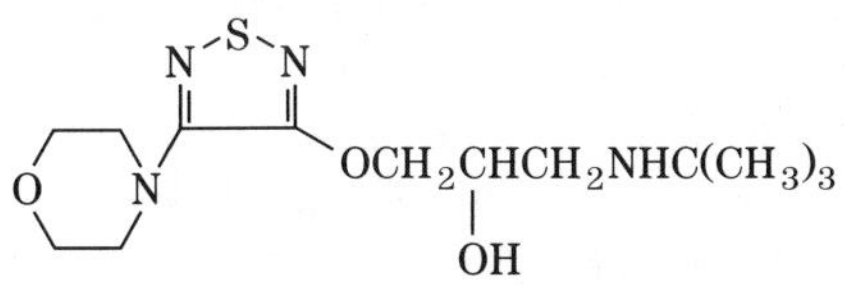 | 346 |

$$N-S-N$$
$$O \; N \;\; OCH_2CHCH_2NHC(CH_3)_3$$
$$|$$
$$OH$$

evaluated in the context of their potential application to pharmaceutical assay and *in vivo* or *in vitro* drug monitoring.

Incorporation of dinonylnaphthalene sulfonic acid (DNNS) alone with the β-adrenergic blockers into a PVC membrane resulted in an acebutolol-selective coated-wire sensor that displayed nearly Nernstian response in the concentration range 10^{-3} to 10^{-5} M[300] as well as in metoprolol-, propranolol-, and timolol-selective sensors of classical type (with internal solution), all displaying nearly Nernstian responses within 10^{-1} to 10^{-5} M concentration.[346] The acebutolol-membrane sensor was not tested in solutions more concentrated than 10^{-3} M, but certainly it responds also to higher concentrations.

For metoprolol, propranolol, and timolol, tetra-(m-chlorophenyl) borate (potassium salt, ClTPB) was also found an adequate site carrier in a PVC membrane (its concentration in the membrane was varied from 2 to 8% [m/m] and no significant differences were observed in the sensors' behavior). For all investigations and applications, membrane compositions were 5.7% electroactive site carrier (DNNS or ClTPB), 63.0% plasticizer (o-NPOE), and 31.3% PVC (m/m). These sensors, containing 10^{-2} M of respective drug hydrochloride as internal solution (in pH 5.0, acetate buffer) showed fast responses ranging from a few seconds for concentrations greater than 10^{-4} M to a few minutes for lower concentrations (10^{-5} or 10^{-6} M).

The critical response characteristics of the DNNS- as well as ClTPB-based β-blocker drug membrane sensors are summarized in Table 5.40. Calibrations were done at constant pH and ionic strength, provided by using acetate buffer, pH 5.0. The highest concentrations of metoprolol and propranolol used for calibrations were 10^{-1} M, whereas for timolol the highest concentration was 10^{-2} M (its solubility in water is lower than for metoprolol and propranolol, respectively).

The metoprolol, propranolol, and timolol sensors were not affected in their response by any inorganic and organic compounds tested (e.g., amino acids, caffeine, vitamins B_1 and B_6, nicotinamide, etc.) nor by neutral fillers used for pharmaceutical preparations. They proved useful for pharmaceutical analysis by direct potentiometry (standard-addition method). In all cases the standard deviation was $< 2.0\%$. Propranolol was also determined in tablets with good precision (standard deviation 1.1%) and an average recovery of 99.4% ($n = 6$) from the nominal value.

Previously, Selinger and Staroscik[344] prepared a working propranolol membrane sensor using tetraphenylborate as counterion and observed a relatively low slope (50 mV decade^{-1}) and short lifetime. They found that determination by direct potentiometry do not give satisfactory results. For potentiometric titrations it was found that only the potassium-ion-selective membrane electrode (membrane composition 15% TPB–K, 30% PVC, and 55% plasticizer [usually an alkylphosphate]) behaved well and was suitable for successive titrations.

Table 5.40 Response Characteristics for β-Blocker-Drug Membrane Sensors[346]

Parameter	Metoprolol sensor		Propranolol sensor		Timolol sensor	
	DNNS	CITPB	DNNS	CITPB	DNNS	CITPB
Slope (mV decade^{-1})^a	54.0 ± 0.4	55.6 ± 0.8	57.4 ± 0.5	55.4 ± 1.2	55.2 ± 0.7	56.6 ± 0.5
Linear range (M)	10^{-1}–10^{-5}	10^{-1}–10^{-5}	10^{-1}–10^{-5}	10^{-1}–10^{-5}	10^{-2}–1.5×10^{-5}	10^{-2}–10^{-5}
Detection limit						
(M)	4.5×10^{-5}	5.6×10^{-6}	4.0×10^{-6}	2.5×10^{-6}	6.3×10^{-6}	4.0×10^{-6}
(μg cm^{-3})	3.1	3.8	1.2	0.7	2.7	1.7

^{a}Standard deviation of average slope value for multiple calibration in 10^{-2} to 10^{-4} M range.

Propranolol-sensitive membrane sensors with improved characteristics of both the coated-wire and conventional types (with internal filling solution) based on didodecylnaphthalene sulfonic acid (DDNS) salt were constructed and characterized by Yamada and Freiser.[345] Both types exhibited near-Nernstian responses down to 10^{-5} M with a lower limit of detection at 10^{-6} M (in solution of propranolol hydrochloride buffered at pH 5.0 with 10^{-2} M acetate buffers). It should be mentioned that at the lower concentrations the sensors should be equilibrated for 30 min to obtain a proper response. Otherwise, both sensors demonstrated a high stability and reproducibility over a period of a month, their lifetime being longer than six months. The sensors showed negligible interference from common inorganic cations (e.g., Ca^{2+}, Mg^{2+}, K^+, and NH_4^+); this is a good indication that such a sensor can be used for determination of propranolol in biological materials. The selectivity coefficient for 1-isoproterenol, a dihydroxylated analogue of isopropranolol is small (2.6×10^{-3} and 1.5×10^{-3} for coated-wire and conventional membrane sensors, respectively) and significant interference does not occur. Some organic cations, mainly tetraalkylammonium, interfere in the electrode response.

Analytical Procedures

i. *For all drugs listed in Table 5.39:*
   Standard solutions of 10^{-2} to 10^{-4} M (I = constant, pH = 5.0; both adjusted with acetate buffer solution) are prepared by serial dilution of 10^{-2} M of the respective drug hydrochloride. Aliquots of the standard solutions are transferred into 100-cm^3 beakers containing Teflon-coated stirring bars. The appropriate drug membrane sensor in conjunction with a double-junction reference electrode is placed successively in standard solutions, and the EMF values are recorded. Graphs of E vs. log C_{drug} are plotted and the unknown concentrations of the samples are determined from these graphs.

ii. *For metoprolol, propranolol, and timolol in the range of micrograms per cubic centimeter:*
   The appropriate β-blocker-drug sensor and SCE are immersed into the aqueous sample solution (50 cm^3) at pH 5.0 (acetate buffer). After potential equilibration by stirring, the EMF value is recorded; 2.5 cm^3 of a 10^{-2} M standard solution of the β-blocker drug is added and the change in millivolt reading (accuracy ± 0.1 mV) is recorded and used to calculate the concentration of the drug.

iii. *For propranolol and oxprenolol:*
   The electrode pair (potassium-ion-selective indicator with SCE as reference) is introduced into the sample solution (30 to 40 cm^3, approximately 5×10^{-3} M) and titrated with 5×10^{-2} M sodium

tetraphenylborate solution. The EMF values are recorded as a function of titrant volume and the curve E vs. V (in cubic centimeters) is plotted. The end point corresponds to the maximum slope on this curve.

iv. *For propranolol in tablets:*
Ten tablets from the same lot are finely powdered and a portion of the powder equivalent to about 5 mg propranolol is transferred to a 50-cm^3 volumetric flask; 5.0 cm^3 acetate buffer solution of pH 5.0 is added and the solution is made up to volume with distilled water. This solution is divided into two 25-cm^3 portions, in which both the indicator and reference electrodes are immersed. After potential equilibration by stirring, and after recording the EMF, 2.5 cm^3 of 10^{-2} M propranolol hydrochloride standard solution (pH 5.0, acetate buffer) is added; the change in EMF is recorded and used to calculate the propranolol content of the tablets.

5.88   Quaternary Ammonium Compounds

Some quaternary ammonium compounds of pharmaceutical interest (topical antiseptics; disinfectants) have been investigated for quantitative analysis with membrane sensors by many scientists during the last twenty years (see Table 5.41).

For the determination of benzalkonium, cetrimide, and cetylpyridinium cations, Diamandis and Hadjiioannou[352] have used a picrate-ion-selective membrane sensor whose membrane contained tetrapentylammonium picrate in 2-nitrotoluene.[355] During their titration of cetylpyridinium with sodium picrate solution, they observed that there is a response of the picrate sensor to this cation. A calibration curve for the cetylpyridinium cation was linear in the range 10^{-3} to 10^{-4} M but with a super-Nernstian (non-equilibrium) slope of 86 mV decade^{-1}. The potentiometric titration method was used with satisfactory results (relative errors and relative standard deviations of 1 to 2%) for determination of these compounds in pharmaceutical preparations. A typical titration curve of this cation is shown in Figure 5.19, along with the first- and second-derivative curves. The most accurate results were obtained when the first-derivative curve was used.

A 3,5-dinitrosalicylate (DNS) selective sensor[353] with a liquid membrane of either tetraphenylphosphonium$^+$–DNSH$^-$ dissolved in p-nitrocumene or dimethyldioctadecylammonium$^+$–DNSH$^-$ dissolved in 1-decanol has been successfully applied in titration of cetrimonium and cetylpyridinium cations. Amounts in the range 0.5 to 2.5×10^{-5} mol were determined with an average error of about 1%. For the determination of the same compounds, a coated-wire sensor with a PVC membrane

Table 5.41 Quaternary Ammonium Compounds Assayed by Membrane Sensors

Compound	Formula (MM)	Refs.
Benzalkonium chloride (bromide)	CH_3 $\overset{\overset{\displaystyle +|}{}}{}$ Ph—CH_2—N—R X$^-$ $|$ CH_3 R = C_8H_{17}—$C_{18}H_{37}$	347, 349, 352, 356–360
Cethexonium bromide	$C_{24}H_{50}BrNO$ (448.6) cyclohexyl—$\overset{+}{N}(CH_3)_2Br^-$ with OH and C_6H_{13} substituents	349, 358
Cetrimide (hexadecyltrimethyl-ammonium bromide)	$R(CH_3)_3N^+Br^-$ R = $C_{12}H_{25}, C_{14}H_{29}, C_{16}H_{33}$ Contains not less than 96% alkyltrimethylammonium bromides calculated as $C_{17}H_{38}BrN$ (MM = 336.4)	347, 349, 352
Cetrimonium bromide (cetyltrimethylammonium bromide)	$C_{19}H_{42}BrN$ (364.5) $[CH_3(CH_2)_{15}\overset{+}{N}(CH_3)_3]Br^-$	14, 347, 349, 351, 353, 354, 358, 359
Cetylpyridinium chloride (bromide) (hexadecylpyridinium)	$C_{21}H_{38}ClN$ (340.0) $C_{21}H_{38}BrN$ (384.5) pyridinium—$\overset{+}{N}$—$(CH_2)_{15}CH_3 X^-$	14, 347, 349, 350, 352–354, 358

plasticizer with o-nitrophenyl octyl ether proved useful.[354] Overall potential jumps of 540 to 590 mV and 450 to 460 mV were recorded for cetylpyridinium bromide and cetrimonium bromide, respectively.

Tetraphenylborate solution was also used as titrant for these compounds when a liquid-membrane tetraphenylborate-selective sensor was employed as indicator.[14] Average errors and relative standard deviations of 1% were reported when 5 to 20 μmol of sample was assayed.

Selig[350] used the following sensing electrodes for the titration of cetylpyridinium chloride vs. tetraphenylborate solution: PVC/DOP-coated graphite, PVC-coated graphite (no plasticizer), and graphite only.

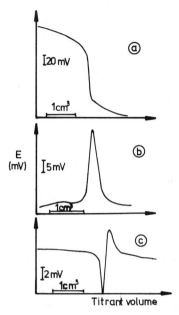

Figure 5.19 (a) Recorded curve for the semiautomatic potentiometric titration of 0.05 mmol cetylpyridinium cation with 2.5×10^{-2} M sodium picrate; (b) first-derivative curve; (c) second-derivative curve. (Reproduced from Diamandis, E. P. and Hadjiioannou, T. P., *Mikrochim. Acta*, 2, 27, 1980. With permission.)

Although end points were obtained with all variants, the quality of the potential breaks and the precision was best with the first sensor; standard deviations were 0.14, 0.32, and 0.58%, respectively.

A *o*-nitrophenyloctyl ether plasticized PVC membrane sensor, showing a Nernst-type response for tetraphenylborate anion within the range 10^{-2} to 10^{-5} M (slope 50 mV decade^{-1}), proved suitable as the end-point detector in the potentiometric titration of various quaternary ammonium salts.[359]

Benzalkonium cation was determined by direct potentiometry in benzalkonium bromide raw material with an ISFET–perchlorate sensor, containing perchlorate–cetylpyridinium as an electroactive material.[360] The purity of the drug substance was found to be 97.6% ($n = 8$, standard deviation 0.57%).

A PVC-membrane selective sensor for benzalkonium bromide, based on the use of benzalkonium–tetraphenylborate ion-pair complex as the electroactive material, showed near-Nernstian response over the concentration range from 10^{-3} to 5×10^{-6} M (detection limit 3×10^{-6} M) with a slope varying between 53.0 and 56.7 mV decade^{-1}, as a function of membrane composition (maximum slope for the ratio [15 to 18] : 2 : 21, PVC : electroactive material : dibutylphthalate, m/m). The membrane sensor showed rapid response (less than 30 s in 10^{-2} to 10^{-5} M range and less than 1 min in 10^{-6} M) and a very good reproducibility in the pH range of 2 to 12.[360] The selectivity coefficients, determined by the separate-solution method, showed no interference from many organic compounds, such as salicylic acid, vitamin B_1, caffeine, novocaine, cinchonine, filcilin, tetramethylammonium, quinine, etc.

Using direct potentiometry, the recovery of pure benzalkonium bromide was 101.8% (standard deviation 4.4%).

A cetrimonium-selective sensor based on a tetrazole ion pair and silicone rubber membranes was found to be useful in analysis of some pharmaceutical formulations.[351] N-(3-Acetyl-5-ethyl-2-hydroxyphenyl)-5-carboxamido-1H-tetrazole and cetyltrimethylammonium bromide were used to prepare the ion pair, and 1,2-dichloroethane was selected as a sensor solvent. The membrane sensor exhibited a linear response to cetrimonium cation over a 10^{-6} to 10^{-4} M concentration range, the upper value being limited by micellization of surfactant. It was observed that the response of the membrane sensor decreased gradually with time, but measurements at any particular time showed a high degree of stability and reproducibility.[351] The sensor responded to other quaternary ammonium ions, the selectivity coefficients decreasing with decreasing chain length. For inorganic ions exhibiting no lipoidal behavior of a change in pH, insignificant interference was observed.

Except for the paper of Shoukry et al.,[361] the literature lacks investigations on cetylpyridinium (hexadecylpyridinium, HDP) selective membrane sensors; all previously reported sensors based on ion pairs of HDP with some counteranions[88, 362-365] were not successful for its determination. They were sensitive only to the counteranions or to some simple anions and cations. This reflects the weak exchange of the HDP cation via the test-solution–membrane interface responsible for the electrode potential.[361] Phosphotungstic acid (PTA) forms an ion association with HDP cation having a mole ratio of $1:3$ ($PT^{3-}:3\,HDP^+$). A PVC-membrane selective sensor for HDP was constructed, based on incorporation of the $PT(HDP)_3$ ion association in the plastic film. The sensor with a membrane composed of 20% $PT(HDP)_3$ (prepared by precipitation of PTA with an excess solution of HDPBr), 40% dioctylphthalate, and 40% PVC showed the best Nernstian behavior (59 mV decade^{-1}, after 2 h of soaking in 10^{-3} M HDP) over a relatively wide range of HDP concentrations (6.3×10^{-6} to 3.1×10^{-3} M) and pH (2.0 to 8.5).

The membrane sensor is highly selective for HDP^+. The inorganic cations did not interfere due to the differences in their ionic size, mobility, and permeability as compared to those of HDP^+. In the case of amino acids, sugars, and amines, the high selectivity is mainly attributed to the differences in polarity and lipophilic nature of their molecules relative to those of HDP^+. Cetrimonium (cetyltrimethylammonium cation) interferes ($k^{pot}_{HDP^+, Cet^+} = 0.35$) because its molecule is nearly as surface-active as the HDP cation.[361] The sensor proved to be useful for the determination of HDP by the standard-addition method and by potentiometric titration in pure solutions and in the disinfection and antiseptic preparations (standard deviations varied from 1.1 to 2.8% in standard-addition method and from 0.18 to 0.82% in potentiometric titration method).

Analytical Procedures

i. *Potentiometric titration with 2.5×10^{-2} M sodium picrate:*
Aliquots (30.0 cm^3) of benzalkonium, cetrimide, and cetylpyridinium cation solutions, respectively, in the range 6.7×10^{-4} to 6.7×10^{-3} M, are pipetted into 50-cm^3 beakers. A volume of 10.0 cm^3 of 1.0×10^{-3} M picric acid solution and 5.0 cm^3 of 1.0×10^{-2} M thiourea solution are added to the sample; after potential equilibration under stirring, increments of titrant are added and EMF values are recorded. The calibration curve is plotted and the end point is evaluated from the first-derivative curve (see also Figure 5.19). (A picrate-ion-selective membrane sensor is used as indicator.)

ii. *Potentiometric titration with 1.0×10^{-2} M monosodium 3,5-dinitrosalicylate:*
Aliquots (25 cm^3) of cetrimonium and cetylpyridinium cation solutions, respectively, in the range 2.0×10^{-4} to 1.2×10^{-3} M, are pipetted into 50-cm^3 beakers. After potential equilibration, under stirring, increments of titrant are added and EMF values recorded vs. volume. The end point corresponds to the maximum slope on the titration curve. (A 3,5-dinitrosalicylate liquid-membrane sensor is used as indicator.)

iii. *Potentiometric titration with 1.0×10^{-2} M sodium tetraphenylborate:*
Aliquots (25 cm^3) of cetrimonium and cetylpyridinium cation solutions, respectively, in the range 2.0×10^{-4} to 1.0×10^{-3} M, are pipetted into 50-cm^3 beakers. After potential equilibration, under stirring, increments of titrant are added and EMF values are recorded vs. volume. The end point corresponds to the maximum slope on the titration curve. (A tetraphenylborate liquid-membrane sensor is used as indicator; a PVC/DOP-coated graphite sensor as well as a PVC membrane plasticized with 2-NPOE can also be used.)

iv. *Benzalkonium determination by direct potentiometry with a benzalkonium membrane sensor:*
Standard solutions of 10^{-3} to 10^{-5} M benzalkonium cation are prepared by keeping the ionic strength at a constant value (e.g., $I = 0.1$ M, KNO$_3$). Aliquots of the standard solutions are transferred into 100-cm^3 beakers containing Teflon-coated stirring bars. The

benzalkonium membrane sensor in conjunction with a reference electrode is placed successively in standard solutions and EMF values recorded. The graph of E vs. log[benzalkonium] is plotted and the unknown concentration of the sample is determined from this graph.

v. *Cetrimonium determination by direct potentiometry with a cetrimonium membrane sensor:*
The same procedure as that described under (iv) is followed, except that standard solutions of cetrimonium cation are prepared for the range 10^{-4} to 10^{-6} M.

vi. *Cetylpyridinium determination by standard-addition method with a cetylpyridinium (HDP) membrane sensor:*
Small increments of 10^{-2} M cetylpyridinium bromide solution are added to 50-cm^3 samples of various concentrations. The change in EMF is recorded after each addition and used to calculate the concentration of the HDP sample solution. For the analysis of pharmaceutical preparations (solutions or powder), adequate portions are quantitatively transferred to 100-cm^3 beakers containing 50 cm^3 of distilled water. The solution is stirred vigorously, and the standard-addition technique is applied as previously described.

vii. *Cetylpyridinium determination by potentiometric titration with 5 × 10^{-5} M phosphotungstic acid:*
An aliquot of cetylpyridinium cation solution containing 0.2 to 2 mg is transferred into a 150-cm^3 beaker, and the solution is diluted to 100 cm^3 with distilled water. After potential equilibration under stirring, increments of titrant are added and EMF values are recorded vs. volume. The end point corresponds to the maximum slope on the titration curve. (For calculation of unknown concentration, a 1-to-3 mole ratio, PT^{3-} : HDP^+ has to be taken into account.)

5.89   Radiopaque Substances

If a combustion method such as the Schöniger combustion[366, 367] is used to mineralize the organic matter, such as X-ray contrast products (see Table 5.42), the iodine content can be determined by measuring the iodate or iodide content. After combustion, the iodine is present as iodine and iodate. These can be reduced to iodide ions, which can be directly determined with an iodide-selective membrane sensor (e.g., Orion Model 94-53).[368] Several reductors (Devarda's alloy at room temperature, Raney nickel at 55°C, and aluminum wire at 60°C) were compared and the best results were obtained with Devarda's alloy.

Table 5.42 Radiopaque Substances Assayed by an Iodide-Membrane Sensor

Compound	Formula	MM
Iobenzamic acid	$C_{16}H_{13}I_3N_2O_3$	662.0

$$I-\!\!\!\bigcirc\!\!\!-CO-N-(CH_2)_2COOH$$
$$H_2N \quad I \quad | \quad \bigcirc$$

| Iodipamide | $C_{20}H_{14}I_6N_2O_6$ | 1139.8 |

COOH ... $NHCO(CH_2)_4CONH$ ... COOH

| Ioglycamic acid | $C_{18}H_{10}I_6N_2O_7$ | 620.1 |

COOH ... $NHCOCH_2OCH_2CONH$ ... COOH

| Iothalamic acid | $C_{11}H_9I_3N_2O_4$ | 613.9 |

COOH ... CH_3CONH ... $CONHCH_3$

| Ipodate | $C_{12}H_{13}I_3N_2O_2$ | 598.0 |

CH_2CH_2COOH ... $N=CH-N(CH_3)_2$

| Sodium salt: $C_{12}H_{12}I_3N_2NaO_2$ | | 619.9 |
| Calcium salt: $C_{24}H_{24}CaI_6N_4O_4$ | | 1234.1 |

The method gave satisfactory results for both the determination of pure drug substances and many pharmaceutical preparations such as tablets, suspensions or solutions, and ampules.

Analytical Procedure

An amount of sample containing about 5 mg of iodine is accurately weighed on ashless paper. A 20-cm^3 volume of 5 M sodium hydroxide solution is used as the absorption liquid. The combustion is carried out in a Schöniger flask filled with oxygen. After combustion, the content of the flask is quantitatively transferred to a conical flask. After 1 g of Devarda's alloy is added, the mixture is shaken for 30 min at room temperature, filtered into a 100-cm^3 volumetric flask, and made up to volume with water. This solution is diluted $1:1$ with 2 M potassium nitrate solution. The iodide content is determined by direct potentiometry, using a calibration curve obtained by standards treated in the same manner, with an iodide-selective membrane sensor as indicator.

5.90   Saccharin (Sodium Salt)

$$C_7H_4NNaO_3S \ (MM = 205.2)$$

Therapeutic category: pharmaceutic aid (flavor); non-nutritive sweetener

Discussion and Comments

An ion-selective sensor sensitive to saccharin was made by using as electroactive membrane the ion-association complex that exists between iron(II)bathophenanthroline chelate and saccharin dissolved in nitrobenzene.[389] The sample and the internal reference solution were separated by the liquid membrane that was placed in the bottom of a U-shaped glass tube. An aqueous solution of 10^{-2} M sodium saccharin was used as the reference solution. The membrane sensor exhibited a linear response over the concentration range 10^{-1} to 10^{-4} M. The measured potentials were independent of the concentration of the ion-association complex in the membrane phase, when this was varied from 5×10^{-5} to 10^{-3} M. The electrode potential was not affected appreciably by pH variation between 3 and 10. The presence of glucose, saccharose, or sorbitol in the sample solution containing saccharin did not interfere with the sensor response. The selectivity coefficient for sodium cyclamate was 2×10^{-2}, whereas selectivity coefficients for citric acid, sodium chloride, lactose, fructose, and sodium phosphate varied between 10^{-3} and 10^{-4}. The presence of salicylic acid or a large amount of benzoic acid interfered with the electrode response. The interference of salicylic acid can be eliminated by addition of aluminum sulfate to the sample solution. In this way, the selectivity coefficient for salicylic acid was decreased from 0.6 to 5×10^{-2}.

In a few Chinese papers the construction and application of saccharin-selective membrane sensors were also discussed.[370-372]

A general anion-selective sensor of PVC membrane based on ion-association complexes as electroactive materials was also proposed for saccharin determination.[371] Ethyl Violet, triheptyldodecylammonium iodide, and tetradodecylammonium iodide as well as tetradodecylammonium chloride have been studied as electroactive materials. Membranes containing either tetradodecylammonium and triheptyldodecylammonium in a PVC matrix (2.0% electroactive material, 73.5% diisooctylphthalate as plasticizer, and 24.5% PVC) displayed Nernstian responses to saccharin over the range of about 10^{-2} to 10^{-6} M.[372] Many inorganic and organic ions as well as carbohydrates, usually present together with saccharin in various drinks (e.g., carbonate, phosphate, tartrate, citrate, sugar, and glucose) do not interfere. For salicylate a selectivity coefficient of 0.56 was reported.

Analytical Procedure

Standard solutions of 10^{-2} to 10^{-4} M sodium saccharin are prepared by serial dilution from a 10^{-1} M stock solution of sodium saccharin. In all standards the ionic strength is maintained at a constant value (e.g., 0.1 M) with sodium chloride. The pH of all standards is also kept constant (pH = 7.0). Aliquots of the standard solutions are transferred into 100-cm^3 beakers containing Teflon-coated stirring bars. A saccharin membrane sensor in conjunction with a reference electrode is placed successively in the standard solutions, and the EMF values are recorded. The graph of E vs. log[saccharin] is plotted and the unknown concentration of the sample is determined from this graph.

5.91   Salicylic Acid (Sodium Salt)

$$C_7H_5NaO_3 \ (MM = 160.1)$$

COONa

—OH

Therapeutic category: topical keratolytic

Discussion and Comments

Some salicylate liquid-membrane sensors containing salicylate salt of methyltricaprylylammonium (Aliquat 336),[104, 105] tetraheptylammonium,[373] and methyltrioctylammonium[106] dissolved in 1-decanol or cetylpyridinium bromide in nitrobenzene[88] have been discussed previously (Coşofreţ,[98] pp. 161–162). All of them show a short linear response range, usually within 10^{-1} to 10^{-3} M.

The construction and analytical applications of an improved liquid-membrane sensor for salicylate that contains tetraoctylammonium salicylate dissolved in p-nitrocumene (10^{-2} M) as the ion exchanger was described by Hadjiioannou and co-workers.[374] Other compounds tested as electroactive materials were tetraheptylammonium, tetradecylammonium, tetradodecylammonium, or tetraoctadecylammonium salicylate dissolved in 4-nitro-m-xylene, 2-nitro-m-xylene, p-nitrocumene, 3-nitrotoluene, or 2-nitrotoluene. The membrane sensor was constructed by using an Orion liquid-membrane electrode body (Model 92) with Orion perchlorate membranes. The internal solution was 10^{-2} M sodium salicylate per 0.1 M sodium chloride.

The calibration data for 21 sensors with salicylates of various symmetric tetraalkylammonium cations dissolved in different organic solvents showed that the limit of linear response was markedly affected when tetraheptylammonium was replaced by tetraoctylammonium in the liquid ion exchanger, but remained practically unchanged when tetraoctylammonium was replaced by bulkier quaternary compounds up to tetraoctyldecylammonium.[374]

The sensor containing tetraoctylammonium salicylate as ion exchanger presents a Nernstian response down to 2×10^{-5} M in the pH range 6 to 9, major interferences being perchlorate and periodate ($k_{Sal, B}^{pot} = 20$ and 28, respectively; mixed-solution method). This membrane sensor was applied in the direct potentiometric determination of salicylic acid in some pharmaceuticals after a previous extraction with water and pH adjustment to 7.0 (phosphate buffer).

A salicylate liquid-membrane sensor with nitron–salicylate ion-pair complex in nitrobenzene[375] provides at pH 4.5 to 9.5 a rapid near-Nernstian response to salicylate in the concentration range 10^{-1} to 10^{-4} M. The selectivity coefficients obtained for 40 different carboxylate, phenolate, and inorganic anions showed that there was negligible interference by most of these ions. The results obtained for the direct potentiometric determination of salicylate in the range 10 μg cm^{-3} to 10 mg cm^{-3} showed an average recovery of 98.9% and a mean standard deviation of 1.8%. Procedures for the determination of salicylate in some keratolytic solutions, ointments, powders and tablet dosage forms containing salicylic acid, methylsalicylate, and acetylsalicylic acid were also described. The average recovery was 98.3% of the nominal values (standard deviation 2.1%), which compared favorably with results given by BP method.[375]

Salicylate-membrane sensors of PVC type were also constructed and characterized[86, 376, 377]; that containing Aliquat 336S–salicylate ion-pair as electroactive material dispersed in PVC (di-n-butylphthalate as plasticizer)[86] showed a linear response in the salicylate concentration range 4×10^{-5} to 1×10^{-1} M (slope 56 mV decade^{-1}) and was used for aspirin assay (see also Section 5.9).

A sensor prepared by incorporating 5,10,15,20-tetraphenyl (porphyrinato) tin(IV)dichloride ($Sn[TPP]Cl_2$) into a plasticized PVC membrane exhibits an anti-Hofmeister selectivity pattern, with high specificity for salicylate over lipophillic inorganic anions (perchlorate, periodate, thiocyanate, iodide, etc.) and biological organic anions (citrate, lactate, and acetate). Moderate selectivity over structural analogues of salicylate (3- and 4-hydroxybenzoate and benzoate) was also observed (log $k_{Sal, B}^{pot}$ = -1.1, -1.0, and -1.4, respectively). Radiotracer uptake experiments using [^{14}C]salicylate clearly showed that the metal center of the metalloporphyrin is critical for selective salicylate transport in the membrane phase.[377] Minimal response to chloride ions (log $k_{Sal, Cl}^{pot}$ = -3.8) made the new sensor potentially useful for estimating salicylate levels in biological samples (serum and urine).

The polymeric membrane composition was typically 1% porphyrin carrier ($Sn[TPP]Cl_2$), 66% dibutylsebacate, and 33% PVC (m/m).

Salicylate was also determined in aqueous solution and pooled serum samples with a carbon dioxide gas sensor (Orion, Model 95-02).[378] The method utilizes the enzyme-catalyzed reaction in which salicylate is stoichiometrically converted to catechol and carbon dioxide:

$$\text{salicylate} + NADH + H^+ + O_2 \xrightarrow[\text{hydroxylase}]{\text{salicylate}} \text{catechol} + NAD^+ + H_2O + CO_2 \quad (5.87)$$

The carbon dioxide produced is sensed with the potentiometric pCO_2 membrane sensor.

The enzyme was immobilized by physically entrapping 1.0 mg (4 units) at the tip of the carbon dioxide sensor with a Technicon Type C dialysis membrane. The slope of the calibration graph (E vs. [salicylic acid]) was found to be 38 mV decade^{-1} for data between 7.5×10^{-5} and 7.3×10^{-4} M (pH 6.0, phosphate buffer). The response time was 8 min for the lowest salicylate concentration (1.08×10^{-5} M) and 2 min for the highest salicylate concentration (8.23×10^{-4} M).

Analytical Procedures

i. *Direct potentiometric method for salicylate determination:*
   Standard solutions of 10^{-2} to 10^{-4} M sodium salicylate are prepared by serial dilution from a 10^{-1} M stock solution of sodium salicylate. In all standards the ionic strength is maintained at a constant value (e.g, 0.1 M, acetate buffer solution of pH 5.5). Aliquots of the standard solutions are transferred into 100-cm^3 beakers containing Teflon-coated stirring bars. An appropriate salicylate-membrane sensor in conjunction with a reference electrode is immersed successively in standard solutions and the EMF values are recorded. The graph of E vs. log[salicylate] is plotted and the unknown concentration of the sample is determined from this graph.

ii. *Determination of salicylate in pharmaceutical preparations:*
   a. *Preparations containing salicylic acid*—An aliquot of solution (e.g., Pyralvex, Norgan, France) (1.0 to 5.0 cm^3) is transferred into a 100-cm^3 beaker and evaporated continuously on a hot plate to near dryness. A 5.0-cm^3 aliquot of a 0.5 M NaOH solution is added to dissolve the residue followed by 45.0 cm^3 of 0.1 M KH$_2$PO$_4$ buffer solution (pH 6.0). The solution is stirred, the electrode pair immersed in the solution, the potential reading recorded and the salicylate content determined from the calibration graph (prepared as before but using phosphate buffer of pH 6.0).

   b. *Preparations containing acetylsalicylic acid*—Five tablets containing acetylsalicylic acid as active principle (e.g., aspirin) are finely powdered and transferred quantitatively into a 250-cm^3 round-bottomed flask followed by 100 cm^3 of 0.5 M NaOH solution. The flask is attached to a condenser and the mixture is heated under reflux for 20 min on a hot plate. The contents of the flask are cooled to ambient temperature and a 5.0-cm^3 aliquot of the hydrolysate is transferred into a 100-cm^3 beaker containing 45.0 cm^3 of 0.1 M KH$_2$PO$_4$ buffer solution (pH 6.0). The salicylate content is determined as previously described.

   c. *Preparations containing methyl salicylate*—A 1.5-g sample of lotion or ointment containing methyl salicylate (e.g., Deep Heat lotion, Mentholatum, U.K., M + W cream, Cupal, U.K.) is transferred quantitatively into a 250-cm^3 round-bottomed flask and 100 cm^3 of a 0.5 M NaOH solution is added. Hydrolysis and measurement are carried out as previously described.

5.92   Steroids

Synthetic glucocorticosteroids (see Table 5.43) were assayed on the basis on their fluoride content, released from $C-F$ bond of such compounds, after a previous Schöniger combustion step.[379] In the method of Mertens et al.,[380] the combustion of the sample was also performed in a nickel Parr microbomb with sodium peroxide and sucrose. In the Schöniger combustion method a polyethylene flask was used. An addition of inorganic oxidizing agents, such as potassium nitrate, potassium chlorate, or sodium peroxide, failed and resulted in low and unreproducible fluorine amounts. Satisfactory results were obtained by using a piece of thin polyethylene foil to wrap the sample before its combustion. Just water was used for the absorption of combustion products. The potentiometric methods with fluoride-selective membrane sensor (Radiometer, F 1052 F) (direct potentiometry at pH 5.2, phthalate buffer and $I = 1.5$ M, KNO_3 and containing 0.01 M trans-1,2-diaminocyclohexane-N,N,N',N'-tetraacetic acid, or potentiometric titration with 0.01 M lanthanum nitrate ethanolic–aqueous solution) were compared with a conventional spectrometric method. Among these methods, the most precise is potentiometric titration, whereas the direct potentiometric method based on a calibration curve procedure provides precision and reproducibility similar to the spectrometric method, but the latter is much more time-consuming.[379]

Analytical Procedures

i. *Mineralization:*
   The accurately weighed samples in amounts of 5 to 17 mg are placed on a filter paper, Whatman No. 1, covered with a 10×15-mm polyethylene foil and wrapped up. They are combustioned according to Schöniger procedure using a 1-dm^3 polyethylene flask containing 15 cm^3 of distilled water. The combustion products are absorbed for 30 to 40 min.

ii. *Potentiometric titration:*
   The solution after mineralization of the sample is quantitatively transferred into a polyethylene beaker, diluted with ethanol to 80% ethanol content, and titrated with 10^{-2} M ethanolic–aqueous solution of lanthanum nitrate (10 cm^3 aqueous 0.1 M solution of lanthanum nitrate made up with 95% ethanol to 100 cm^3) in the presence of a fluoride-selective sensor as indicator and SCE as reference. The graph of E vs. volume is plotted and the equivalence point corresponds to the maximum slope on this curve.

Table 5.43   Some Steroids Assayed by Fluoride Membrane Sensor

Compound	Formula (MM)	$\%F_{theor}$
Dexamethasone acetate	$C_{24}H_{28}FO_7$ (449.5) 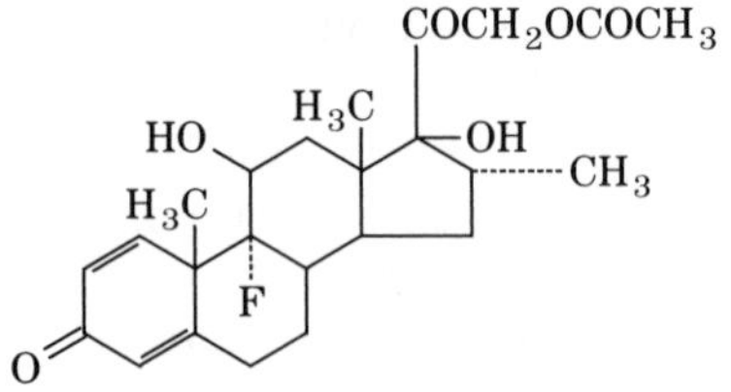	4.23

$COCH_2OCOCH_3$
H_3C
HO OH
H_3C CH_3
F
O

| Fludrocortisone acetate | $C_{23}H_{31}FO_6$ (422.5) 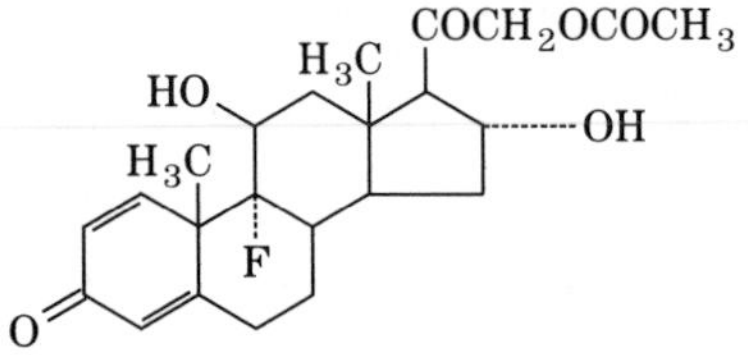 | 4.50 |

$COCH_2OCOCH_3$
H_3C
HO OH
H_3C
F
O

| Flumethasone pivalate | $C_{27}H_{36}F_2O_6$ (494.6) 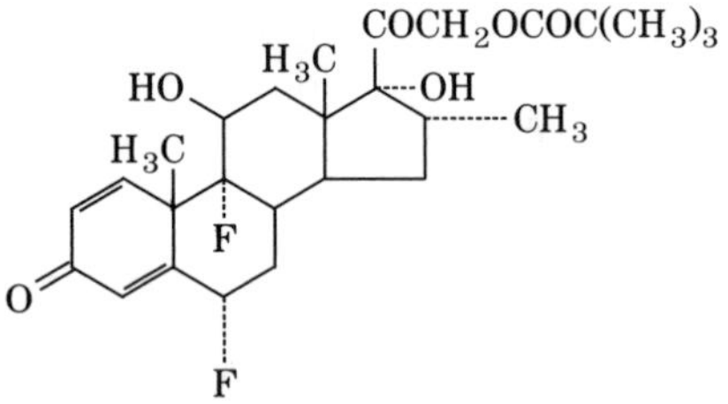 | 7.68 |

$COCH_2OCOC(CH_3)_3$
H_3C OH
HO CH_3
H_3C
F
O
F

| Fluocinolone acetonide | $C_{24}H_{30}F_2O_6$ (452.5) 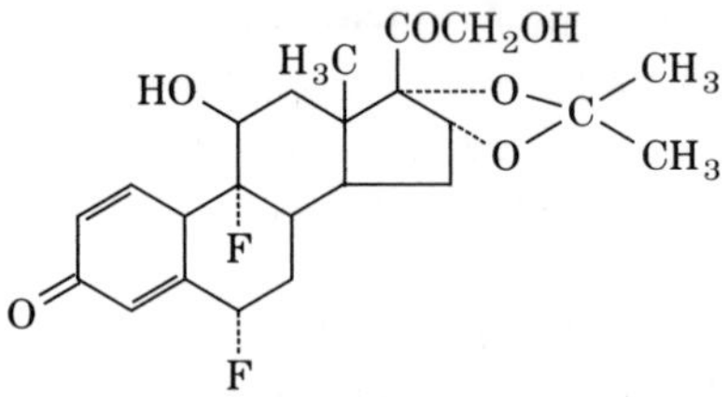 | 8.40 |

$COCH_2OH$
H_3C CH_3
HO O—C
O CH_3
F
O
F

| Triamcinolone acetonide | $C_{23}H_{31}FO_6$ (434.5) 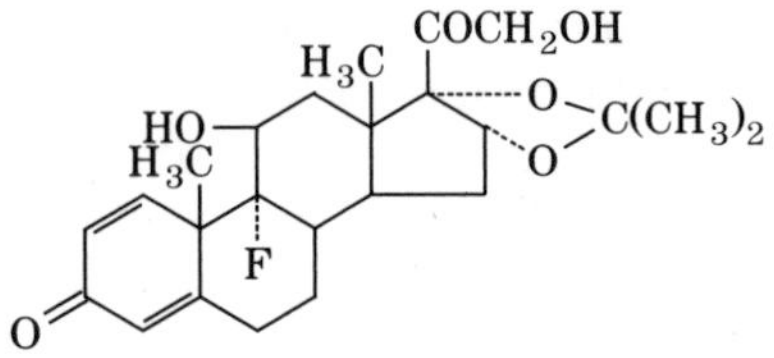 | 4.37 |

$COCH_2OH$
H_3C
HO O
H_3C $C(CH_3)_2$
O
F
O

iii. *Direct potentiometry:*

Into each 100-cm^3 volumetric flask are pipetted 10 cm^3 phthalate buffer of pH 5.2 ($I = 1.5$, KNO$_3$) with complexing agent (see the preceding text) and various volumes of standard sodium fluoride solution, to contain 10 to 500 μg F$^-$ in a flask, made up to the mark with distilled water, and transferred into polyethylene beakers, where the EMF values are measured. These are used to construct the calibration graph E vs. log[F$^-$]. The combustion sample is measured in the same manner and the unknown concentration is calculated from the calibration graph.

5.93   Sulfonamides

A large number of sulfonamides (see Tables 5.44 and 5.45) have been determined with various membrane sensors[381–391] by adequate techniques. Only in one method[383] was a destructive technique used to convert divalent sulfur from the respective sulfonamide into lead sulfide. In this technique is used the reaction with solid potassium hydroxide at 250 to 280°C for 5 to 10 min, followed by addition of alkali plumbite. The excess lead(II) was potentiometrically titrated with 0.01 M EDTA–Na$_2$ at pH 4.6 using a lead-ion-selective membrane sensor (Orion 94-82). The method was applied to acetazolamide and sulfathiazole determination and the results agreed with those obtained by the flask combustion method, with the difference that both sulfonamides showed sulfur figures to be exactly double of those obtained by the plumbite method, because both compounds contain two sulfur atoms per mole and only one of these is in the divalent state.

A simple, rapid, and accurate method[302] for determination of some sulfonamides listed in Tables 5.44 and 5.45 is based on the well-known ability of these compounds to form highly insoluble derivatives with mercury(II) ions:

$$2\,R\!-\!SO_2\!-\!NH\!-\!R' + Hg^{2+} \longrightarrow (R\!-\!SO_2\!-\!\underset{\underset{R'}{|}}{N}\!-\!)_2Hg + 2\,H^+$$

$$(5.88)$$

A mercury(II)-ion-selective electrode with a liquid membrane[177] as well as a commercial Ag$^+$/S^{2-}-crystal membrane electrode (Orion, Model 94-16) were used to monitor Reaction 5.88. Experiments for elucidating the mechanism of interaction of mercury(II) ions with the silver sulfide membrane sensor led to the conclusion that this sensor has a linear response to mercury(II) with a slope of 60 mV decade^{-1} through the

Table 5.44   Sulfa Drugs (Antibacterials) Assayed by Membrane Sensors

$$H_2N-\!\!\!\bigcirc\!\!\!-SO_2NH-R$$

Sulfa drug	R	Formula (MM)
Sulfacetamide	$-COCH_3$	$C_8H_{10}N_2O_3S$ (214.2)
Sulfadiazine	(pyrimidin-2-yl)	$C_{10}H_{10}N_4O_2S$ (250.3)
Sulfadimethoxine	(2,6-dimethoxypyrimidin-4-yl, OCH_3 and OCH_3)	$C_{12}H_{14}N_4O_4S$ (310.3)
Sulfadimidine	(4,6-dimethylpyrimidin-2-yl, CH_3 and CH_3)	$C_{12}H_{14}N_4O_2S$ (278.3)
Sulfadoxine	(pyrimidinyl, H_3CO and OCH_3)	$C_{12}H_{14}N_4O_4S$ (310.3)
Sulfafurazole	(isoxazolyl, H_3C and CH_3, $O-N$)	$C_{11}H_{13}N_3O_3S$ (267.3)
Sulfamerazine	(pyrimidinyl, CH_3)	$C_{11}H_{12}N_4O_2S$ (264.3)
Sulfamethoxazole	(isoxazolyl, $N-O$, CH_3)	$C_{10}H_{11}N_3O_3S$ (253.3)
Sulfamethoxydiazine	(pyrimidinyl, $-OCH_3$)	$C_{11}H_{12}N_4O_3S$ (280.3)
Sulfanilamide	$-H$	$C_6H_8N_2O_2S$ (172.2)
Sulfapyridine	(pyridin-2-yl)	$C_{11}H_{11}N_3O_2S$ (249.3)
Sulfathiazole	(thiazol-2-yl, S, N)	$C_9H_9N_3O_2S_2$ (255.3)

Table 5.45 Other Sulfa Drugs Assayed by Membrane Sensors

Sulfa drug	Formula (MM)	Therapeutic category
Acetazolamide	$C_4H_6N_4O_3S$ (222.3)	Carbonic, anhydrase inhibitor; diuretic; used in treatment of glaucoma
Chlorpropamide	$C_{10}H_{13}ClN_2O_3S$ (276.8)	Oral hypoglycemic
Furosemide	$C_{12}H_{11}ClN_2O_5S$ (330.7)	Diuretic
Hydrochlorothiazide	$C_7H_8ClN_3O_4S_2$ (297.7)	Diuretic
Tolbutamide	$C_{12}H_{18}N_2O_3S$ (270.3)	Oral hypoglycemic

following ion-exchange process taking place on the surface of the silver-crystal membrane:

$$Hg^{2+} + Ag_2S_{crystal} \rightleftharpoons [AgHgS^+]_{crystal} + Ag^+ \qquad (5.89)$$

However, the use of an excess of mercury(II) for completely precipitating the sulfa drugs and titration of the excess mercury(II) with EDTA or NTA produced more satisfactory results.

Sulfadiazine, sulfadimidine, sulfamerazine, sulfamethoxydiazine, and sulfapyridine (all as sodium salts) could be directly titrated with copper(II) sulfate and/or silver nitrate solutions.[387] The potentiometric jumps at

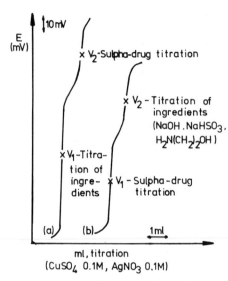

Figure 5.20 Potentiometric titration of an injectable solution of sulfamethoxydiazine with (a) the Cu^{2+}-selective membrane sensor and (b) the Ag^{+}-selective membrane sensor. (Reproduced from Baiulescu, G. E., Kandemir, G., Ionescu, M. S., and Cristescu, C., *Talanta*, 32, 295, 1985. With permission.)

the equivalence point are big enough for the maximum error of determination to be 0.5% with the Ag^{+}-membrane sensor and 1.9% with the Cu^{2+}-membrane sensor (both sensors were laboratory-made). The method is applicable to sulfa drugs with $pK_a \geq 6$. Sulfafurazole ($pK_a = 5.5$) does not form insoluble salts with Cu(II) or Ag(I). The determination of sulfamethoxydiazine in injectable solutions for veterinary use was examined. These solutions contain the sulfa drug dissolved in relatively concentrated sodium hydroxide solution together with sodium bisulfite and ethanolamine. All these ingredients form salts or complexes with both Cu(II) and Ag(I). The potentiometric titration curves obtained are presented in Figure 5.20; two significant potential jumps are obtained.

In the titration with Cu(II), all the ingredients are titrated together in the first step of the titration curve, whereas, in the titration with Ag(I), the first potential jump corresponds to the sulfa drug and the second to the other ingredients. For this reason the titration with silver nitrate was recommended for determination of sulfamethoxydiazine in injectable solutions. The method is not applicable to formulations containing more than one titrable sulfa drug.

For sulfamethoxazole and other sulfa-drug assays, a back-titration of Ag(I) excess with potassium iodide solution[388] or sodium chloride solution[389] was preferred.

A potentiometric sensor based on the precipitation of silver diethyl-dithiocarbamate within a graphite rod[385] was used as indicator for the potentiometric titration of sulfonamides at pH 8.0 with 10^{-2} *M* silver nitrate. In this case one equivalent of silver is consumed per mole of sulfonamide. Another modified graphite electrode (containing a mixture of Ag_2S–CuS), which is sensitive for both silver and copper ions[384] could be used as indicator electrode in potentiometric titrations of sulfonamides with Ag^+ or Cu^{2+} at pH 8.0. Average recoveries for certified sulfonamide standards in the concentration range 50 μmol to 0.3 mmol (equivalent to 10 to 50 mg) were 99.7% (mean standard deviation 0.2%) with silver(I) nitrate and copper(II) nitrate titrants. The determination of the same samples under the same conditions but using the commercially available solid-state silver- and copper-ion-selective membrane sensors gave comparable results. The influence of some excipients and diluents commonly used in the preparation of tablets, syrups, injections, drops, and ointments (e.g., magnesium stearate, talc powder, gum arabic, carboxymethylcellulose, cocoa butter, vanillin, Tween-80, polyvinylpyrolidone, glucose, lactose, sucrose, ethylene glycol, and glycerol) in potentiometric determinations were negligible, even when they were added in a very large excess over the respective sulfonamide.

The response of silver sulfide membrane sensors[392] to sulfonamide drugs was investigated for sulfonamides forming soluble complexes with silver (sulfacetamide, acetazolamide, furosemide, and hydrochlorothiazide)[386] and for those precipitating silver ions (e.g., sulfathiazole and sulfadimethoxine).[382, 386] It is known that when a silver sulfide membrane sensor is used in solutions free from silver ions, but containing a ligand forming relatively weak complexes with silver, the ligand does not dissolve the membrane but may react with adsorbed silver ions.[47, 386] The equation describing the response of the silver sulfide sensor to ligands is given by

$$E = E_{Ag^+}^0 - \frac{RT}{F} \ln\left(\frac{\alpha_{Ag^+}}{\beta_p}\right) - p\frac{RT}{F} \ln(L - p\alpha_{Ag^+}) + \frac{RT}{F} \ln \gamma_{Ag^+}$$

$$(5.90)$$

where α_{Ag^+} is the silver ion activity due to grain boundaries and adsorbed silver ions, β_p is the apparent formation constant of the silver complex formed, L is the total ligand activity, γ_{Ag^+} is the activity coefficient of the Ag^+, and p is the coordination number. Plots of the measured potentials vs. expected response were usually obtained within 1 to 2 min and steady potentials were established within 5 min, except for sulfacetamide, when about 10 min were required. As can be seen in Figure 5.21, the slope of linear portion of the calibration curve as well as the linear range depend on the respective sulfonamide.

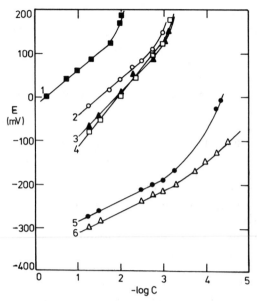

Figure 5.21 Response of the Ag_2S membrane sensor to different sulfonamides: (1) sulfacetamide; (2) furosemide; (3) acetazolamide; (4) hydrochlorothiazide; (5) sulfadimethoxine; (6) sulfathiazole. pH = 9.5; $I = 1$ M (KNO_3); 25°C. (Reproduced from Malecki, F. and Staroscik, R., *Anal. Chim. Acta*, 139, 353, 1982, Elsevier Science Publishers, Physical Sciences and Engineering Division. With permission.)

The data in Figure 5.21 were used to calculate the complex formation constants (log β_2) of the silver–sulfonamide complexes. The values obtained were 3.31, 5.34, 6.04, and 6.47 for sulfacetamide, furosemide, acetazolamide, and hydrochlorothiazide, respectively. The E_L^0 values (obtained by extrapolation to 1 M activity of the sulfonamide) depend linearly on log$(\beta_2)^{1/2}$ with a slope of 118 mV, which is in the close agreement with theoretical prediction.[393] It can be concluded that the electrode potential is controlled mainly by the reaction of non-lattice silver ions with the respective sulfonamide, resulting in a 1 : 2 complex.[386] The slopes for sulfathiazole and sulfadimethoxine, which form 1 : 1 precipitates with silver(I), were found to be near-Nernstian, because of the relatively high pK_{so} values of the precipitates.

The silver sulfide membrane sensor proved useful for drug analysis in pharmaceutical preparations both by potentiometric titrations and direct potentiometry. In the latter case a relative standard deviation of 2.9% for amounts of 25 mg hydrochlorothiazide was reported. Sensors sensitive to sulpha drugs were first described for the determination of sulfamerazine and sulfisodimine.[381] These sensors were constructed by dissolving ion associates formed between an iron(II)–bathophenanthroline chelate and a sulpha drug in nitrobenzene to form a liquid membrane for the following electrochemical cell:

$$\text{SCE} \left| \begin{array}{c} \text{Reference} \\ \text{solution} \end{array} \right| \text{Liquid membrane} \left| \begin{array}{c} \text{Sample} \\ \text{solution} \end{array} \right| \text{SCE} \qquad \text{(5.IV)}$$

An aqueous solution (10^{-2} M) of the sodium salt of the sulfa drug was used as reference solution. The liquid membrane, located in the bottom of a glass U-tube, separated sample and reference solutions.

The response time of both membrane sensors was only a few seconds and the potential values were reproducible to ± 1 mV. The sensors gave linear responses in the 10^{-1} to 10^{-3} M range (slope 57 mV decade^{-1}). EMF values were not affected by pH between 8.5 and 11.0 for sulfisodamine and between 2.3 and 10.0 for sulfamerazine. Hence all determinations were made at pH 9.0, adjusted with TRIS and sulfuric acid buffer.

Some sulfa-drug-sensitive membrane sensors were prepared by coating a copper disk (electroplated with platinum or gold) with ion-pair complexes such as cetyltrioctylammonium sulfathiazole, sulfadoxine, and sulfadimethoxine.[390, 391] Among the PVC-coated copper disk sensors, the sensor based on the ion-pair complex of cetyltrioctylammonium with sulfathiazole or sulfadoxine shows the best response characteristics, giving a near-Nernstian anionic response ranging from 10^{-5} to 10^{-2} M with a slope of 58 mV decade^{-1}, the detection limit being 2.5×10^{-6} M. The sensor made with cetyltrioctylammonium sulfadimethoxine as electroactive material gives nearly the same electrode performances. However, the sensor made with the ion-pair complex with cetyltributylammonium gives only a sub-Nernstian slope of 46 to 50 mV decade^{-1} with a shorter linearity range down to only about 10^{-4} M. The sensors made with cetyltrimethylammonium ion pairs cannot be used successfully for sulfonamide determinations.

Sulfathiazole, sulfadoxine, and sulfadimethoxine sensors respond to many sulfa drugs tested, e.g., sulfamethoxazole, sulfadoxine, sulfadimethoxine, sulfacetamide, sulfadiazine, and sulfanilamide, except sulfaguanidine and trimethoprim, which are only slightly soluble in the basic medium used. Sulfathiazole sensor shows the best response characteristics (with respect to response slope, linear response range and detection limit).

Yao et al.[391] studied in detail the influence of substrate metal on the electrode characteristics and found that the gold- or platinum-plated metal substrates exhibit better electrode response and can be used for at least three months. Concerning the pH effect on the CTOA–sulfathiazole sensor response, it was observed that the potential readings were not affected by pH in the range from 8.0 to 11.5. For the CTOA–sulfadoxine and CTOA–sulfadimethoxine sensor, the pH ranges in which the potential remains constant are 8.0 to 12.0 and 8.2 to 12.0, respectively.

The selectivity coefficients were, in the main, of the same order of magnitude for the sensors tested and decreased in the order sulfathiazole, sulfamethoxazole > sulfadoxine, sulfadimethoxine > sulfacetamide, sulfadiazine > sulfanilamide.

The CTOA–sulfathiazole sensor was used in the direct potentiometric determination of sulfa drugs at or lower than concentration levels of milligrams per cubic centimeter. The results for sulfadiazole, sulfamethoxazole, sulfadiazine, and sulfacetamide determinations gave an average recovery of 99.3% with standard deviation of 1.3% (standard-addition and/or calibration method). The sensor has also been applied to the determination of sulpha drugs in pharmaceutical tablets and ophthalmic solutions. The sulpha-drug content can generally be determined without preliminary filtration or other separation treatment and the results were in good agreement with those obtained by the pharmacopoeial method (the mean relative standard deviations were between 0.9 and 5.7%).[391]

Analytical Procedures

i. *Sulfa-drug membrane sensors:*
   A sample (drug substance or homogeneous powder with active principle) containing 0.2000 to 0.3000 g of sulfa drug is taken in a 100-cm^3 beaker and diluted with a minimum volume of methanol; 10 cm^3 of 1 M KNO$_3$ and approximately 50 cm^3 distilled water are successively added and the pH is adjusted to 9.0 with diluted sodium hydroxide solution. The sample is quantitatively transferred to a 100-cm^3 volumetric flask and diluted to volume with distilled water. The EMF of this solution is measured and compared to those of the standards prepared in the same manner.

ii. *Mercury(II)-ion-selective or Ag$^+$/S^{2-}-crystal membrane sensor:*
   The weighed sample (10 to 20 mg) is dissolved in about 5 cm^3 methanol in a 100-cm^3 beaker and 5 cm^3 10^{-2} M mercury(II) nitrate solution is added. After precipitation, 10 cm^3 distilled water is added and the pH is adjusted to 6.0 with hexamine. The pair of electrodes (mercury(II)- or Ag$^+$/S^{2-}-ion-selective membrane sensor, with SCE as reference) is introduced into the solution containing mercury(II)–sulpha-drug precipitate, and titration is carried out under stirring with 10^{-2} M EDTA–Na$_2$ solution. The EMF is recorded as a function of the titrant volume, and the end point corresponds to the maximum slope of the titration curve.

iii. *Ag$^+$/S^{2-}-crystal membrane sensor (for sulfonamides that form sparingly soluble precipitates with Ag(I) solution):*

The solid sample (substance or tablet powder) containing 100 to 200 mg of sulfonamide is mixed with 50 cm^3 of 0.1 M NaOH and the solution or suspension is potentiometrically titrated with standard silver nitrate solution (10^{-1} or 5×10^{-2} M). The EMF is recorded as a function of titration volume, and the end point corresponds to the maximum slope of the titration curve.

5.94   Thiambutosine

$$C_{19}H_{25}N_3OS \ (MM = 343.5)$$

$$(CH_3)_2N \!-\!\!\!\bigcirc\!\!\!-\!NH\!-\!\underset{\underset{S}{\|}}{C}\!-\!NH\!-\!\!\!\bigcirc\!\!\!-\!OC_4H_9$$

Therapeutic category: antileprotic

Discussion and Comments

As thiambutosine contains a divalent sulfur, the method developed by Hassan and Eldesouki[383] can be applied with satisfactory results for its determination. The method is based on decomposition of sample (5 to 10 mg) in a Pyrex test tube, with solid potassium hydroxide at 250 to 280°C for 5 to 10 min, followed by addition of alkali plumbite whereby lead sulfide is stoichiometrically formed. The excess lead(II) ions can be determined by potentiometric titration with EDTA solution at pH 4.6 using a lead(II)-selective membrane sensor. An average recovery of 99.0% and a mean standard deviation of 0.9% were reported. However, divalent sulfur attached to an aromatic moiety (e.g., Methylene Blue, promazine, thioridazine, etc.) is not desulfurized under these conditions. This is probably due to difficulties encountered in the cleavage of C—S bond owing to the interaction of unshared p-electron pair of sulfur with the π electrons of the benzene ring. Tetravalent and hexavalent sulfur in both aliphatic and aromatic compounds also do not decompose into sulfide.[383]

Analytical Procedure

An amount of 5 to 10 mg of the sample is transferred to the bottom of a Pyrex test tube (10 × 1 cm). Three pellets (about 0.2 to 0.3 g) of potassium hydroxide and one drop of water are added and the test tube is placed in a sand bath at 250 to 280°C for 8 to 10 min. The content is cooled at room temperature and then 5 cm^3 of potassium plumbite solution (8 g of lead(II) nitrate is dissolved in 1 dm^3 of 0.5 M potassium hydroxide) is added. The mixture is shaken and the tube is placed in a boiling water bath for 2 to 3 min. The content is cooled, centrifuged, and decanted; 2.5 cm^3 of supernatant is transferred into a 150-cm^3 beaker. A volume of 25 cm^3 of buffer solution of pH 4.6 is added. The electrode pair (lead(II)-ion-selective in conjunction with a double-junction reference electrode) is immersed into the solution and this is potentiometrically titrated with 0.01 M EDTA solution. A blank is carried out and the sulfur content is calculated in the usual way (1 cm^3 of 0.01 M EDTA ≡ 0.322 mg sulfur).

5.95   Thyroid and Antithyroid Agents

Table 5.46 lists some thyroid and antithyroid agents that can be determined by membrane sensors. Some of them have been discussed in the previous monograph (Coşofreţ,[98] pp. 297–299 and 303–307) and as a consequence, just a little attention will be paid here for their assays.

Thiourea and phenylthiourea could be potentiometrically titrated with silver nitrate standard solution in the concentration range 10^{-1} to 10^{-3} M in the presence of 1 or 0.1 M sodium hydroxide. The reactions that take place were easily followed with a sulfide membrane sensor[394-397] as well as with a silver(I) liquid-membrane sensor.[37] The titration curves of thiourea gave two potential breaks and both may be used for evaluating the thiourea concentration (Equations 5.91 and 5.92) whereas the titration curves of phenylthiourea gave only a potential jump (Equation 5.93):

$$\begin{array}{c} H_2N \\ \diagdown \\ H_2N \diagup \end{array} C{=}S + 2\,AgNO_3 \longrightarrow Ag_2S + H_2N{-}C{\equiv}N + 2\,HNO_3$$
$$(cyanamide)$$

$$(5.91)$$

$$H_2N{-}C{\equiv}N + 2\,AgNO_3 \longrightarrow Ag_2N{-}C{\equiv}N + 2\,HNO_3 \quad (5.92)$$

$$\begin{array}{c} H_2N \\ \diagdown \\ C_6H_5{-}NH \diagup \end{array} C{=}S + 2\,AgNO_3 \longrightarrow Ag_2S + \begin{array}{c} H_2N \\ \diagdown \\ C_6H_5NH \diagup \end{array} C{=}O + 2\,HNO_3 \quad (5.93)$$

Thiouracil and methyl thiouracil could be potentiometrically titrated with silver nitrate solution only when acetate buffer solutions (pH 5.6) were used.[397] In 0.1 M sodium hydroxide solutions, the appearance of the breaks in the titration curves was impeded by co-precipitation of silver oxide, which started before the theoretical end point was reached. Both potentiometric and infrared investigations support the reaction

Table 5.46 Thyroid and Antithyroid Agents Assayed by Membrane Sensors

Compound	Formula (MM)	Therapeutic category	Ref.
Thiourea	CH_4NS (62.1)	Thyroid inhibitor	37, 394–397, 400
Phenylthiourea	$C_7H_8N_2S$ (155.2)	Thyroid inhibitor used in medical genetics	396, 400
Thiouracil	$C_4H_4N_2OS$ (128.2)	Has been used as antithyroid agent in angina pectoris and in congestive heart failure	397
Methylthiouracil	$C_5H_6N_2OS$ (142.2)	Antithyroid agent	397
Methimazole	$C_4H_6N_2S$ (114.2)	Antithyroid agent which depresses the formation of thyroid hormone; its main effect is to reduce the formation of diiodotyrosine and hence of thyroxine.	398
Liothyronine sodium	$C_{15}H_{11}I_3NNaO_4$ (673.4)	Thyroid replacement therapy	368

Table 5.46   Continued

Compound	Formula (MM)	Therapeutic category	Ref.
Levothyroxine sodium	$C_{15}H_{10}I_4NNaO_4$ (798.9)	Thyroid hormone	399

Compound	Formula (MM)	Therapeutic category	Ref.
Carbimazole	$C_7H_{10}N_2O_2S$ (186.2)	Thyroid inhibitor	383

Compound	Formula (MM)	Therapeutic category	Ref.
Thyroid (thyroid hormone)	Contains not less than 0.17% and not over 0.23% iodine in thyroid combination	Thyroid hormone	399

mechanisms shown in Equation 5.94,[397]

$$(5.94)$$

where R = H or CH_3.

Przyborowski[398] used a copper(II)-ion-selective membrane sensor (Radiometer, F 111 Cu) for the potentiometric titration of methimazole with standard copper(II) sulfate solution at pH 5.6 (acetate buffer). The equivalence point corresponds to a molar ratio of 2 : 1 (drug : Cu(II)). The method was not affected by various excipients from pharmaceutical preparations.

Divalent sulfur in carbimazole and other structurally related aliphatic compounds was selectively determined by reaction with solid potassium hydroxide at 250 to 280°C for 5 to 10 min, followed by addition of alkali plumbite whereby lead(II) sulfide is stoichiometrically formed. The excess Pb(II) is measured by potentiometric titration with EDTA at pH 4.6 (acetate buffer) using a lead(II)-selective membrane sensor[383] (see also

Section 5.94). The analytical results showed an average recovery of 98.9% and a standard deviation of 0.9%.

A simple procedure for the determination of thyroid in pharmaceutical preparations by potentiometric titration with 10^{-3} M silver nitrate solution using an iodide-selective or silver sulfide–selective membrane sensor was developed by Richheimer and Schachet.[399] First of all, the tablet samples were combusted in a muffle furnace at 675 to 700°C with anhydrous potassium carbonate, after previously being ground to a fine powder. The results obtained by this method compared well with those obtained by the official methods. The method has been successfully applied to routine quality control work for content uniformity determinations of tablets ranging in potency from 16 to 324 mg of thyroid, and to the analysis of organically bound iodine in other pharmaceutical preparations such as sodium levothyroxine tablets (recovery 99.8%, relative standard deviation 1.8% for 0.1-mg preparation).

Analytical Procedures

  i. *Thiourea and phenylthiourea by potentiometric titration with 10^{-2} M silver nitrate solution:*
     The pair of electrodes (silver(I)- or sulfide-ion-selective membrane as indicator with SCE as reference) is introduced into the sample solution (30 to 40 cm³, approximately 10^{-3} M concentration; 1 M alkaline medium), which is titrated under stirring with 10^{-2} M silver nitrate solution. The electrode potential is recorded as a function of titrant volume, and the end point corresponds to the maximum slope of the titration curve. (For thiourea the second potential jump is used.)

 ii. *Thiourea and methylthiouracil by potentiometric titration with 10^{-2} M silver nitrate:*
     The pair of electrodes (sulfide-ion-selective membrane and SCE as reference) is introduced into the sample solution (30 to 40 cm³, approximately 10^{-3} M in pH 5.6 acetate buffer), which is titrated under stirring with 10^{-2} M silver nitrate solution, as previously described.

iii. *Methimazole by potentiometric titration with 10^{-1} M copper(II) sulfate:*
     The pair of electrodes (copper(II)-ion-selective membrane as indicator and SCE as reference) is introduced into the sample solution (30 to 40 cm³, approximately 2×10^{-2} M at pH 5.6 adjusted with acetate buffer) and titration performed under stirring with 0.1 M copper(II) sulfate solution. The EMF is recorded as a function of titrant volume, and the end point corresponds to the maximum slope of the titration curve.

 iv. *Liothyronine sodium:*
     See the procedure from Section 5.89.

v. *Carbimazole:*
See the procedure from Section 5.94.

vi. *Thyroid assay:*

a. *Uncoated tablets and bulk material*—Twenty tablets are finely powdered and a portion equivalent to approximately 635 mg of thyroid (proportionately less should be used if the iodine content is > 0.2%) is weighed into a large crucible; this portion is mixed with approximately 8 g of anhydrous potassium carbonate, compressed, and then overlayed with 16 g of carbonate. The mixture is ignited at 675 to 700°C in a preheated muffle furnace for 25 min. The cooled char is transferred to a 600-cm^3 beaker, using water to facilitate the transfer, and is then acidified to pH 2.5 ± 1 with dilute phosphoric acid while stirring vigorously with a magnetic stirrer. Water is added to bring the volume to 400 cm^3, and the mixture is titrated immediately with 10^{-3} M silver nitrate solution using an iodide-selective membrane sensor as indicator electrode and a suitable reference electrode (1 cm^3 of 10^{-3} M silver nitrate is equivalent to 0.1269 mg of iodine).

b. *Coated tablets*—The coated sample is prepared in the same way as uncoated tablets except that the char is treated with hot water, filtered, and then acidified with dilute phosphoric acid, and the volume is brought to 400 cm^3 with water (if calcium sulfate is used in formulation, the acidified solution is boiled for at least 30 min). The solution is cooled to room temperature and titrated with 10^{-3} M silver nitrate solution, as previously described.

vii. *Levothyroxine sodium assay in tablets:*
Twenty tablets are finely powdered and an amount equivalent to 1 to 2 mg of levothyroxine sodium is weighed into a crucible. The procedure for thyroid assay in uncoated tablets is followed except that 1 cm^3 of 10^{-3} M silver nitrate is equivalent to 0.1997 mg of levothyroxine sodium.

5.96   Trimethoprim

$$C_{14}H_{18}N_4O_3 \; (MM = 290.3)$$

Therapeutic category: antibacterial

Discussion and Comments

A membrane sensor selective to trimethoprim is based on the ion-pair complex of trimethoprim with silicotungstate and it was constructed by coating a copper disk (electroplated with platinum or gold) with electroactive material in PVC.[401] Dipicrylamine as well as phosphomolybdate have also been tested as counterions for the sensor preparations, but the trimethoprim–silicotungstate sensor displayed the best response to trimethoprim, the Nernstian-type response range being from 2×10^{-3} to 3×10^{-5} M with a slope of 51 ± 1.5 mV decade^{-1} and a detection limit of 10^{-5} M.

PVC-membrane sensors (classical type) were also prepared in order to compare their performances, and no significant differences were found between the linearity ranges and the response slopes of the two types.[401]

The trimethoprim membrane sensor was not affected by pH changes within the range 3.5 to 6.5; at pH > 7, trimethoprim tends to precipitate, causing the potential to decrease progressively. Selectivity coefficients determined for 23 organic and inorganic substances (mixed-solution method) showed that only chlorpheniramine, atropine, and glycopyrolate are likely to cause interference, but these substances will rarely be contained in the same pharmaceutical preparations as trimethoprim.

It was found that during three months of use, the response time of the sensor in 10^{-6} M solution is about 1 min, in 10^{-4} M solutions 15 to 30 s and in 10^{-2} to 10^{-3} M solutions < 10 s.

The trimethoprim–silicotungstate membrane sensor was applied to the potentiometric determination of trimethoprim from aqueous solutions and tablets by the standard-addition method and the "single-drop" method, both with good results.[401] In the second method, only one or two drops of sample solution can be used for analysis by keeping the sensor inverted and injecting the sample into the limited space between the indicator sensor and reference electrode, to form a very small cell.

Trimethoprim in pharmaceuticals (tablets of trimethoprim and trimethoprim + sulfamethoxazole) were also determined by a trimethoprim membrane sensor based on trimethoprim–tetraphenylborate as electroactive material in a PVC matrix (Nernstian-type response from 4.4×10^{-5} to 2.2×10^{-3} M, slope 57.3 mV decade^{-1}, and detection limit 4.6×10^{-6} M). The analysis results compared well with those obtained by differential pulse polarography.[402]

Analytical Procedure

Three standard solutions of trimethoprim (in the range 2×10^{-2} to 3×10^{-5} M) are prepared by accurately weighed drug substance and subsequent dilutions, respectively. In all cases the pH is kept constant with acetate buffer solution of pH 5.0. Adequate volumes of standard solutions are transferred into 100 cm^3 beakers containing magnetic stirring bars. The EMF of these solutions are recorded with a trimetho-prim membrane sensor in conjunction with a reference electrode (e.g., SCE) and plotted vs. log C. The unknown sample concentration is determined from this graph.

For tablets assay as well as for drug substance, the standard-addition method can also be used.

5.97   Vanillin

$$C_8H_8O_3 \ (MM = 152.1)$$

Therapeutic category: pharmaceutical aid (flavor)

Discussion and Comments

A vanillate membrane sensor based on tetraheptylammonium vanillate in a PVC matrix was constructed for the determination of vanillin after its *in situ* oxidation to vanillate.[403] The sensor exhibits a Nernstian response (slope 59.07 mV decade^{-1}) in the range 10^{-1} to 4×10^{-4} M vanillate over a reasonably wide working pH range (7.5 to 10.2), a fast response time (less than 30 s) and is stable for at least one month.

For the determination of vanillin, this was first oxidized to vanillic acid, which was then determined by the vanillate-selective membrane sensor at pH 10.0. The oxidation was carried out in aqueous solution when vanillin was treated with two equivalents of sulfamic acid and 1.5 equivalents of sodium chlorite at room temperature for 1 h. The chemical yield of the oxidation was found to be consistently 75%. Longer reaction times or larger amounts of oxidant did not increase the yield of the reaction.[403]

For the determination of a vanillin sample, adjusting the amount of vanillin found by the vanillate-selective membrane sensor method by a factor of 0.75 will give the equivalent amount of vanillin in the sample. The results obtained were found to be fairly accurate and in good agreement with those found by an official method.

Analytical Procedure

i. *Calibration of the vanillate sensor:*
A 0.1 M sodium vanillate solution is prepared by dissolving 0.025 mol of vanillic acid and 1.1 equivalents of sodium hydrogen carbonate solution in distilled water. By appropriate dilution, a series of standard solutions with concentrations of 0.04, 0.01, 0.004, 0.001, 0.0004, and 0.0001 M are obtained. Aliquots of 50 cm^3 of 10^{-1} to 10^{-4} M sodium vanillate solution are transferred into 100-cm^3 beakers containing magnetic stirring bars; the pH is adjusted to 10.0 by the addition of few drops of 1 M sodium carbonate solution. The vanillate–PVC matrix sensor and the SCE are immersed in the solutions, and the measured EMF values are plotted as a function of the logarithm of the vanillate concentration.

ii. *Oxidation of vanillin:*
Vanillin (2 mmol) and sulfonic acid (4 mmol) are dissolved in 40 cm^3 of distilled water. Sodium chlorite (3 mmol) is then added and the solution is stirred at room temperature for 1 h. Vanillic acid separates out as fine crystals.

iii. *Determination of vanillin:*
Three samples of vanillin of about 200 mg are oxidized to vanillic acid by the preceding procedure. The crude reaction mixture is taken up in 1.1 equivalents of sodium carbonate solution. With appropriate dilution, the amount of vanillic acid is determined by the vanillate sensor at pH 10.0. After correcting the oxidation yield factor (0.75), the equivalent amount of vanillin in the sample is calculated and the results are compared with the original sample masses of vanillin.

5.98   Vitamins

Considerable research has been done in the field of the assay of vitamins by potentiometric techniques with sensitive membrane sensors. Table 5.47 gives some references for some vitamins from the B group as well as for vitamin H (an antirickettsial and a sunscreen agent) and vitamin C, assayed by membrane electrode potentiometry. Some comments and analytical procedure for vitamin PP (nicotinamide) assay can be found in Section 5.46.

A platinum electrode modified with a covalently bound monolayer of ferrocene molecules[404] acts as a potentiometric sensor for vitamin C.[405] The electroactive site in ferrocene is iron(II), which can be oxidized to iron(III). The oxidation of ascorbic acid is known to be catalyzed by various iron(III) compounds.[423–425] A stable potential is developed at the modified electrode surface when it is immersed in a solution of L-ascorbic

Table 5.47   Some Vitamins Assayed by Membrane Sensors

Vitamin	Formula (MM)	Ref.
C (ascorbic acid)	$C_4H_8O_6$ (176.1)	136, 282, 404–407

$$CH_2-OH$$
$$H-C-OH$$

H (aminobenzoic acid)	$C_7H_7NO_2$ (137.1)	408

$$H_2N-\!\!\langle\ \rangle\!\!-COOH$$

B$_1$ (thiamine hydrochloride)	$C_{12}H_{18}Cl_2N_4OS$ (337.3)	9, 14, 150, 409–416

$$Cl^-$$

$$(C_{12}H_{22}N_4O_9P_2S, MM = 460.3,$$
is thiamine pyrophosphate salt
or vitamin B$_1$ diphosphate salt)

B$_2$ (riboflavin)	$C_{17}H_{20}N_4O_6$ (376.4)	417

$$CH_2-CH-CH-CH-CH_2-OH$$
OH OH OH

B$_6$ (pyridoxine hydrochloride)	$C_8H_{12}ClNO_3$ (205.6)	409, 416, 418–420

B$_{12}$ (cyanocobalamin)	$C_{63}H_{88}CoN_{14}O_{14}P$ (1355.4)	421, 422

acid and the potential response is linear over the range 10^{-3} to 10^{-6} M (slopes of 50 ± 8.8 mV decade^{-1} were obtained with 15 examined sensors). The potentiometric response at the ferrocene-modified electrode is probably due to a mixed potential of the ferrocene–ferrocinum couple and the L-ascorbic acid–dehydroascorbic acid couple. This could explain the slope of the one-electron reaction and the 28.5 mV required for a two-electron reaction with a Nernstian response at $19 \pm 0.5°C$.[404] Two sensors were applied for the determination of L-ascorbic acid in the range 10^{-4} to 10^{-5} M (glycine buffer, pH 2.2, 0.1 M KCl) by the standard-addition method. The mean value of recovery was 99.7% with a relative standard deviation of 1.9%.

Koupparis and Hadjiioannou[136] used a chloramine-T-selective membrane sensor with a liquid membrane of nickel bathophenanthroline–chloramine-T dissolved in nitro-p-cymene for the determination of ascorbic acid by direct potentiometry or by indirect potentiometric titration with chloramine-T. Ascorbic acid in the 4 to 40 mg range can be determined with relative errors and relative standard deviation of about 1%.

Ascorbic acid was also determined by direct potentiometry in injection solutions or in tablets with relative standard deviation of 1 to 2% for 0.7 to 35 mg of ascorbic acid.

Christova et al.[406] proposed, for the determination of ascorbic acid and other reductants that react stoichiometrically with iodine, a method based on quantitative oxidation in solutions of pH 2.0 with 0.1 M iodide-free ethanolic solution of iodine. The activity of the iodide ions formed were determined with an iodide-ion-selective membrane sensor (Radiometer, F 1032 I or Crytur iodide-selective).

A direct semi-automatic potentiometric titration method of ascorbic acid with potassium iodate, using an iodide-ion-selective membrane sensor (Orion, Model 94-53A) to monitor the reaction, was proposed by Calokerinos and Hadjiioannou.[407] The method is based on the overall reactions before and after the equivalence point:

$$3 C_6H_8O_6 + IO_3^- \rightarrow 3 C_6H_6O_6 + I^- + 3 H_2O \qquad (5.95)$$

$$IO_3^- + 6 H^+ + 5 I^- \rightarrow 3 I_2 + 3 H_2O \qquad (5.96)$$

A sharp increase in potential at the end point (Reaction 5.96) makes very precise the location of the titrant volume corresponding to this point. A 1 M HCl solution was chosen as a reaction medium. Results for the determination of ascorbic acid in the range 0.1 to 1.0 mg cm^{-3} (in aqueous solutions) showed a relative error of $\pm 1.5\%$. The method has been applied for the determination of ascorbic acid in tablets, and the results checked closely with those obtained by the standard method.

It was also found that in 10^{-3} M H_2SO_4, ascorbic acid is totally oxidized by potassium iodate. The time needed for a 40-mV potential change was inversely proportional to concentration of ascorbic acid. This is the basic principle for a kinetic determination method proposed recently.[282] A calibration graph obtained by plotting the reciprocal time $(1000/t, \text{s}^{-1})$ vs. concentration was linear in the range 5×10^{-4} M with a very high slope (382 dm^3 mol^{-1} s^{-1}).

The method used for the potentiometric determination of arene–diazonium salts based on ion-pair formation between the diazonium cation and tetraphenylborate[408] can be applied to the determination of 4-aminobenzoic acid (vitamin H), too. First of all, the aromatic amine must be converted into respective arenediazonium salt by using 1 M sodium nitrite solution under cooling at $0°C$ with ice; then, an aliquot of solution is potentiometrically titrated with sodium tetraphenylborate solution. Titration is done under cooling with ice and is followed potentiometrically with organic ion-selective membrane sensors comprising PVC membranes plasticized with polar solvents (e.g., o-nitrophenyloctyl ether, didecylphthalate, dimethoxybenzene, etc.) and coated on aluminum wires. Most of the ion-pair compounds ($Ar—N^+ \equiv N \ AB^-$) investigated by Vytras et al.[408] are practically insoluble in water. Both the steepness of the break in potentiometric titration curve and the overall size of the potential break are governed by the solubility product of the precipitate formed. It was observed that, in the case of 4-carboxybenzene diazonium cation as well as for other compounds containing hydrophilic groups such as $—COOH$ and $—OH$, the potentiometric titration curves have no well-defined end points, presumably because the respective diazonium salts tend to form zwitterions.

Ion-selective membrane sensors sensitive to vitamins B_1 and B_6 were first described by Ishibashi et al.[409] They are based on the ion-association complexes of respective vitamins with tetraphenylborate or dipicrylamine dissolved in organic solvents such as 1,2-dichloroethane and nitrobenzene (concentration 10^{-4} M). Both sensors displayed useful range between 10^{-2} and 10^{-5} M with no significant interference from common inorganic cations.

Vitamin B_1 liquid-membrane sensors containing either picrolonate[412] and tetra(m-methylphenyl)borate[415] as ion-site carriers in the respective membrane displayed relatively wide Nernst-type ranges. For the second type, the electrode function covered the 10^{-1} to 10^{-6} M range, whereas in the first type the range was within 10^{-5} to 5×10^{-5} M (nitrobenzene being used as solvent in both cases).

Yao et al.[415] showed that in citrate buffered media, vitamin B_1 was found to be stable enough at pH 8, and the sensor showed a twofold sensitivity in the determination, with a slope of 54 mV decade^{-1}. In the pH range 2 to 4, the slope of the sensor varied between 28 and 29 mV decade^{-1}. For the sensor based on thiamine–picrolonate,[412] a slope of 40 mV decade^{-1} was found (pH 6.0 to 6.5). In this pH range a mixture of monoprotonated and diprotonated vitamin B_1 is present in the solution.

Although for the sensor based on thiamine–picrolonate negligible interferences were observed for vitamin B_2, vitamin B_6, vitamin B_{12}, nicotinamide, folic acid, and calcium pantothenate, the sensor based on thiamine–tetra(m-methylphenyl)borate showed interferences by vitamins B_2 and B_6 at pH 2 to 3 ($k_{B_1,B}^{pot} = 158$ and 19.9, respectively) and showed slight interference at pH 7.0 ($k_{B_1,B}^{pot} = 0.4$ and 0.33, respectively).

Both dinonylnaphthalenesulfonate and tetra(m-chlorophenyl)borate were used as ionic site carriers to prepare PVC-type membrane sensors sensitive to vitamin B_1 as well as vitamin B_6.[416] When the electroactive material in the membrane varied from 2 to 8% (o-NPOE as plasticizer) no significant changes in sensor behavior were observed. Vitamin B_6 membrane sensors showed near-Nernstian responses to diprotonated cation (pH 3.5, acetate buffer) over the range 10^{-1} to 10^{-5} M with fast response and good reproducibility of potential measurements. Vitamin B_6, nicotinamide, caffeine, various amino acids, etc. did not show any interference in the sensor response.

A vitamin B_1–selective coated graphite sensor with thiamine–tetraphenylborate as the electroactive material in PVC, with DBP as the plasticizer, at pH 2 to 5 provides also a rapid, nearly Nernstian response to vitamin B_1 divalent cation in the range 10^{-1} to 10^{-5} M.[413, 414] For both PVC-type vitamin B_1 membrane sensors the internal reference solution was 10^{-2} M thiamine hydrochloride, whereas for the coated graphite sensor this solution was eliminated.

All vitamin B_1 membrane sensors were successfully applied to vitamin B_1 assay in aqueous solutions as well as in pharmaceutical preparations, such as tablets and injections. This was performed by direct potentiometry or by potentiometric titrations. By using the calibration graph method with the picrolonate-based liquid-membrane sensor, the determination of vitamin B_1 in pure solutions showed an average recovery of 98% ($n = 16$, each sample in replicate and containing 30 to 1000 μg cm^{-3} of vitamin B_1) and a standard deviation of 1.8%. An average recovery of nominal value of 98.5% and a standard deviation of 1.7% was found for vitamin B_1 determination—after a simple dilution step—in some tablets and injections. A standard deviation of less than 1% was found with the tetra(m-methylphenyl)borate-based liquid-membrane sensor by the potentiometric titration method.

When a liquid-membrane tetraphenylborate sensor was used as indicator in potentiometric titrations with sodium tetraphenylborate solution at pH 7.0, a potential break of only 80 mV was recorded for an amount of

15 to 25 μmol of thiamine. Even in these conditions relative standard deviations were 0.4 and 0.5% for potentiometric titrations of pure aqueous solutions and tablets, respectively.[14]

A potentiometric break of 135 mV was recorded with a fluoroborate-selective membrane sensor, in solutions buffered at pH 9.0 with 2 M sodium acetate[9]; standard deviation in this case was 0.32%.

Hassan et al.[416] applied the desulfurization procedure using solid potassium hydroxide (250 to 280°C for 5 to 8 min) and alkali plumbite solution (1.0 mg cm^{-3} Pb(II) in 0.2 M KOH). The excess lead(II) was determined at pH 4.5 (acetate buffer) with EDTA solution, using the lead(II)-selective sensor as an end-point detector. The titration procedure was simplified by using the Gran's plot technique, three aliquots of the titrant being sufficient for locating the equivalence point in the titration. The results obtained for determination of pure vitamin B$_1$ in amounts ranging from 3 to 5 mg showed an average recovery of 99% and a mean relative standard deviation of 0.7%. An average assay of 98.0% and a mean standard deviation of $\pm 0.6\%$ were obtained for the determination of vitamin B$_1$ in some pharmaceutical preparations. The effects of vitamin B$_2$, B$_6$, and B$_{12}$ and nicotinamide, as well as the effect of excipients usually present in preparations containing these vitamins are negligible. Citrate, magnesium, and calcium also did not affect the accuracy of the method. The citrate ion is decomposed by the alkali during the reaction step and magnesium and calcium react with EDTA at higher pH values.

It is well known that thiamine exists in blood and tissue both in the free form and as phosphate esters. Thiamine pyrophosphate or vitamin B$_1$ diphosphate salt is the metabolically active co-enzyme for thiamine in a large number of enzymes catalyzing acyl group transfer reactions, for example, decarboxylation of α-keto acids and the formation of α-hydroxycarbonyl linkages.[411] A potentiometric method for the determination of thiamine pyrophosphate was described by Seegopaul and Rechnitz.[411] The method is based on measuring the initial rate of carbon dioxide formation from a reaction sequence involving the recombination of thiamine pyrophosphate using pyruvate decarboxylase apoenzyme with the haloenzyme. The proposed method is highly selective and, with optimum conditions experimentally established, the calibration graph of thiamine pyrophosphate concentration vs. reaction rate gave a linear range of up to 30 ng cm^{-3}, without any separation procedures of secondary products.[411]

Electroactive materials such as those containing ion-pair complexes of pyridoxine with dipicrylamine,[418, 419] tetraphenylborate,[418] dinonylnaphthalenesulfonate,[416] or tetra(m-chlorophenyl)borate[416] proved adequate for construction of various types of vitamin B$_6$ membrane sensors. Nernstian-type response ranges varied from 10^{-1} to 10^{-5} M to 10^{-1} to 10^{-4} M as a function of membrane composition. The sensor prepared with 0.5% pyridoxine–dipicrylaminate as the electroactive material in a PVC matrix (DBP as plasticizer) displayed a near-Nernstian response over 10^{-1} to 10^{-5} M with a very low detection limit (1.9×10^{-6} M).

Not many organic substances have been tested as potent intereferences for the vitamin B_6 membrane sensors, but from the values reported recently[416], it can be concluded that these sensors are highly affected by the more hydrophobic drug substances such as alkaloids or β-adrenergic blockers.

All vitamin B_6 membrane sensors were applied to the potentiometric assay of pyridoxine hydrochloride in aqueous sample solutions or pharmaceuticals by using standard-addition or substraction methods as well as Gran's technique or potentiometric titration with sodium tetraphenylborate solution. Recoveries of 100.2 and 101.2% with standard deviations of 1.6 and 1.5% were reported for vitamin B_6 determination from tablets and injectable solutions, respectively, with a PVC-matrix B_6 membrane sensor.[416]

A pK_a value of 4.91, in good agreement with the literature data, was found for vitamin B_6 by using a potentiometric method with a pyridoxine membrane sensor.[418]

A new potentiometric biosensor using binding protein was proposed for the selective determination of riboflavin (vitamin B_2).[417] It is known that riboflavin-binding protein (RBP) binds riboflavin, through non-covalent interaction, more tightly than it binds riboflavin analogues (see references 7 and 8 in Yao and Rechnitz[417]). The binding of RBP to riboflavin is reversible and RBP dissociates into apo-RBP and free riboflavin at low pH. The new potentiometric sensor for riboflavin utilizes this effect. The response principle of the sensor depends on the change in membrane charge induced by displacement of apo-RBP from the riboflavin to produce a shift in membrane potential. Both negatively charged flavin adenine dinucleotide (FAD) and positively charged acriflavin were used as riboflavin analogues for immobilization at the membrane surface.

The apo-RBP-complexed membrane was prepared by immersing the acriflavin or FAD bound membrane (obtained as described in detail in Yao and Rechnitz[417]) into an apo-RBP solution (2 mg of apo-RBP in 2×10^{-2} M TRIS–HCl buffer at pH 7.7) for 1 h. Membrane disks of 3 mm were cut for placement into the Orion 92 series electrode bodies (phosphate buffer of pH 7.7 as an internal filling solution). Prior to measurements each sensor was immersed for about 30 min in the background electrolytes until the sensor achieved a constant potential.

From the structural formulae in Figure 5.22 as well as from the fact that there are significant differences in bioaffinity constants of various flavin derivatives by aporiboflavin and FAD possess the necessary charge characteristics for attachment to the membrane surfaces and for selective binding with aporiboflavin binding properties. These properties were

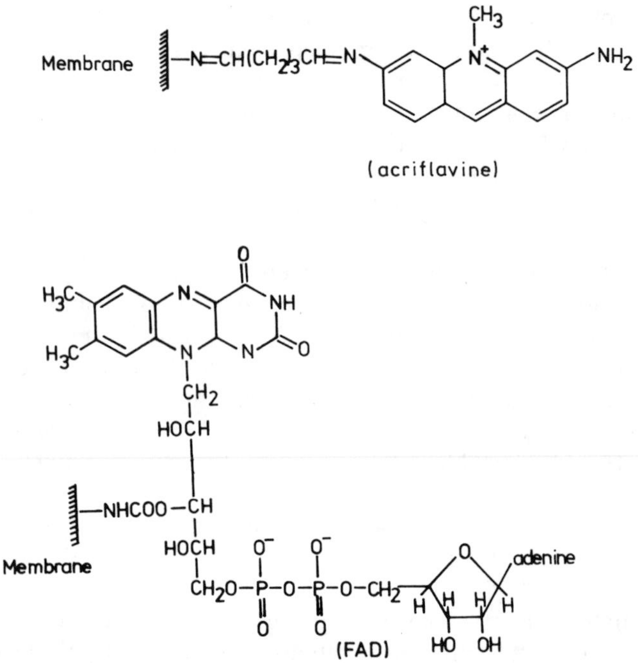

(acriflavine)

(FAD)

Figure 5.22 Structural characteristics of acriflavin and FAD bound on the electrode membrane. (Reprinted with permission from Yao, T. and Rechnitz, G. A., *Anal. Chem.*, 59, 2115, 1987. Copyright 1987 American Chemical Society.)

utilized for the construction and operation of potentiometric sensor according to the scheme represented in Figure 5.23.

The first step is the preparation of apo-RBP-complexed membranes. The interaction between riboflavin analogue (acriflavin or FAD) and apo-RBP depends on non-covalent binding of these compounds.[417] This interaction blocks the positively or negatively charged group of the riboflavin analogue bound membrane, because apo-RBP is a macromolecule compared to riboflavin analogues. Apo-RBP binds riboflavin with a bioaffinity constant of about 1.3 nM, e.g., by several orders of magnitude more tightly than it binds FAD, acriflavin, or related flavins. The addition of riboflavin, therefore, causes the displacement of apo-RBP from the membrane surface to form stable halo-RBP (second step). The displacement of apo-RBP–riboflavin complex causes a shift of the Donnan equilibrium in the direction of increasing charge of the membrane surface accompanied by a potential change.[417]

Figure 5.24 illustrates the potentiometric behavior resulting from the two-step processes previously described. In the case of acriflavin-bound membrane electrode (Figure 5.24A) the potential changed in the positive

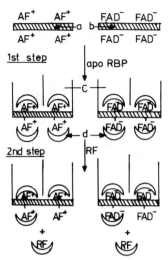

Figure 5.23 Schematic diagram of apo-RBP–riboflavin analogue bound membrane electrodes: (a) acriflavin (AF$^+$) bound membrane; (b) flavinadenine dinucleotide (FAD$^-$) bound membrane; (c) internal solution; (d) aporiboflavin-binding protein (apo-RBP); RF = riboflavin. The first step is the preparation of apo-RBP-complexed membranes and the second is the sensing step for riboflavin. (Reprinted with permission from Yao, T. and Rechnitz, G. A., *Anal. Chem.*, 59, 2115, 1987. Copyright 1987 American Chemical Society.)

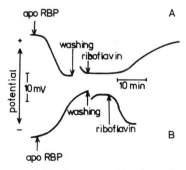

Figure 5.24 Responses of riboflavin analogue bound membrane electrodes to apo-RBP (90 μg) and riboflavin (1.3 μM) additions: (A) acriflavin-bound membrane electrode; (B) FAD-bound membrane electrode. (Reprinted with permission from Yao, T. and Rechnitz, G. A., *Anal. Chem.*, 59, 2115, 1987. Copyright 1987 American Chemical Society.)

direction when riboflavin was added (long response time because of small difference in the bioaffinity constant for riboflavin, 1.3×10^{-9}, and acriflavin, 1.8×10^{-7}). As shown in Figure 5.24B, the apo-RBP–FAD bound membrane electrode gave a well-defined response for the addition of riboflavin. Because the FAD has a negatively charged site, the direction of the potential change was opposite to that of the apo-RBP–acriflavin bound membrane electrode. The response time of this sensor was shorter than that of the first one.

The calibration curves for both sensors have shown that the sensitivity was almost the same, but the apo-RBP–FAD bound membrane electrode surpasses the apo-RBP–acriflavin bound membrane electrode with regard to its rapid-response characteristics and good reproducibility of measurements. The relative standard deviations (for $n = 6$) were 6.8% for the apo-RBP–FAD electrode and 11.8% for the apo-RBP–acriflavin electrode at a concentration of 1.0 μM riboflavin for experiments carried out with freshly conditioned sensors.

Other vitamins (e.g., vitamins A, B, and C) did not interfere with the measurements of riboflavin even in 50-fold excess. It was observed that if the freshly prepared bound membrane was stored in 0.02 M phosphate buffer (pH 7.7) in the refrigerator it retained binding activity for apo-RBP for 20 to 25 days.[417]

Since the papers of Dessouky and Pungor[421] and Goldstein and Duca,[422] which described methods for the determination of cyanocobalamin (vitamin B_{12}) based on the use of cyanide membrane sensors, no other potentiometric method for this drug has been developed. In the first method, the cyano group is released following reduction with different reduction agents (ascorbic acid, tin(II) chloride in hydrochloric acid, calcium hypophosphite in hydrochloric acid or in sulfuric acid) under reflux or by exposure to a strong source of visible light at room temperature (two 500-W lamps) for 30 min.[421] By heating at 140 to 145°C in sodium hydroxide solution (pH 12 to 13), cyano group from vitamin B_{12} can be converted into cyanide and then quantitatively determined by a cyanide-ion-selective membrane sensor.[422] This method has the advantage of the rapid determination of the cyano group in vitamin B_{12}. Moreover, it is selective and accurate and the error does not exceed $\pm 3\%$.

Analytical Procedures

i. *Vitamin C determination with a chloramine-T membrane sensor:*

5.0 cm^3 of 5×10^{-2} M chloramine-T (CAT) solution (250 μmol), 20.0 cm^3 of sample solution, and 2.0 cm^3 of phosphate buffer solution (pH 7.0) are pipetted into the reaction cell. When the potential has stabilized to within ± 0.1 mV (in about 1 to 2 min), the potential E is recorded. Four ascorbic acid standards in the 10^{-3} to 10^{-2} M range, containing 2.0 cm³ of buffer per 20.0 cm³ of standard solution are included with each series of unknown samples. The excess of chloramine-T is found from a plot of E (in millivolts) vs. log[CAT] in excess. The amount of ascorbic acid in the sample is equivalent to the amount of chloramine-T consumed, the stoichiometry of the reaction being 1 : 1.

ii. *Vitamin C determination with an iodide-selective membrane sensor:*
10.0 cm³ of potassium iodate working solution (2.0×10^{-4} M) and 10.0 cm³ of 0.002 M sulfuric acid solution are transferred into the measuring cell and the potential is recorded under stirring. After potential equilibration, 0.10 cm³ of sample solution is injected and the potential change with time is recorded. The cell is emptied and the procedure repeated for each sample solution. The time required for the recorded cell potential to increase from 5 to 45 mV is measured.

For calibration, aliquots of ascorbic acid are introduced into the cell and change in potential is measured as before. Calibration graphs are constructed by plotting the reciprocal time ($1000/t$, s^{-1}) vs. concentration.

iii. *Vitamin H determination with a coated-wire ion-selective sensor (aluminum wire dipped into a solution of PVC (0.09 g) and o-nitrophenyl octyl ether plasticizer (0.2 cm³) in 3 cm³ of tetrahydrofuran, and the solvent allowed to evaporate):*
The respective arenediazonium salt solution is prepared by titration of precisely measured volume of the vitamin H (p-aminobenzoic acid) solution with sodium nitrate under cooling to 0°C with ice; the end point of the diazotization is checked both potentiometrically (with a Pt electrode vs. SCE) and with iodide–starch papers. The solution of the diazonium salt (in 0.005 M HCl) is transferred to a 100-cm³ beaker, the volume is adjusted to 50 to 75 cm³ (to obtain an approximately 5×10^{-3} M solution of the titrated substance), and the tetraphenylborate solution (2.5%) is added from an automatic burette. The titration vessel is cooled with ice to 0°C externally. The change in EMF is recorded and plotted vs. titrant volume. The equivalence point is evaluated from the maximum slope of the titration curve and used to calculate the concentration of the unknown sample concentration.

iv. *Vitamin B_1 and vitamin B_6 determination by direct potentiometry:*
Stock solutions (0.1 M) of vitamin B_1 hydrochloride and vitamin B_6 hydrochloride are prepared in distilled water and by keeping both the pH and ionic strength at constant values with acetate buffer

solution (pH 3.5); 10^{-2}, 10^{-3}, and 10^{-4} M vitamin solutions are obtained from the respective stock solution by successive dilutions. The sensors (vitamin-sensitive and SCE) are placed in the standard solutions, and EMF readings (linear axis) are plotted against concentration (logarithmic axis). The sample concentration is determined from this graph.

v. *Vitamin B_1 and vitamin B_6 determination from pharmaceutical preparations with vitamin membrane sensors:*

a. *Vitamin B_1 and vitamin B_6 for injections*—1.00 cm^3 of the commercial product is diluted with distilled water to a final volume of 50 cm^3; 2.5 cm^3 of this solution is diluted with distilled water and acetate buffer of pH 3.5 (10% buffer solution, v/v) to a 25-cm^3 volumetric flask. This solution (V_x) is used for analysis. The appropriate vitamin membrane sensor and SCE are immersed in this solution. After potential equilibration by stirring, the EMF value is recorded; 2.5 cm^3 of a 10^{-2} M standard solution of the respective vitamin hydrochloride solution (pH 3.5) is added and the change in millivolt reading (accuracy ± 0.1 mV) is recorded and used to calculate the vitamin concentration of the respective injectable solution.

b. *Vitamin B_1 and vitamin B_6 tablets*—At least 10 tablets are made into a powder. An appropriate amount of the powder, equivalent to approximately 5 mg of vitamin, is weighed and dissolved in a 50-cm^3 volumetric flask; 5.0 cm^3 of acetate buffer solution of pH 3.5 is added and the solution is made up to volume with distilled water. This solution is divided into two 25-cm^3 portions, in which both indicator and reference electrodes are immersed. After electrode equilibration by stirring and after recording the EMF, 2.5 cm^3 of 10^{-2} M standard solution of the respective vitamin hydrochloride solution (pH 3.5) is added and the change in millivolt reading (accuracy ± 0.1 mV) is recorded and used to calculate the vitamin content of the tablets.

vi. *Vitamin B_1 determination from pharmaceutical preparations with lead(II)-ion-selective membrane sensor:*

For formulated vitamin B_1, 20 tablets are ground or the contents of 5 capsules are mixed. An aliquot of the powder equivalent to 5 to 10 mg of vitamin is accurately weighed and transferred to the bottom of a Pyrex test tube (10 × 1 cm). Three pellets (about 0.2 to 0.3 g) of potassium hydroxide are added and the tube is placed in a 250 to 280°C sand bath for 5 to 8 min. Then the contents are cooled at room temperature; 5 cm^3 of 0.02 M potassium plumbite solution (prepared by dissolving 6.624 g lead nitrate in 1 dm^3 0.2 M KOH) is added to alkali reaction product in the test tube and the tube is shaken and placed in a boiling water bath for 2 min. The reaction mixture is quantitatively transferred to a 150-cm^3 beaker and the tube is washed with 30 cm^3 of pH 4.5 acetate buffer solution. The solution is diluted to 50 cm^3 with buffer, and a lead-ion-selective membrane sensor in conjunction with a double-junction reference electrode is immersed into it. While stirring, three to five 1-cm^3 portions of standardized 0.005 M EDTA are added; after each addition the potential is recorded when it attains constant value (after about 20 s). Using Gran's plot paper, the potential vs. EDTA volume is plotted. The intercept on the horizontal axis indicates the equivalence point. A blank experiment is carried out under identical conditions. The corrected equivalence point is used to calculate the content of vitamin B$_1$ in the respective pharmaceutical product.

vii. *Vitamin B$_2$ determination with a potentiometric biosensor for riboflavin, based on the use of aporiboflavin-binding protein:*
Three standards of riboflavin solutions (in 0.1 M TRIS–HCl buffer, pH 7.7) in the range 0.1 to 2 μM concentration are freshly prepared from a 0.1 mM riboflavin stock solution (in 0.01 M acetate, pH 5.0; stored at 4°C and protected from light). Aliquots of these standards are transferred into 100-cm^3 beakers containing magnetic stirring bars. The potentiometric biosensor in conjunction with a saturated calomel electrode is immersed successively in these solutions, and the EMF values are recorded vs. concentration. The unknown concentration of riboflavin is determined from this calibration graph. (Prior to measurements the biosensor is immersed for about 30 min in the background electrolytes until it achieves a constant potential.)

viii. *Vitamin B$_{12}$ determination with a cyanide-ion-selective membrane sensor:*
About 1 to 10 mg of the sample is weighed directly into a vial and, avoiding moisture contact, the vial is carefully introduced into a quartz test tube, which is placed in an electric oven; the nitrogen flow to the quartz tube is adjusted to 45 to 50 bubbles per minute as measured in the receiving test tube, which contains 10 cm^3 of 0.1 M potassium nitrate, adjusted to pH 12 to 13 with 0.1 M sodium hydroxide. By heating the oven to about 140 to 145°C (15 to 20 min), the cyanide content of the sample in the quartz tube is distilled as hydrogen cyanide and trapped in the 0.1 M potassium nitrate set at pH 12 to 13, where it is determined by means of a cyanide membrane sensor (direct potentiometry).

References

1. D. S. Papastathopoulos and G. A. Rechnitz, *Anal. Chem.*, **48**, 862 (1976).
2. M. A. Arnold and G. A. Rechnitz, *Anal. Chem.*, **53**, 515 (1981).
3. Y. Umezawa, M. Kataoka, W. Takami, E. Kimura, T. Koike, and H. Nada, Proc. 2nd Beijing Conf. and Exhib. on Instrum. Anal., Beijing, Oct., 1987, p. 1361.

4. Y. Umezawa, M. Kataoka, W. Takami, E. Kimura, T. Koike, and H. Nada, *Anal. Chem.*, **60**, 2392 (1988).

5. K. Vytras, *Int. Lab.*, Sept., 35 (1979).

6. K. Vytras, *Amer. Lab.*, Nov., 93 (1979).

7. K. Vytras, *Ion-Selective Electrode Rev.*, **7**, 77 (1985).

8. K. Vytras and V. Riha, *Česk. Farm.*, **26**, 9 (1977).

9. W. Selig, *Mikrochim Acta 2*, 138 (1980).

10. J. Kálmán, K. Tóth, and D. Küttel, *Acta Pharm. Hung.*, **41**, 267 (1971).

11. J. Kálmán, K. Tóth, and D. Küttel, *Acta Pharm. Hung.*, **42**, 152 (1972).

12. S. Yao, G. Shen, and G. Dai, *Kexue Tongbao*, **29**, 1416 (1984).

13. V. V. Coşofreţ, C. Ştefănescu, and A. A. Bunaciu, *Talanta*, **26**, 1035 (1979).

14. T. K. Christopoulos, E. P. Diamandis, and T. P. Hadjiioannou, *Anal. Chim. Acta*, **143**, 143 (1982).

15. E. P. Diamandis and T. P. Hadjiioannou, *Anal. Chim. Acta*, **123**, 341 (1981).

16. B. Li, Z. Zhang, X. You, T. Lu, and G. Yin, *Analyst*, **113**, 57 (1988).

17. G.-L. Shen, Y. Zhang, and R.-Q. Yu, *Yaoxue Xuebao*, **20**, 151 (1985).

18. E. P. Diamandis, E. Athanasiou-Malaki, D. S. Papastathopoulos, and T. P. Hadjiioannou, *Anal. Chim. Acta*, **128**, 239 (1981).

19. H.-X. Lin, H. Liu, and G.-L. Sheng, *Yaowu Fenxi Zazhi*, **5**, 150 (1985).

20. S. S. M. Hassan and F. S. Tadros, *Anal. Chem.*, **56**, 542 (1984).

21. S. S. M. Hassan, M. A. Ahmed, and F. S. Tadros, *Talanta*, **34**, 723 (1987).

22. G. Shen, S. Yao, Y. Jiang, and W. Wu, *Fenxi Huaxue*, **11**, 481 (1983).

23. D.-P. Huang, L.-Q. Zhang, X.-Y. Lu, Z. Nan, Z.-H. Lu, and G.-J. Song, *Yaoxue Xuebao*, **22**, 545 (1987).

24. H. Chi and T.-H. Zhou, *Yaoxue Xuebao*, **18**, 278 (1983).

25. J. Huang, P.-Z. Pao, and S.-S. Qian, *Yaowu Fenxi Zazhi*, **7**, 31 (1987).

26. S.-Z. Yao, A.-X. Zhu, and L.-H. Nie, *Yaowu Fenxi Zazhi*, **6**, 205 (1986).

27. S. S. M. Hassan, M. A. Ahmed, and M. S. Saoudi, *Anal. Chem.*, **57**, 1126 (1985).

28. S. Yao, G. Shen, and G. Dai, *Kexue Tongbao*, **29**, 1416 (1984).

29. V. V. Coşofreţ and R. P. Buck, *J. Pharm. Biomed. Anal.*, **3**, 123 (1985).

30. L. Cunningham and H. Freiser, *Anal. Chim. Acta*, **139**, 97 (1982).

31. C. E. Efstathiou, E. P. Diamandis, and T. P. Hadjiioannou, *Anal. Chim. Acta*, **127**, 173 (1981).

32. Z. Dai, J. Xie, X. Su, and W. Li, *Gaodeng Xuexiao Huaxue Xuebao*, **8**, 882 (1987).

33. S. Yao, G. Shen, and H. Wu, *Yaowu Fenxi Zazhi*, **4**, 281 (1984).

34. J. I. Anzai, C. Isomura, and T. Osa, *Chem. Pharm. Bull.*, **33**, 236 (1985).

35. S.-Z. Yao and J.-H. Liu, *Huaxue Xuebao*, **43**, 611 (1985).

36. M. S. Ionescu, D. Negoiu, and V. V. Coşofreţ, *Anal. Lett.*, **16**, 553 (1983).

37. V. V. Coşofreţ, *Rev. Roum. Chim.*, **23**, 1489 (1978).

38. S. S. M. Hassan and M. B. Elsayes, *Anal. Chem.*, **51**, 1651 (1979).

39. G. Shen, S. Yao, and S. Sun, *Gaodeng Xuexiao Huaxue Xuebao*, **5**, 136 (1984).

40. H.-M. Zhong, C.-Y. Wang, L.-Y. Wang, and F. Lu, *Yaowu Fenxi Zazhi*, **6**, 169 (1986).

41. M. S. Ionescu, A. A. Abrutis, M. Lăzărescu, E. Pascu, G. E. Baiulescu, and V. V. Coşofreţ, *J. Pharm. Biomed. Anal.*, **5**, 59 (1987).

42. G. A. Rechnitz, R. K. Kobos, S. J. Riechel, and C. R. Gebauer, *Anal. Chim. Acta*, **94**, 357 (1977).

43. S. R. Grobler, N. Basson, and C. W. Van Wyk, *Talanta*, **29**, 49 (1982).

44. D. P. Nikolelis and T. P. Hadjiioannou, *Anal. Chim. Acta*, **147**, 33 (1983).

45. P. W. Alexander and C. Maitra, *Anal. Chem.*, **53**, 1590 (1981).

46. E. M. Athanasiou-Malaki and M. A. Koupparis, *Analyst*, **112**, 757 (1987).

47. P. K. C. Tseng and W. F. Gutknecht, *Anal. Chem.*, **47**, 2316 (1975).

48. L. C. Gruen and B. S. Harrap, *Anal. Biochem.*, **42**, 377 (1971).

49. W. Selig, *Mikrochim Acta*, 453 (1973).

50. M. A. Jensen and G. A. Rechnitz, *Anal. Chim. Acta*, **101**, 125 (1978).

51. C. Liteanu, E. Hopîrtean, and R. Vlad, *Rev. Roum. Chim.*, **21**, 988 (1976).

52. E. M. Athanasiou-Malaki and M. A. Koupparis, *Anal. Chim. Acta*, **161**, 348 (1984).

53. E. G. Sarantonis, E. P. Diamandis, and M. Y. Karayannis, *Anal. Biochem.*, **155**, 129 (1986).

54. N. Smit and G. A. Rechnitz, *Biotechnol. Lett.*, **6**, 209 (1984).

55. M. Matsui and H. Freiser, *Anal. Lett.*, **3**, 161 (1970).

56. S. Kuriyama and G. A. Rechnitz, *Anal. Chim. Acta*, **131**, 91 (1981).

57. G. G. Guilbault and F.-R. Shu, *Anal. Chim. Acta*, **56**, 333 (1971).

58. M. A. Arnold and G. A. Rechnitz, *Anal. Chem.*, **52**, 1170 (1980).

59. G. A. Rechnitz, M. A. Arnold, and M. E. Meyerhoff, *Nature*, **278**, 466 (1979).

60. M. A. Arnold and G. A. Rechnitz, *Anal. Chim. Acta*, **113**, 351 (1980).

61. R. R. Walters, B. E. Moriarty, and R. P. Buck, *Anal. Chem.*, **52**, 1680 (1980).

62. R. R. Walters, P. A. Johnson, and R. P. Buck, *Anal. Chem.*, **52**, 1684 (1980).

63. P. M. Kovach and M. E. Meyerhoff, *Anal. Chem.*, **54**, 217 (1982).

64. R. M. Ianniello and A. M. Yacynych, *Anal. Chim. Acta*, **131**, 123 (1981).

65. W. C. White and G. G. Guilbault, *Anal. Chem.*, **50**, 1481 (1978).

66. K.-W. Fung, S.-S. Kuan, H.-Y. Sung, and G. G. Guilbault, *Anal. Chem.*, **51**, 2319 (1979).

67. D. P. Nikolelis and T. P. Hadjiioannou, *Anal. Lett.*, **16**, 401 (1983).

68. G. G. Guilbault and G. Nagy, *Anal. Lett.*, **6**, 301 (1973).

69. C.-P. Hsiung, S.-S. Kuan, and G. G. Guilbault, *Anal. Chim. Acta*, **90**, 45 (1977).

70. T. Matsunaga, I. Karube, N. Teraoka, and S. Suzuki, *Anal. Chim. Acta*, **127**, 245 (1981).

71. G. G. Guilbault and F.-R. Shu, *Anal. Chem.*, **44**, 2161 (1972).

72. C. L. Di Paolantonio and G. A. Rechnitz, *Anal. Chim. Acta*, **141**, 1 (1982).

73. S. R. Grobler and C. W. Van Wyk, *Talanta*, **27**, 602 (1980).

74. W. E. Morf, G. Kahr, and W. Simon, *Anal. Chem.*, **46**, 1538 (1974).

75. D. Tse, T. Kuwana, and G. Royer, *J. Electroanal. Chem.*, **98**, 345 (1979).

76. R. J. H. Wilson, G. Kay, and M. Lilly, *Biochem. J.*, **109**, 137 (1968).

77. E. G. Sarantonis and M. I. Karayannis, *Anal. Biochem.*, **130**, 177 (1983).

78. A. M. Berjonneau, C. T. Minh, and G. Broun, *C.R. Hebd. Scéances Acad. Sci., Sér. D*, **275**, 121 (1972).

79. E. Hopîrtean and E. Stefăniga, *Rev. Roum. Chim.*, **21**, 305 (1976).

80. E. Hopîrtean, C. Liteanu, and E. Stefăniga, *Rev. Roum. Chim.*, **19**, 1651 (1974).

81. L. Z. Cai, H. J. Pan, and L. S. Cui, *Anal. Proc. (London)*, **24**, 339 (1987).

82. E. P. Diamandis and T. K. Christopoulos, *Anal. Chim. Acta*, **152**, 281 (1983).

83. A. Mitsana-Papazoglou, T. K. Christopoulos, E. P. Diamandis, and T. P. Hadjiioannou, *Analyst*, **110**, 1091 (1985).

84. C. Luca, C. Baloescu, G. Semenescu, and T. Țolea, *Rev. Chim. (Bucharest)*, **30**, 72 (1975).

85. S. S. Badawy, H. F. Shoukry, and M. M. Omar, *Anal. Chem.*, **60**, 758 (1988).

86. K.-K. Choi and K.-W. Fung, *Anal. Chim. Acta*, **138**, 385 (1982).

87. Y.-Y. Zhao and C.-Y. Wang, *Yaowu Fenxi Zazhi*, **7**, 223 (1987).

88. E. Hopîrtean and E. Veress, *Rev. Roum. Chim.*, **23**, 273 (1978).

89. G. D. Carmack and H. Freiser, *Anal. Chem.*, **49**, 1577 (1977).

90. A. K. Covington, T. R. Harbinson, and A. Sibbald, *Anal. Lett.*, **15**, 1423 (1982).

91. G.-L. Shen, X.-Y. Shi, and R.-Q. Yu, *Yaoxue Xuebao*, **22**, 841 (1987).

92. V. V. Coşofreţ and A. A. Bunaciu, *Anal. Lett.*, **12**, 617 (1979).

93. G.-L. Shen, L.-J. Cai, and R.-Q. Yu, *Yaoxue Xuebao*, **21**, 226 (1986).

94. H. Hopkala, *Acta Pol. Pharm.*, **43**, 577 (1986).

95. R. W. Cattrall and H. Freiser, *Anal. Chem.*, **43**, 1905 (1971).

96. H. J. James, G. D. Carmack, and H. Freiser, *Anal. Chem.*, **44**, 856 (1972).

97. B. M. Kneebone and H. Freiser, *Anal. Chem.*, **45**, 449 (1973).

98. V. V. Coşofreţ, *Membrane Electrodes in Drug-Substances Analysis*, Pergamon Press, Oxford, 1982, p. 141.

99. F. Peter and R. Rosset, *Anal. Chim. Acta*, **64**, 397 (1973).

100. P. D'Orazio and G. A. Rechnitz, *Anal. Chem.*, **49**, 41 (1977).

101. G. E. Baiulescu, V. V. Coşofreţ, and M. Blasnic, in *Ion-Selective Electrodes*, E. Pungor and I. Buzás, eds., Akad. Kiadó, Budapest, 1978, p. 207.

102. R. Staroscik and T. Blaskiewicz, *Pharmazie*, **40**, 248 (1985).

103. M. T. Benignetti, L. Campanella, and T. Ferri, *Z. Anal. Chem.*, **296**, 412 (1979).

104. G. J. Coetzee and H. Freiser, *Anal. Chem.*, **40**, 2071 (1968).

105. G. J. Coetzee and H. Freiser, *Anal. Chem.*, **41**, 1128 (1969).

106. I. Shigematsu, A. Ota, and M. Matsui, *Bull. Inst. Chem. Res., Kyoto Univ.*, **51**, 268 (1973).

107. H. Hara, S. Okazaki, and T. Fujinaga, *Anal. Chim. Acta*, **121**, 119 (1980).

108. P. Amoroso, L. Campanella, G. De Angelis, T. Ferri, and R. Morabita, *J. Membr. Sci.*, **16**, 259 (1983).

109. L. Campanella, T. Ferri, and M. Gabrielli, *Rev. Roum. Chim.*, **27**, 681 (1982).

110. Y. Zhu and S. Huang, *Hunan Daxue Xuebao*, **11**, 139 (1984).

111. V. V. Coşofreţ and R. P. Buck, *Anal. Chim. Acta*, **162**, 357 (1984).

112. T. Goina, Şt. Hobai, and A. Rodeanu, *Farmacia (Bucharest)*, **24**, 89 (1976).

113. V. V. Coşofreţ, P. G. Zugrăvescu, and G. E. Baiulescu, *Talanta*, **24**, 461 (1977).

114. H. Soep and P. Demoen, *Microchem. J.*, **4**, 77 (1960).

115. E. Debal and R. Levy, *Mikrochim. Acta*, 285 (1964).

116. R. Bennewitz, *Mikrochim. Acta*, 54 (1960).

117. W. J. Kirsten, *Mikrochim. Acta*, 369 (1960).

118. E. Pella, *Mikrochim. Acta*, 472 (1961).

119. E. Pella, *Mikrochim. Acta*, 369 (1965).

120. W. Potman and E. A. M. F. Dahmen, *Mikrochim. Acta*, 303 (1972).

121. M. S. Ionescu, S. Sitaru, G. E. Baiulescu, and V. V. Coşofreţ, *Rev. Chim. (Bucharest)*, **38**, 256 (1987).

122. K. Shirahama, H. Kamaya, and I. Ueda, *Anal. Lett.*, **16**, 1485 (1983).

123. M. S. Ionescu, A. A. Abrutis, N. Rădulescu, G. E. Baiulescu, and V. V. Coşofreţ, *Analyst*, **110**, 929 (1985).

124. S. S. M. Hassan and M. A. Ahmed, *J. Assoc. Off. Anal. Chem.*, **69**, 618 (1986).

125. M. S. Ionescu, H. Scurei, V. Tamas, A. Voiculescu, G. E. Baiulescu, and V. V. Coşofreţ, *Rev. Chim. (Bucharest)*, **38**, 1138 (1987).

126. G. Shen, S. Yao, and X. Liu, *Yaowu Fenxi Zazhi*, **3**, 257 (1983).

127. S.-Z. Yao, W.-L. Ma, A.-X. Zhu, and L.-H. Nie, *Yaoxue Xuebao*, **21**, 285 (1986).

128. A. G. Fogg, M. A. Abdalla, and H. P. Henriques, *Analyst*, **107**, 449 (1982).

129. M. A. Abdalla, A. G. Fogg, and C. Burgess, *Analyst*, **107**, 213 (1982).

130. Y. Asano and S. Ito, *Nippon Kagaku Kaishi*, 1494 (1980).

131. K. Matsumoto, H. Seijo, T. Watanabe, I. Karube, I. Satoh, and S. Suzuki, *Anal. Chim. Acta*, **105**, 429 (1979).

132. I. Satoh, I. Karube, and S. Suzuki, *J. Soiid-Phase Biochem.*, **2**, 1 (1977).

133. I. Karube, S. Suzuki, S. Kinoshita, and J. Mizuguchi, *Ind. Eng. Chem. Prod. Res. Dev.*, **10**, 160 (1971).

134. M. T. M. Zaki, *Anal. Lett.*, **18**, 1697 (1985).

135. Y. M. Dessouky, K. Tóth, and E. Pungor, *Analyst*, **95**, 1027 (1970).

136. M. A. Koupparis and T. P. Hadjiioannou, *Anal. Chim. Acta*, **94**, 367 (1977).

137. M. A. Koupparis and T. P. Hadjiioannou, *Mikrochim. Acta 2*, 267 (1978).

138. M. A. Koupparis and T. P. Hadjiioannou, *Anal. Chim. Acta*, **96**, 31 (1978).

139. S. S. M. Hassan and M. H. Eldessouki, *Talanta*, **26**, 531 (1979).

140. J. G. Pentari and C. E. Efstathiou, *Anal. Chim. Acta*, **153**, 161 (1983).

141. A. Campiglio and G. Traverso, *Mikrochim. Acta 1*, 485 (1980).

142. V. V. Coşofreţ and R. P. Buck, *Anal. Chim. Acta*, **174**, 299 (1985).

143. Q.-X. Zhu and Z.-J. Tian, *Yaowu Fenxi Zazhi*, **6**, 104 (1986).

144. K. Kina, N. Maekawa, and N. Ishibashi, *Bull. Chem. Soc. Jpn.*, **46**, 2772 (1973).

145. K. Fukamaki and N. Ishibashi, *Bunseki Kagaku*, **27**, 152 (1978).

146. Y.-L. Liang and Q.-M. Huang, *Yaoxue Xuebao*, **20**, 628 (1985).

147. G. Shen and X. Li, *Hunan Daxue Xuebao*, **11**, 86 (1984).

148. A. F. Shoukry, S. S. Badawi, and I. M. Issa, *J. Electroanal. Chem.*, **233**, 29 (1987).

149. G.-L. Shen, Y.-P. Jian, and B.-Q. Yu, *Yaowu Fenxi Zazhi*, **5**, 75 (1985).

150. W. Selig, *Z. Anal. Chem.*, **308**, 21 (1981).

151. S. Tagami and Y. Muramoto, *Chem. Pharm. Bull.*, **32**, 1018 (1984).

152. D. S. Papastathopoulos and G. A. Rechnitz, *Anal. Chem.*, **47**, 1792 (1975).

153. L. Campanella, L. Sorrentino, and M. Tomassetti, *Anal. Lett.*, **15**, 1515 (1982).

154. L. Campanella, L. Sorrentino, and M. Tomassetti, *Analyst*, **108**, 1490 (1983).

155. U. Biader Ceipidor, R. Curini, G. D'Ascenzo, M. Tomassetti, A. Alessandrini, and C. Montesani, *G. Ital. Chim. Clin.*, **5**, 127 (1980).

156. L. Campanella, M. Tomassetti, and M. Cordatore, *J. Pharm. Biomed. Anal.*, **4**, 155 (1986).

157. L. Campanella, F. Mazzei, M. Tomassetti, and R. Sbrilli, *Analyst*, **13**, 325 (1988).

158. G. Baum, *Anal. Lett.*, **3**, 105 (1970).

159. G. Baum, F. B. Ward, and S. Yaverbaum, *Clin. Chim. Acta*, **36**, 405 (1972).

160. G. Baum, M. Lynn, and F. B. Ward, *Anal. Chim. Acta*, **65**, 385 (1973).

161. E. Hopîrtean and M. Miklos, *Rev. Chim. (Bucharest)*, **29**, 1178 (1978).

162. C. T. Minh, R. Guyonnet, and J. Beaux, *C.R. Acad. Sci. Paris, Sér. C*, **286**, 115 (1978).

163. K. Kina, N. Maekawa, and N. Ishibashi, *Bull. Chem. Soc. Jpn.*, **46**, 2772 (1973).

164. A. Jaramillo, S. Lopez, J. B. Justice, Jr., J. D. Salamone, and D. B. Neill, *Anal. Chim. Acta*, **146**, 149 (1983).

165. M. F. Suaud-Chagny and J. F. Pujol, *Analusis*, **13**, 25 (1985).

166. R. Tor and A. Freeman, *Anal. Chem.*, **58**, 1042 (1986).

167. M. Gotoh, E. Tamiya, M. Momoi, Y. Kagawa, and I. Karube, *Anal. Lett.*, **20**, 857 (1987).

168. T. Sullivan, Master's thesis, Emory University, Atlanta, GA, 1981.

169. N. Ishibashi and H. Kohara, *Anal. Lett.*, **4**, 785 (1971).

170. A. Mitsana-Papazoglou, E. P. Diamandis, and T. P. Hadjiioannou, *J. Pharm. Sci.*, **76**, 485 (1987).

171. N. Ishibashi, K. Kina, and N. Maekawa, *Chem. Lett.*, 119 (1973).

172. K. Fukamachi and N. Ishibashi, *Yakugaku Zasshi*, **99**, 126 (1979).

173. X.-Y. Hu and Z.-Z. Leng, *Yigao Gongye*, **19**, 313 (1988); *Chem. Abstr.*, **109**, 197,29ot (1988).

174. S.-Z. Yao, W.-L. Ma, A.-X. Zhu, and L.-H. Ni, *Yaowu Fenxi Zazhi*, **6**, 274 (1986).

175. M. J. M. Campbell, B. Demetriou, and R. Jones, *Analyst*, **105**, 605 (1980).

176. A. A. Bunaciu, M. S. Ionescu, I. Enăchescu, G. E. Baiulescu, and V. V. Coşofreţ, *Analusis*, **16**, 131 (1988).

177. M. S. Ionescu, V. V. Coşofreţ, T. Panaitescu, and M. Costescu, *Anal. Lett.*, **13**, 715 (1980).

178. G. E. Baiulescu and V. V. Coşofreţ, *Applications of Ion-Selective Membrane Electrodes in Organic Analysis* (Ellis Horwood Series in Anal. Chem.), John Wiley & Sons, Chichester, 1977.

179. V. V. Coşofreţ, *Ion-Selective Electrode Rev.*, **2**, 159 (1980).

180. N. E. Nashed and S. Lindenbaum, *Analyst*, **112**, 205 (1987).

181. G. J. Moody, R. B. Oke, and J. D. R. Thomas, *Analyst*, **95**, 910 (1970).

182. T. Higuchi, C. R. Illian, and J. L. Tossounian, *Anal. Chem.*, **42**, 1674 (1970).

183. S.-Z. Yao, *J. Pharm. Biomed. Anal.*, **5**, 325 (1987).

184. S. S. M. Hassan and M. B. Habib, *Anal. Chem.*, **53**, 508 (1981).

185. M. Y. Keating and G. A. Rechnitz, *Anal. Lett.*, **18**, 1 (1985).

186. M. Hashimoto and K. Kawano, in *Enzyme Immunoassay*, E. Ishikawa, T. Kawai, and K. Miyai, eds., Igaku-Shoin Ltd., Tokyo, 1981.

187. K. Vytras, *Coll. Czech. Chem. Commun.*, **42**, 3168 (1977).

188. W. Selig, *Z. Anal. Chem.*, **308**, 21 (1981).

189. I. A. Gurev and T. S. Vyatchanina, *Zh. Anal. Khim.*, **34**, 976 (1979).

190. A. A. Kulugin and I. A. Gurev, *Zh. Anal. Khim.*, **35**, 2424 (1980).

191. K. Vytras and M. Dajkova, *Anal. Chim. Acta*, **141**, 377 (1982).

192. A. G. Fogg, A. A. Al-Sibbai, and K. S. Yoo, *Anal. Lett.*, **10**, 173 (1977).

193. C. Brugges, A. G. Fogg, and D. T. Burns, *Lab. Pract.*, **22**, 472 (1973).

194. A. G. Fogg and K. S. Yoo, in *Ion Selective Electrodes*, E. Pungor and I. Buzás, eds., Akad. Kiadó, Budapest, 1978, p. 9.

195. A. G. Fogg and K. S. Yoo, *Anal. Chim. Acta*, **113**, 165 (1980).

196. K. Fukamaki, R. Nakagawa, M. Morimoto, and N. Ishibashi, *Bunseki Kagaku*, **24**, 428 (1975).

197. A. Jyo, K. Fukamaki, W. Koga, and N. Ishibashi, *Bull. Chem. Soc. Jpn.*, **50**, 670 (1977).

198. Y. Yamamoto, T. Tarumoto, and E. Iwamoto, *Anal. Chim. Acta*, **64**, 1 (1973).

199. M.-C. Ni, *Zhongcaoyao*, **16**, 56 (1985).

200. S. S. M. Hassan and M. M. Saoudi, *Analyst*, **111**, 1367 (1986).

201. S. S. M. Hassan and G. A. Rechnitz, *Anal. Chem.*, **58**, 1052 (1986).

202. S. Tagami and M. Fujita, *J. Pharm. Sci.*, **71**, 523 (1987).

203. *United States Pharmacopeia*, 20th rev., U.S. Pharmacopeial Convention, Inc., Rockville, MD, 1980, p. 548.

204. *The Japanese Pharmacopoeia*, 9th rev., Hirokawa, Tokyo, 1976, p. 436.

205. L. Campanello and D. Gozzi, *Anal. Chim.* (*Rome*), **67**, 345 (1977).

206. L. Campanella, G. De Angelis, T. Ferri, and D. Gozzi, *Analyst*, **102**, 723 (1977).

207. L. Campanella, T. Ferri, D. Gozzi, and T. Scorcelletti, in *Ion-Selective Electrodes*, E. Pungor and I. Buzás, eds., Akad. Kiadó, Budapest, 1978, p. 307.

208. L. Campanella, T. Ferri, and D. Gozzi, *Rev. Roum. Chim.*, **23**, 281 (1978).

209. D. P. Nikolelis, C. E. Efstathiou, and T. P. Hadjiioannou, *Analyst*, **104**, 1181 (1979).

210. C. E. Efstathiou, D. P. Nikolelis, and T. P. Hadjiioannou, *Anal. Lett.*, **15**, 1179 (1982).

211. B. J. Wincke, M. J. Devleeschouwer, J. Dony, and G. J. Patriarche, *Intern. J. Pharm.*, **21**, 265 (1984).

212. T. Tagami and H. Maeda, *J. Pharm. Sci.*, **72**, 988 (1983).

213. T. Goina, S. Hobai, and H. Rodeanu, *Farmacia* (*Bucharest*), **24**, 89 (1976).

214. K. Vytras, J. Kalous, Z. Kalabova, and M. Remes, *Anal. Chim. Acta*, **141**, 163 (1982).

215. A. Campiglio, *Mikrochim. Acta 2*, 347 (1982).

216. W. Selig, *Mikrochim. Acta 2*, 141 (1982).

217. M. S. Ionescu, M. Lăzărescu, A. Ionescu, and G. E. Baiulescu, *Talanta*, **34**, 887 (1987).

218. L. Macholan and L. Schanel, *Coll. Czech. Chem. Commun.*, **42**, 3667 (1977).

219. I. A. Gur'ev, E. A. Gushchina, and E. N. Mitina, *Zh. Anal. Khim.*, **34**, 1184 (1979).

220. S. G. Back, *Anal. Lett.*, **7**, 793 (1971).

221. T. P. Hadjiioannou and E. P. Diamandis, *Anal. Chim. Acta*, **94**, 443 (1977).

222. S. S. M. Hassan, *Anal. Chem.*, **49**, 45 (1977).

223. M. E. El-Taras, E. Pungor, and G. Nagy, *Anal. Chim. Acta*, **82**, 285 (1976).

224. H. Gao, D. Guo, and X. Song, *Fenxi Huaxue*, **15**, 554 (1987).

225. S.-M. Zhu, J.-H. Jiang, and C.-Z. Peng, *Yaoxue Xuebao*, **21**, 198 (1986).

226. B. C. Jones, J. E. Heveran, and B. Z. Senkowski, *J. Pharm. Sci.*, **60**, 1036 (1971).

227. W. I. Rogers and J. A. Wilson, *Anal. Biochem.*, **32**, 31 (1969).

228. T.-S. Ma, *Anal. Chem.*, **30**, 1557 (1958).

229. S. Ikeda, *Anal. Lett.*, **7**, 343 (1974).

230. M. A. Koupparis, C. E. Efstathiou, and T. P. Hadjiioannou, *Anal. Chim. Acta*, **107**, 91 (1979).

231. C. E. Efstathiou and T. P. Hadjiioannou, *Anal. Chim. Acta*, **89**, 55 (1977).

232. I. Karube, T. Matsunaga, and S. Suzuki, *Anal. Chim. Acta*, **109**, 39 (1979).

233. R. F. Cosgrove and A. E. Beezer, *Anal. Chim. Acta*, **105**, 77 (1979).

234. R. F. Cosgrove, *J. Appl. Bacteriol.*, **44**, 199 (1978).

235. R. F. Cosgrove, *Anal. Proc.*, **21**, 295 (1984).

236. M. Thompson, P. Y. Worsfold, J. M. Holuk, and E. A. Stubley, *Anal. Chim. Acta*, **104**, 195 (1979).

237. D. L. Simpson and R. K. Kobos, *Anal. Lett.*, **15**, 1345 (1982).

238. D. L. Simpson and R. K. Kobos, *Anal. Chem.*, **55**, 1974 (1983).

239. D. L. Simpson and R. K. Kobos, *Anal. Chim. Acta*, **164**, 273 (1984).

240. A. F. Shoukry and S. S. Badawi, *Microchem. J.*, **36**, 107 (1987).

241. M. Mascini, A. Memoli, and F. Olana, *Anal. Chim. Acta*, **200**, 237 (1987).

242. S.-Z. Yao and M.-H. Cheng, *Yaoxue Xuebao*, **20**, 450 (1985).

243. P. T. Veltsistas and M. I. Karayannis, *Analyst*, **112**, 1579 (1987).

244. S.-Z. Yao, J. Xiao, and L.-H. Nie, *Yaoxue Xuebao*, **23**, 281 (1988).

245. S.-Z. Yao, J. Shian, and L.-H. Nie, *Gaodeng Xuexiao Huaxue Xuebao*, **9**, 637 (1988).

246. *United States Pharmacopeia*, 20th rev. U.S. Pharmacopeial Conv. Inc., Rockville, MD, 1980, p. 956.

247. S. J. Updike and G. P. Hicks, *Nature*, **214**, 986 (1967).

248. S. P. Bessman and R. D. Schultz, *Trans. Am. Soc. Artif. Internal Organs*, **19**, 361 (1973).

249. M. K. Weibch, N. Dritschilo, H. J. Bright, and A. E. Humphrey, *Anal. Biochem.*, **52**, 402 (1973).

250. H. J. Kunz and M. Stasny, *Clin. Chem.*, **20**, 1018 (1974).

251. G. G. Guilbault and G. J. Lubrano, *Anal. Chim. Acta*, **44**, 439 (1973).

252. G. G. Guilbault and G. J. Lubrano, *Anal. Chim. Acta*, **60**, 254 (1972).

253. M. Nanjo and G. G. Guilbault, *Anal. Chim. Acta*, **73**, 367 (1974).

254. D. R. Thévenot, R. Sternberg, P. R. Coulet, and D. C. Gautheron, *Bioelectrochem. Bioenerg.*, **5**, 541 (1978).

255. R. A. Llenado and G. A. Rechnitz, *Anal. Chem.*, **45**, 826 (1973).

256. R. A. Llenado and G. A. Rechnitz, *Anal. Chem.*, **45**, 2165 (1973).

257. G. Nagy, L. H. von Storp and G. G. Guilbault, *Anal. Chim. Acta*, **66**, 443 (1973).

258. D. R. Thévenot, R. Sternberg, P. R. Coulet, J. Laurent, and D. C. Gautheron, *Anal. Chem.*, **51**, 96 (1979).

259. C.-C. Liu, L. B. Wingard, Jr., S. C. Wolfson, Jr., S.-J. Yao, A. L. Drash, and J. G. Schiller, *Bioelectrochem. Bioenerg.*, **6**, 19 (1979).

260. S. R. Grobler and G. A. Rechnitz, *Talanta*, **27**, 283 (1980).

261. M. Koyama, Y. Sato, M. Aizawa, and S. Suzuki, *Anal. Chim. Acta*, **116**, 307 (1980).

262. D. S. Papastathopoulos, D. P. Nikolelis, and D. P. Hadjiioannou, *Analyst*, **102**, 852 (1977).

263. M. Mascini, M. Iannello, and G. Palleschi, *Anal. Chim. Acta*, **146**, 135 (1983).

264. C. Bertrand, P. R. Coulet, and D. G. Gautheron, *Anal. Chim. Acta*, **126**, 23 (1981).

265. I. K. Al-Hitti, G. J. Moody, and J. D. R. Thomas, *Analyst*, **109**, 1205 (1984).

266. M. Mascini and A. Memoli, *Anal. Chim. Acta*, **182**, 113 (1986).

267. S. S. M. Hassan and M. M. Elsaied, *Analyst*, **112**, 545 (1987).

268. R. Sternberg, D. S. Bindra, G. S. Wilson, and D. R. Thévenot, *Anal. Chem.*, **60**, 2781 (1988).

269. C. E. Efstathiou and T. P. Hadjiioannou, *Anal. Chem.*, **47**, 864 (1975).

270. C. E. Efstathiou, T. P. Hadjiioannou, and E. McNelis, *Anal. Chem.*, **49**, 410 (1977).

271. M. Kudoh, M. Kataoka and T. Kambara, *Talanta*, **27**, 495 (1980).

272. S.-Z. Yao and G.-H. Liu, *Talanta*, **32**, 1113 (1985).

273. M. Bochenska and J. F. Biernat, *Anal. Chim. Acta*, **162**, 369 (1984).

274. F. N. Assubail, G. J. Moody, and J. D. R. Thomas, *Analyst*, **113**, 61 (1988).

275. G. E. Baiulescu and V. V. Coşofreţ, *Rev. Chim. (Bucharest)*, **26**, 1051 (1975).

276. G. E. Baiulescu, V. V. Coşofreţ, and C. Cristescu, *Rev. Chim. (Bucharest)*, **26**, 429 (1975).

277. G. E. Baiulescu, V. V. Coşofreţ, and F. G. Cocu, *Talanta*, **23**, 329 (1976).

278. H. Hopkala and L. Przyborowski, *Pharmazie*, **44**, 65 (1989).

279. S. S. Badawi, A. F. Shoukry, M. S. Rizk and M. M. Omar, *Talanta*, **35**, 487 (1988).

280. S. Uchiyama, Y. Sato, Y. Tofuku, and S. Suzuki, *Anal. Chim. Acta*, **209**, 351 (1988).

281. M. A. Koupparis and T. P. Hadjiioannou, *Talanta*, **25**, 477 (1978).

282. I. I. Koukli and A. C. Calokerinos, *Anal. Chim. Acta*, **192**, 333 (1987).

283. S.-Z. Yao and G.-L. Dai, *Yaowu Fenxi Zazhi*, **4**, 284 (1984).

284. G.-L. Shen and D.-Q. Liao, *Yaowu Fenxi Zazhi*, **4**, 200 (1984).

285. G. J. Patriarche and J. R. Sepulchre, *Analusis*, **14**, 351 (1986).

286. D. W. Mendenhall, T. Higuchi, and L. A. Sternson, *J. Pharm. Sci.*, **68**, 746 (1979).

287. S. Srianujata, W. R. White, T. Higuchi, and L. A. Sternson, *Anal. Chem.*, **50**, 232 (1978).

288. S. Tagami, *Chem. Pharm. Bull.*, **27**, 1820 (1979).

289. S. Tagami, *Chem. Pharm. Bull.*, **28**, 2642 (1980).

290. Y. Michotte, D. L. Massart, and L. Dryon, *Pharm. Acta Helv.*, **52**, 152 (1977).

291. A. Campiglio, *Mikrochim. Acta 2*, 71 (1977).

292. R. W. Wood and H. L. Welles, *J. Pharm. Sci.*, **68**, 1272 (1979).

293. P. Seegopaul and G. A. Rechnitz, *Anal. Chem.*, **56**, 852 (1984).

294. S. S. Badawy, A. F. Shoukry, and I. M. Issa, *Analyst*, **111**, 1363 (1986).

295. F. Rakiás, K. Tóth, and E. Pungor, *Anal. Chim. Acta*, **121**, 93 (1980).

296. S.-Z. Yao and G. Gao, *Yaoxue Xuebao*, **19**, 550 (1984).

297. M. S. Ionescu, V. Badea, G. E. Baiulescu, and V. V. Coşofreţ, *Talanta*, **33**, 101 (1986).

298. Y.-L. Liang and Q.-M. Huang, *Yaowu Fenxi Zazhi*, **5**, 81 (1985).

299. E. R. Hogue and W. C. Landgraf, *Anal. Lett.*, **14**, 1757 (1981).

300. L. Cunningham and H. Freiser, *Anal. Chim. Acta*, **157**, 157 (1984).

301. S.-M. Zhu, Q. Cheng and W.-F. Lu, *Yaowu Fenxi Zazhi*, **7**, 340 (1987).

302. C. R. Martin and H. Freiser, *J. Chem. Educ.*, **57**, 512 (1980).

303. G. J. Papariello, A. K. Mukherji, and C. M. Shearer, *Anal. Chem.*, **45**, 790 (1973).

304. L. F. Cullen, J. F. Rusling, A. Schleifer, and G. J. Papariello, *Anal. Chem.*, **46**, 1955 (1974).

305. H. Nilsson, A. C. Akerlund, and K. Mosbach, *Biochim. Biophys. Acta*, **320**, 529 (1973).

306. H. Nilsson, K. Mosbach, S. O. Enfors, and N. Molin, *Biotechnol. Bioeng.*, **20**, 527 (1978).

307. S. O. Enfors and N. Molin, *Process Biochem.*, **13**, 9 (1978).

308. G. J. Olliff, R. T. Williams, and J. M. Wright, *J. Pharm. Pharmacol.*, **30**, 45 (1978).

309. S. O. Enfors and H. Nilsson, *Enzyme Microb. Technol.*, **1**, 260 (1979).

310. J. Kulys, V. Gureniciene, and V. Laurinavicius, *Antibiotiki* (*Moscow*), **25**, 655 (1980).

311. J. Pittou and J. W. Poole, *J. Pharm. Sci.*, **61**, 1594 (1972).

312. A. G. Dobrolyubov, Yu. G. Emel'yanov, D. D. Grinshpan, S. B. Itsygin, and F. N. Kaputskii, *Antibiot. Med. Biotechnol.*, **32**, 894 (1987).

313. S. A. Dolidze, *Soobshch. Akad, Nauk Gruz. SSR*, **128**, 601 (1987); *Chem. Abstr.*, **108**, 148,823q (1988).

314. J. Anzai, M. Shimada, T. Osa, and C. Chen, *Bull. Chem. Soc. Jpn.*, **60**, 4133 (1987).

315. J. Janata and S. Moss, *Biomed. Eng.*, **11**, 241 (1976).

316. B. Danielsson, I. Lundstrom, K. Mosbach, and L. Stiblert, *Anal. Lett.*, **12**, 1189 (1979).

317. S. Caras and J. Janata, *Anal. Chem.*, **52**, 1935 (1980).

318. S. Caras and J. Janata, *Anal. Chem.*, **57**, 1924 (1985).

319. L. Campanella, M. Tomassetti, and R. Sbrilli, *Ann. Chim.* (*Rome*), **77**, 483 (1986).

320. L. Campanella, F. Mazzei, R. Sbrilli, and M. Tomassetti, *J. Pharm. Biomed. Anal.*, **6**, 299 (1988).

321. B. Karlberg and U. Forsman, *Anal. Chim. Acta*, **83**, 309 (1976).

322. A. Blazsek-Bodó, A. Varga, and I. Kiss, *Rev. Chim.* (*Bucharest*), **29**, 464 (1978).

323. S. S. M. Hassan, *Talanta*, **23**, 738 (1976).

324. C. R. Martin and H. Freiser, *Anal. Chem.*, **52**, 562 (1980).

325. C. R. Martin and H. Freiser, *Anal. Chem.*, **52**, 1772 (1980).

326. C. Kaiser and P. E. Settler in *Burger's Medicinal Chemistry*, 4th ed., Part III, M. E. Wolff, ed., John Wiley & Sons, New York, 1981, Chapter 56.

327. V. V. Coşofreţ and R. P. Buck, *Analyst*, **109**, 1321 (1984).

328. S. Yao, G. Shen, and Z. Ruo, *Yaowu Fenxi Zazhi*, **4**, 100 (1984).

329. S.-Z. Yao and D.-Z. Liu, *Scientia Sinica*, **29**, 954 (1986).

330. A. F. Shoukry and S. S. Badawy, *J. Assoc. Off. Anal. Chem.*, **71**, 1042 (1988).

331. *United States Pharmacopeia*, 20th rev., U.S. Pharmacopeial Convention Inc., Rockville, MD, 1980, pp. 142–144 and 699–670.

332. E. I. Issacson and J. N. Delgado, in *Burger's Medicinal Chemistry*, 4th ed. Part III, M. E. Wolff, ed., John Wiley & Sons, New York, 1981, pp. 829–858.

333. G.-L. Shen, X.-Y. Shi, and R.-Q. Yu., *Yaoxue Xuebao*, **22**, 841 (1987).

334. V. V. Coşofreţ and R. P. Buck, *J. Pharm. Biomed. Anal.*, **4**, 45 (1986).

335. S.-Z. Yao and U.-Q. Tang, *Yaoxue Xuebao*, **19**, 455 (1984).

336. *United States Pharmacopeia*, 20th rev., U.S. Pharmacopeial Convential Inc., Rockville, MD, 1980, pp. 620–622.

337. E. Hopîrtean and F. Cörmos, *Stud. Univ. Babes–Bolyai, Ser. Chem.*, **22**, 35 (1977).

338. D. Negoiu, M. S. Ionescu, and V. V. Coşofreţ, *Talanta*, **28**, 377 (1981).

339. G.-L. Sheng, H. Liu, M. Han, and R.-Q. Yu. *Fenxi Huaxue*, **13**, 706 (1985).

340. A. F. Shoukry, J. M. Issa, R. El-Sheik, and M. Zarch, *Microchem. J.*, **37**, 299 (1988).

341. V. V. Coşofreţ, P. G. Zugrăvescu, and G. E. Baiulescu, *Talanta*, **24**, 461 (1977).

342. Y.-C. Shen, J.-H. Fan, D.-P. Huang, and L.-Q. Meng, *Yaoxue Xuebao*, **22**, 785 (1987).

343. T. K. Christopoulos and E. P. Diamandis, *Analyst*, **112**, 1293 (1987).

344. K. Selinger and R. Staroscik, *Pharmazie*, **33**, 208 (1978).

345. T. Yamada and H. Freiser, *Anal. Chim. Acta*, **125**, 179 (1981).

346. Z.-R. Zhang, D.-Y. Mao, Y.-X. Li, and V. V. Coşofreţ, *Talanta*, **37**, 673 (1990).

347. S. Pinzauti and E. La Porta, *Analyst*, **102**, 938 (1977).

348. M. J. Philippe, *Sci. Tech. Pharm.*, **3**, 419 (1974).

349. B. J. Birch, *Ion-Selective Electrode Rev.*, **3**, 3 (1981).

350. W. Selig, *Anal. Lett.*, **15**, 309 (1982).

351. S. S. Davis and O. Olejnik, *Anal. Chim. Acta*, **132**, 51 (1981).

352. E. P. Diamandis and T. P. Hadjiioannou, *Mikrochim. Acta 2*, 27 (1980).

353. T. P. Hadjiioannou and P. C. Gritzapis, *Anal. Chim. Acta*, **126**, 51 (1981).

354. K. Vytras, M. Dajkova, and V. Mach, *Anal. Chim. Acta*, **127**, 165 (1981).

355. T. P. Hadjiioannou and E. P. Diamandis, *Anal. Chim. Acta*, **94**, 443 (1977).

356. Y.-C. Shen, J.-H. Fan, D.-P. Huang, and L.-Q. Meng, *Yaoxue Xuebao*, **22**, 785 (1987).

357. A. S. Suchoruchkina, S. V. Kashcheev, and S. D. Mikhailova, *Vopr. Khim. Technol.*, **81**, 45 (1986); *Chem. Abstr.*, **107**, 51,136j (1987).

358. E. Benoit, P. Leroy, and A. Nicolas, *Ann. Pharm. Fr.*, **43**, 177 (1985).

359. T. Masadome, T. Imato, and N. Ishibashi, *Bunseki Kagaku*, **36**, 508 (1987).

360. S.-M. Zhu, L.-P. Guo, and X.-Z. Dong, *Yaoxue Xuebao*, **22**, 314 (1987).

361. A. F. Shoukry, S. S. Badawy, and R. A. Farghali, *Anal. Chem.*, **60**, 2399 (1988).

362. A. F. Shoukry, S. S. Badawy, and Y. M. Issa, *Anal. Chem.*, **59**, 1078 (1987).

363. C. Luca and G. Semenescu, *Rev. Chim.* (*Bucharest*), **26**, 946 (1975).

364. W. Selig, *Microchem. J.*, **25**, 200 (1980).

365. Z. Hu, X. Qian, and J. Chen, *Fenxi Huaoxue*, **12**, 145 (1984).

366. W. Schöniger, *Mikrochim. Acta*, 123 (1955).

367. W. Schöniger, *Mikrochim. Acta*, 869 (1956).

368. M. Vandeputte, L. Dryon, L. de Hertogh, and D. L. Massart, *J. Pharm. Sci.*, **68**, 1416 (1979).

369. N. Hazemato, N. Kamo, and Y. Kobatake, *J. Assoc. Off. Anal. Chem.*, **57**, 1205 (1974).

370. G.-X. Zhang, H. Fu, and A. Liu, *Fenxi Huaxue*, **13**, 533 (1985).

371. D.-M. Feng, C.-Z. Chen, and X.-L. Chen, *Fenxi Huaxue*, **15**, 114 (1987).

372. C.-Z. Chen, D.-M. Feng, A.-S. Lu, and H. Li, *Huaxue Tongbao*, **8**, 46 (1987).

373. W. M. Haynes and H. J. Wagenknecht, *Anal. Lett.*, **4**, 491 (1971).

374. A. Mitsana-Papazoglou, E. P. Diamandis, and T. P. Hadjiioannou, *Anal. Chim. Acta*, **159**, 393 (1984).

375. S. S. M. Hassan and M. A. Hamada, *Analyst*, **113**, 1709 (1988).

376. J. L. F. C. Lima, A. A. C. C. Machado, M. Montenegro, B. S. M. Conceicao, and A. M. Roque de Silva, *Rev. Port. Farm.*, **37**, 11 (1987); *Chem. Abstr.*, **108**, 156,544q (1988).

377. N. A. Chaniotakis, S. B. Park, and M. E. Meyerhoff, *Anal. Chem.*, **61**, 566 (1989).

378. T. Fonong and G. A. Rechnitz, **158**, 357 (1984).

379. H. Hopkala and L. Przyborowski, *Pharmazie*, **43**, 422 (1988).

380. J. Martens, P. van den Vinkel, A. H. Boeckstijns, and D. L. Massart, *J. Pharm. Belg.*, **29**, 181 (1974).

381. N. Hazemoto, N. Kamo, and Y. Kabatake, *J. Pharm. Sci.*, **65**, 435 (1976).

382. M. S. Ionescu, S. Cilianu, A. A. Bunaciu, and V. V. Coşofreţ, *Talanta*, **28**, 383 (1981).

383. S. S. M. Hassan and M. H. Eldesouki, *Mikrochim. Acta*, 27 (1979).

384. S. S. M. Hassan and M. H. Eldesouki, *J. Assoc. Off. Anal. Chem.*, **64**, 1158 (1981).

385. S. S. M. Hassan and M. M. Habib, *Microchem. J.*, **26**, 181 (1981).

386. F. Malecki and R. Staroscik, *Anal. Chim. Acta*, **139**, 353 (1982).

387. G. E. Baiulescu, G. Kandemir, M. S. Ionescu, and C. Cristescu, *Talanta*, **32**, 295 (1985).

388. C.-Y. Wang and S.-P. Yang, *Yaowu Fenxi Zashi*, **7**, 359 (1987).

389. S. I. Obtemperanskaya, R. Shahid, M. M. Buzlanova, and V. V. Zhmurova, *Vestn. Mosk. Univ. Ser. 2 Khim.*, **26**, 427 (1985); *Chem. Abstr.*, **103**, 183,638b (1985).

390. S.-Z. Yao, J. Xiao, and L.-H. Nie, *Yaoxue Xuebao*, **22**, 889 (1987).

391. S.-Z. Yao, J. Xiao, and L.-H. Nie, *Talanta*, **34**, 977 (1987).

392. R. Staroscik and F. Malecki, *Acta Pol. Pharm.*, **34**, 643 (1977).

393. P. Seegopaul and G. A. Rechnitz, *Anal. Chem.*, **55**, 1929 (1983).

394. M. K. Papay, K. Tóth, and E. Pungor, *Anal. Chim. Acta*, **56**, 291 (1971).

395. E. Pungor, K. Tóth, and M. K. Papay, *Chem. Anal. (Warsaw)*, **17**, 947 (1972).

396. M. K. Papay, V. P. Izvekov, K. Tóth, and E. Pungor, *Anal. Chim. Acta*, **69**, 173 (1974).

397. M. T. Neshkova, V. P. Izvekov, M. K. Papay, K. Tóth, and E. Pungor, *Anal. Chim. Acta*, **75**, 439 (1975).

398. L. W. Przyborowski, in E. Pungor and I. Buzas, eds., *Ion-Selective Electrodes*, Akad. Kiadó, Budapest, 1978, p. 519.

399. S. L. Richheimer and M. S. Schachet, *J. Pharm. Sci.*, **72**, 822 (1983).

400. S. S. M. Hassan, *Mikrochim. Acta*, 405 (1977).

401. S.-J. Yao, J. Shiao, and L.-H. Nie, *Talanta*, **34**, 983 (1987).

402. N.-X. Wang and Z.-B. Xie, *Yaoxue Xuebao*, **22**, 848 (1987).

403. W.-H. Chan, W.-M. Lee, C.-L. Foo, and W.-K. Tang, *Analyst*, **12**, 845 (1987).

404. M. Petersson, *Anal. Chim. Acta*, **147**, 359 (1983).

405. M. Sharp, M. Petersson, and K. Edstrom, *J. Electroanal. Chem.*, **109**, 271 (1980).

406. R. Christova, M. Ivanova, and M. Novkirishka, *Anal. Chim. Acta*, **85**, 301 (1976).

407. A. C. Calokerinos and T. P. Hadjiioannou, *Microchem. J.*, **28**, 464 (1983).

408. K. Vytras, M. Remes, and H. Hubesova-Svobodova, *Anal. Chim. Acta*, **124**, 91 (1981).

409. N. Ishibashi, H. Kina, and N. Maekawa, *Chem. Lett.*, 119 (1973).

410. S. S. M. Hassan, M. T. Zaki, and M. H. Eldesouki, *J. Assoc. Off. Anal. Chem.*, **62**, 315 (1979).

411. P. Seegopaul and G. A. Rechnitz, *Anal. Chem.*, **55**, 1929 (1983).

412. S. S. M. Hassan, M. L. Iskander, and N. E. Nashed, *Z. Anal. Chem.*, **320**, 584 (1985).

413. C.-Y. Wang and Y.-L. Guo, *Microchem. J.*, **35**, 369 (1987).

414. C.-Y. Wang and Y.-L. Guo, *Yaoxue Tongbao*, **21**, 143 (1986).

415. S.-Z. Yao, X.-M. Xu, and G.-L. Shen, *Yaoxue Xuebao*, **18**, 612 (1983).

416. Z.-R. Zhang, Y.-X. Li, D.-Y. Mao, and V. V. Coşofreţ, *J. Pharm. Biomed. Anal.*, **8**, 385 (1990).

417. T. Yao and G. A. Rechnitz, *Anal. Chem.*, **59**, 2115 (1987).

418. G. Shen, M. Han, H. Liu, and R.-Q. Yu, *Gaodeng Xuexiao Huaxue Xuebao*, **6**, 941 (1985).

419. Z.-Q. Dai, *Yaoxue Tongbao*, **21**, 77 (1986).

420. C. Plese, W. Fox, and K. Williams, *Clin. Chem.*, **29**, 407 (1983).

421. Y. M. Dessouki and E. Pungor, *Analyst*, **96**, 442 (1971).

422. S. Goldstein and A. Duca, *J. Pharm. Sci.*, **65**, 1831 (1976).

423. M. F. Dantartas and J. J. Evans, *J. Electroanal. Chem.*, **109**, 301 (1980).

424. K.-N. Kuo and R. W. Murray, *J. Electroanal. Chem.*, **131**, 37 (1982).

425. N. Winograd, H. N. Blount, and T. Kuwana, *J. Phys. Chem.*, **73**, 3456 (1969).

Part III

DRUG-RELEASE MONITORING BY MEMBRANE SENSORS

Chapter 6

IN VITRO MONITORING

One possible application of membrane sensors is to monitor the dissolution tablets, capsules, and suppositories when the pharmaceutical form contains an ion to which the sensor responds selectively in a short time. The advantages of a potentiometric measurement over the commonly employed spectrophotometric methods are twofold. First, the membrane sensor may be placed in the dissolution media, where it continuously measures ion concentration vs. time. This eliminates the need for periodic sample collection or circulation of the sample for measurement outside the dissolution chamber. Second, the potentiometric method presents an alternative analytical technique when the spectrophotometric methods are limited by interferences or when the active principle of the drug does not absorb in the visible or UV spectrum.

6.1   Potassium Membrane Sensor

The safety and efficiency of slow-release formulations must be demonstrated before they may be introduced into clinical use. Potassium chloride tablets are often administered in the form of sustained-release preparations.[1, 2] Slow-release preparations containing potassium chloride present the risk that a large initial dose, or continuing smaller doses, will result in ulceration of the gut and in other side effects.[3] The dissolution rates of 12 different potassium chloride tablets were measured using a potassium-selective glass electrode.[2] The apparatus used by Thomas[2] is essentially a temperature-controlled double-walled vessel, with stirring device (Figure 6.1). The stirring rate was 10 rpm to avoid turbulence and abrasion. The potassium membrane sensor (A. H. Thomas 4923-Q10) in conjunction with a potassium-free liquid junction reference electrode was connected to a pH/mV-meter to measure the EMF.

The apparatus of Marshall and Brook[4] (Figure 6.1) has been modified to include a Beckman pump (flow 20 cm^3 min^{-1}), which pumps the

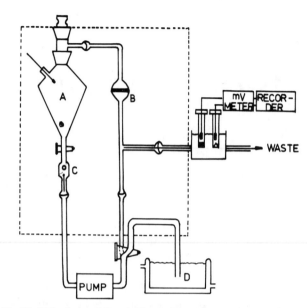

Figure 6.1 Modified Marshal and Brook dissolution rate apparatus. Dissolution fluid was pumped from D through a one-way valve C into the dissolution flask A and then through a sintered-glass filter B to trap any particles carried over. The fluid passed into a small reservoir before going to waste. (Adapted from Thomas, W. H., *J. Pharm. Pharmacol.*, 25, 27, 1973. With permission.)

dissolution fluid from receptacle D (a double-walled temperature-controlled vessel). The fluid was made to pass through a small reservoir before going to waste. This glass reservoir contains the potassium-selective membrane sensor and the reference electrode. The amount of sample dissolving in the agitation zone at any given time was obtained by potentiometric measurement of the potassium concentration in the fluid passing the sensors. The apparatus attains perfect sink conditions if the sampling rate equals the pump rate, with fresh fluid continuously bathing the sample. Sink conditions are claimed to exist when $C \ll C_s$, usually $C \leq 0.1C_s$ (where C_s is the concentration of the solute at saturation and C is the solute concentration at time t).

For the calibration curve in the non-sink method of Levy and Hayes,[5] standard solutions of 3, 30, 300, 3000, and 30,000 mg of potassium chloride were prepared in a buffer solution of the following composition: 0.2 M TRIS (250 cm^3), 0.1 N HCl (450 cm^3), and distilled water (to 1000 cm^3), adjusted to pH 6.8. All measurements were made at 37 $\pm$ 0.5°C. The calibration curve was linear above 30 mg dm^{-3} with a slope of 61.5 mV decade^{-1}. This calibration curve was used to obtain the experimental concentrations of potassium in the simulated biological fluids. The same buffer solution was used as dissolution medium in the modified Marshall and Brook sink method, but in larger volumes because

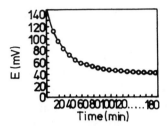

Figure 6.2 A graph of E (in millivolts) readings for a non-sink method of determination of the dissolution rate. (Reproduced from Thomas, W. H., *J. Pharm. Pharmacol.*, 25, 27, 1973. With permission.)

it flows to waste. The apparatus was initially filled with the medium (300 cm^3) and the pump was primed. The medium was circulated for 2 to 3 min, and the volume of effluent was checked with a pump rate, which was set at 20 ± 0.2 cm^3 min^{-1}. The tablets were then introduced into the dissolution vessel A, and EMF readings were recorded. Six runs of each sample were made for a minimum of 180 min each.

Where tablets containing added diuretic were tested, it was established that there was no potassium response from these drugs. (The following tablets were used: hydrochlorothiazide 0.25 mg; cyclopenthiazide 0.50 mg; guanethidine sulfate 10 mg; trichloromethiazide 4.0 mg; methyldopa 250.0 mg; hydrochlorothiazide 15.0 mg; bendrofluazide 2.5 mg).

The performance of the potassium-selective membrane sensor was compared with flame photometric determination of K$^+$ on the same solutions, giving a correlation coefficient of $r = 0.99$. The average percentage recovery values were 100.8% by potentiometry and 99.8% by flame photometry.

Figure 6.2 shows a typical graph of E readings (in millivolts) versus time, using the Levy and Hayes non-sink method.[5] The potential readings were related to potassium concentration by reference to the calibration graph.

For non-sink conditions, a plot of the logarithm of the amount of potassium (log(%K)) remaining undissolved at time t vs. time was linear, i.e.,

$$\log 100 \frac{W_0 - W}{W_0} = \text{constant} - K_1 t \qquad (6.1)$$

where W is the weight of solute or tablet dissolved. Differentiation gives

$$\frac{dW}{dt} = (W_0 - W) K_1 \times 2.303 \qquad (6.2)$$

and

$$\frac{dW}{dt} = K_R (W_0 - W) \qquad (6.3a)$$

where $K_R = K_1 \times 2.303$ is the release constant.

Because the volume V remains constant, Equation 6.3a can be rewritten as

$$\frac{dC}{dt} = K_R(C_0 - C) \qquad (6.3b)$$

i.e., the release of potassium chloride from tablets shows a first-order dependence on the weight of potassium chloride remaining. This may be compared with the Noyes–Whitney equation (Equation 6.4), which requires a first-order dependence on $(W_s - W)$ rather than $(W_0 - W)$ and is in the form

$$\frac{dW}{dt} = \frac{DS}{Vh}(W_s - W) \qquad (6.4)$$

where dW/dt is the total dissolution rate across the dissolving surface, D is the diffusion coefficient, S is the surface area exposed to the dissolution medium, h is the effective thickness of the film of diffusion layer, W_s is the weight of solute in the diffusion layer, and W is the weight of solute in the bulk of the solution at time t.[2]

First-order rate constants obtained from the slopes K_R varied with the composition of the formulation products. The plots indicate that the first portion of potassium chloride was released immediately but at a very slow rate. The remainder was released exponentially.[2] This initial lag time may be attributed to the presence of various thicknesses of the sugar coating on the tablet surface. Owing to this lag time, which varied from 0 to 45 min in most cases, the release-rate constant (K_R) could not be ascertained during this phase. It was possible, however, to make a comparative study of the various release characteristics of the tablets by collating the release constants.[2]

Some of the products tested by Thomas[2] were of the non-disintegrating fat–wax matrix type of dosage form. Subject to the diffusion process being rate-determining, it was assumed reasonable to apply Higuchi's equations (Equations 6.5 and 6.6):

$$Q = [Dt(2A - C_s)C_s]^{1/2} \qquad \text{(for uniform matrix)} \qquad (6.5)$$

$$Q = \left[\frac{DE}{T}(2A - EC_s)C_s t\right]^{1/2} \qquad \text{(for non-homogeneous matrix)} \qquad (6.6)$$

where Q is the amount of drug released after time t per unit exposed area, D is the diffusivity of drug in permeating fluid, A is the total amount of drug present in the matrix per unit volume, E is the porosity of the matrix and refers to the volume fraction that is permeated by the solvent and available for diffusion in the already leached portion of matrix, and T is the tortuosity factor of the capillary system (straight channel $T = 1$). Plots of percentage concentration vs. $t^{1/2}$ should be linear. All were substantially linear over much of the dissolution period, excepting the initial and final stages.

Dissolution rate-limited absorption indicates that there is no increase in drug concentration in the gastrointestinal fluid, i.e., the fluids function as a perfect sink, which is a necessary condition of agreement between *in vitro* and *in vivo* tests. Equation 6.7, which applies to these conditions ($C \ll C_s$), indicates that if the surface area is held constant under sink and nonreactive conditions, then the rate of dissolution is constant, i.e., the kinetics are of zero order:

$$\frac{dW}{dt} = KSC_s \qquad (6.7)$$

Integration leads to

$$W = KSC_s t \qquad (6.8)$$

A plot of W vs. t will yield a straight line with a slope of KSC_s in milligrams per minute per cubic decimeter.

The percentage of original dosage released per time interval was computed from the calibration curve (E (in millivolts) vs. $\log[K^+]$) and is presented in Figure 6.3. The curve shows that a definite maximum dissolution rate was reached.[2]

The most satisfactory products obtained by Thomas[2] appear to be those prepared from a fat–wax potassium chloride matrix, an insoluble wax coat on a non-disintegrating wax core and a combination of potassium chloride, cellulose acetate phthalate, ethyl cellulose, and polyethylene glycol. The methods used to measure the dissolution rates have proved to be of value in assessing the characteristics of slow-release preparations containing potassium chloride.

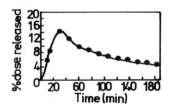

Figure 6.3 Elution rate vs. time using the forced convection sink method. (Reproduced from Thomas, W. H., *J. Pharm. Pharmacol.*, 25, 27, 1973. With permission.)

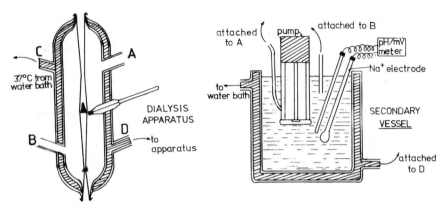

Figure 6.4 Diagram of modified apparatus of Setnikar and Fantelli. See text for details. (Reproduced from Thomas, W. H. and McCormack, R., *J. Pharm. Pharmacol.*, 23, 490, 1971. With permission.)

6.2   Sodium Membrane Sensor

The release of sodium ions from suppositories containing sodium pheno-barbital (200 mg), manufactured with cocoa butter, cocoa butter plus beeswax, Dehydag, and polyethylene glycols, was measured by Thomas and McCormack[6] with a sodium-selective membrane sensor.

It is well known that the release of a drug from a suppository is critically dependent on the physical characteristics of the base. It follows that base and active principle must be considered together. Physical factors that affect drug release from a suppository include (i) particle size of the suspended drug, (ii) the effect of surface-active agents on the mucous fluids secreted over the absorbing surface, and (iii) the binding of the drug to various components of the base.[6] Diffusion of the drug to the surface for absorption is one of the rate-limiting steps.[7]

The modified Setnikar–Fantelli[9] apparatus used by Thomas and McCormack[6] is shown in Figure 6.4. The primary part of it is a double-walled glass vessel with open ends, in the interior of which is supported a length of dialysis tubing (Union Carbide 36/36 membrane) previously tied with thread 5 cm from the lower end. The ends of the dialysis tubing are folded back over the open ends of the apparatus and securely tied. A small immersible electric pump placed in the secondary vessel (Figure 6.4) circulates the eluting fluid outside the dialysis membrane. Water at 37°C is circulated through the walls of the dialysis apparatus and the secondary vessel, which are interconnected.

The sodium membrane sensor (A. H. Thomas, Model 4923-L10) mea-suring the release of the drug was placed in the secondary vessel and a Radiometer pH/mV-meter with expanded scale was used to measure the potential difference between this and the reference electrode.

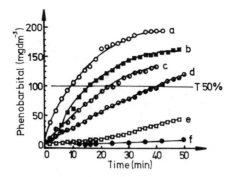

Figure 6.5 The release of sodium phenobarbital as assayed by Na$^+$ release from the cocoa butter- and Dehydag-based suppositories: (a) cocoa butter, T_{50} = 9.2; (b) Dehydag base I, T_{50} = 15.6; (c) cocoa butter +2% beeswax, T_{50} = 23; (d) cocoa butter +3% beeswax, T_{50} = 38.8; (e) Dehydag base II, T_{50} = 92 mm; (f) cocoa butter +5% beeswax, no T_{50} value. (Reproduced from Thomas, W. H. and McCormack, R., *J. Pharm. Pharmacol.*, 23, 490, 1971. With permission.)

For the calibration curve, solutions containing 5, 10, 50, 100, 200, 600, 1000, and 1400 mg of sodium phenobarbital per cubic decimeter were prepared in the same buffer solution as described under Section 6.1. The calibration curve was checked before each series of measurements (slope 61.5 mV decade^{-1} at 37°C). The wet dialysis tubing was secured firmly at the lower end of the apparatus and the upper end was left open until the inner chamber was filled with TRIS buffer solution (representing total body fluids) (one liter of TRIS buffer was used for each run).

Graphs of Na$^+$ (as milligrams of sodium phenobarbital per cubic decimeter vs. time were plotted and the time required for the suppository to release half of its contents to the medium was designated as the T_{50} value.

Figure 6.5 shows the release of drug as assessed by Na$^+$ release from the cocoa butter suppositories and the Dehydag base I and base II suppositories (these are considered to be melting fat-type bases) and Figure 6.6 shows the drug-release characteristics of the dissolving type of suppository—those with polyethylene glycol base.

The drug release rate at 37°C is dependent on the melting characteristics of the fatty suppositories, so that melting-point determination does give some indication of release in this class of suppository. With water-soluble bases, quantitative measurement of release is the most desirable parameter of availability of the drug.[6]

The selective sensor–dialysis membrane method described by Thomas and McCormack[6] furnishes a means of measuring a continuous change in the amount of drug released from a suppository in conditions which approximate those *in vivo*, namely: (i) an average temperature of 36 to

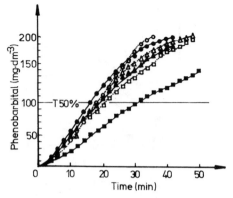

Figure 6.6 The release of sodium phenobarbital as assessed by Na^+ release from the PEG bases: (▲) PEG I, T_{50} = 19; (○) PEG II, T_{50} = 18.5; (△) PEG III, T_{50} = 19; (●) PEG IV, T_{50} = 16.4; (■) PEG V, T_{50} = 31.6; (□) PEG VI, T_{50} = 21.2 min. (Reproduced from Thomas, W. H. and McCormack, R., *J. Pharm. Pharmacol.*, 23, 490, 1971. With permission.)

37°C; (ii) little or no peristaltic movement; (iii) minimal quantity of unbound water present in the liquid state; and (iv) a pressure of 0 to 50 cm of water. Furthermore, this method can be easily modified to measure release under sink conditions.

The utility of a sodium-selective membrane sensor (Beckman No. 39278) in following the dissolution of tablets containing sodium salicylate was demonstrated by Masson et al.[9] The tablet dissolution apparatus employed was the same as described by Levy and Hayes.[5] The sensor was used in its optimal pH range of 7 to 10 and a 10^{-3} M Na^+ concentration was included as a blank in all of the buffered solutions. After the initial calibration, the membrane sensor was stabilized in the dissolution media, a tablet (containing 50 mg sodium salicylate, 100 mg lactose, 345 mg cornstarch, and 5 mg stearic acid) was added, and the sodium salicylate concentration was followed as a function of time.

Figure 6.7 shows the change in electrode potential for the dissolution of a sodium salicylate tablet. Because tableting procedure introduces some inherent variation in the individual characteristics of each tablet, it was necessary to ascertain that the recorded variation was indeed due to the tablet and not to the analytical method. This was accomplished by correlation with results from atomic absorption spectrophotometry and UV spectrophotometric analysis.[9]

Needham et al.[10] studied the use of a continuous flowing stream apparatus to follow tablet dissolution. By using the technique of continuous analysis in flowing streams to monitor dissolution, the entire profile can be recorded and experimental error can be reduced to a minimum. The apparatus used (Figure 6.8A) allows a choice of analytical module

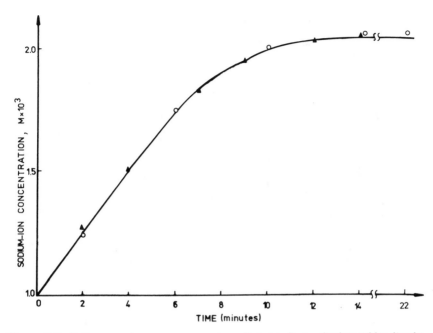

Figure 6.7 Sodium membrane sensor response during sodium salicylate tablet dissolution (—): (▲) spectrophotometric salicylate assay (pH 9, $\lambda = 296$ nm); (○) atomic absorption assay for Na$^+$; $T = 37 \pm 1°$C. (From Masson, W. D., Needham, T. E., and Price, J. C., *J. Pharm. Sci.*, 60, 1756, 1971. Reproduced with permission of the copyright owner, the American Pharmaceutical Association.)

using an inexpensive, universally available dissolution cell. A sodium membrane sensor may be mounted in a flow cell and employed as a sensor in addition to a spectrophotometric sensor. This apparatus provides a quick and accurate means of analyzing *in vitro* tablet dissolution.

Both commercial (sodium warfarin, sodium butabarbital, and sodium bicarbonate) and specially manufactured (sodium salicylate) tablets were used. The reproducibility of the apparatus as well as its ability to detect differences in tablet hardness and to analyze and to differentiate between tablets of different drug potency, were evaluated.[10]

The dissolution medium consisted of a buffer mixture[11] of potassium hydroxide and dibasic potassium phosphate at pH 8 with an adjusted sodium ion concentration of 10^{-3} M.

The dissolution cell of the apparatus is a commercially available filter unit (Swinnex-25, Millipore), using a standard 0.22-μm filter to prevent disintegrated tablet particles from leaving the cell. A peristaltic pump was used to push the solvent at a constant flow rate from the reservoir through the dissolution cell and analytical system (either a spectrophotometer with an adapted time drive to record absorbance as a function of

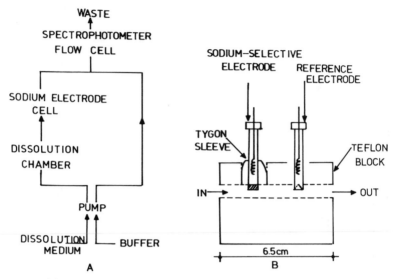

Figure 6.8 (A) Schematic representation of automated flow system; (B) side view of sodium membrane sensor cell. (From Needham, T. E., Jr., Luzzi, L. A., and Masson, W. D., *J. Pharm. Sci.*, 62, 1860, 1973. Reproduced with permission of the copyright owner, the American Pharmaceutical Association.)

time or a sodium membrane sensor with connected recorders (Figure 6.8). The analytical modules were connected singly or in series to the dissolution cell.

At the beginning of each tablet dissolution run, the analytical system was equilibrated to its baseline with the dissolution sodium and the dissolution chamber was disassembled and dried. A tablet was placed in the chamber, a new filter was set in, and the apparatus was closed. The pump was started and the changes in concentration were recorded as a function of time. The solution was discarded after being pumped through the analytical module. Variance in the rate of dissolution between tablets of the same batch was found to be comparable to the USP, Levy beaker[7, 12] and magnetic basket[13, 14] methods of dissolution testing.

6.3   Chloride Membrane Sensor

Chen et al.[15] performed some experiments in order to determine the suitability of using a chloride-ion-selective membrane sensor (Orion, Model 94-17A) for the measurement of pseudoephedrine hydrochloride dissolution from commercially available compressed tablets. Dissolution experiments were carried out in 500 cm^3 of distilled water using the USP paddle method[16] at 100 rpm. Both chloride ion and pseudoephedrine (UV spectrophotometry) were measured at six different sampling times. When a chloride membrane sensor was used to monitor the release of

Table 6.1 Mean[a] ($\pm$ Standard Error of Mean) Percentage of Drug Dissolved from Two Separate Batches of 60-mg Pseudoephedrine Hydrochloride Tablets[15]

Time (minutes)	Batch A		Batch B	
	Chloride membrane sensor method	UV method ($\lambda = 257$ nm)	Chloride membrane sensor method	UV method ($\lambda = 257$ nm)
2	31.1 ± 1.2	31.3 ± 1.3	32.9 ± 3.8	29.5 ± 2.1
5	57.6 ± 1.8	60.9 ± 4.3	62.9 ± 6.3	49.2 ± 1.2
10	88.2 ± 1.7	81.8 ± 2.3	88.8 ± 4.6	76.2 ± 1.7
15	98.3 ± 0.5	98.3 ± 2.0	96.4 ± 1.4	92.1 ± 1.8
30	99.0 ± 0.1	99.0 ± 0.7	99.4 ± 0.2	99.3 ± 0.3
60	99.8 ± 0.2	98.2 ± 0.7	99.8 ± 0.2	99.3 ± 0.4

[a] $n = 6$.

the drug, the dissolution medium (500 cm^3 distilled water) contained also 10 cm^3 of ionic strength adjuster (5 M NaNO$_3$). For these measurements the indicator membrane sensor (Cl$^-$-selective) together with a double-junction reference electrode (Orion, Model 90-02) were placed in the flask 5 cm from the bottom to permit direct measurement without sample withdrawal. Standard solutions of 0.01 to 1.00 mg cm^{-3} pseudoephedrine hydrochloride were prepared for the construction of the calibration curve (ionic strength kept at a constant value with sodium nitrate).

Dissolution test results for both analytical methods are summarized in Table 6.1 for the two batches of 60-mg pseudoephedrine hydrochloride tablets. Although some differences existed at the 5-, 10-, and 15-min sampling times for both batches, no statistically significant differences were found between the dissolution rates as determined by the two methods at the 95% confidence level.

Although the paper of Chen et al.[15] was limited to a single drug, the method may have application for other drugs available as hydrochloride salts.

6.4 Fluoride Membrane Sensor

Some *in vitro* studies performed by Yonese et al.[17] showed that the amount of fluoride incorporated in a remineralization treatment can be increased substantially if the tooth is demineralized carefully prior to the remineralization. This finding suggested that successful remineralization might be attained *in vivo* if the teeth could be demineralized to the same extent as in the *in vitro* experiments; in other papers, Higuchi and co-workers[18, 19] described some studies in the development of a fluoride topical delivery system designed to achieve *in vivo* results similar to the results obtained *in vitro*. The prototype fluoride delivery device involved micronized calcium fluoride maintained at the tooth surface with a cellulose film. Together with salivary calcium and phosphate (or simulated saliva), this system was able to generate and maintain the appropriate thermodynamic activity driving force for significant fluorapatite deposition in a reasonably short time (about 48 h).

In vitro remineralization can be very successful when the ionic activity product $K_{FAP}(a_{Ca^{2+}}^{10}a_{PO_4^{3-}}^6 \cdot a_{F^-}^2)$ is about 10^{-108}.[17] Less-concentrated solutions provide less driving force for remineralization; more-concentrated solutions may result in the rapid precipitation of calcium fluoride (**I**) or dicalcium phosphate dihydrate (**II**) in the prepared solutions themselves or in enamel pores, thereby blocking or retarding remineralization. It was decided by Higuchi and co-workers[18] to control the solution conditions at the enamel surface by supplying fluoride in the form of (**I**) suspended in a film adhering to the enamel surface.

Calculations have shown that mixtures of **I** and **II** or **I** alone in the film should result in solution compositions appropriate for remineralization at the enamel surface. Furthermore, the relatively low solubility of **I** limits the rate at which the suspended particles dissolve, so that fluoride applied in this way is inherently long-acting. This fact made the film design much simpler than when a more soluble fluoride source, such as sodium fluoride, is used in the film and must then control the release rate.[18]

Two delivery devices were proposed by Higuchi and co-workers[18] (see Figure 6.9): mixed polymer film with suspended **I** or **I–II** (device A) and cellulose film containing micronized **I** or **I–II** (device B).

For both delivery devices, diffusivity of **I** or **I–II** in polymer films and the effects of film thickness and particle size of suspended **I** or **I–II** on remineralization as well as the effects of longer remineralization times were studied in detail.[18]

It is well known that, when a solute is suspended in a film, its diffusivity D_e through the film can be obtained by measuring the amount of solute released from the film as a function of time according to[20]

$$Q = (2D_e t A C_s)^{1/2} \tag{6.9}$$

where Q is the amount of solute released per unit area after time t, A is the total amount of solute in the film per unit volume, and C_s is the solubility of the solute in the film.

The diffusivities of **I** through ethylcellulose, ethylcellulose–glycerin, and ethylcellulose–povidone films were calculated from the slopes of the plots, amount of **I** released vs. the square root of time (Figure 6.10) using Equation 6.9. Because the values of D_e in these films demon-

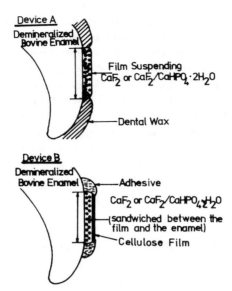

Figure 6.9 Fluoride delivery devices. (From Yonese, M., Iyer, B. V., Fox, J. L., Hefferren, J. J., and Higuchi, W. I., *J. Pharm. Sci.*, 70, 907, 1981. Reproduced with permission of the copyright owner, the American Pharmaceutical Association.)

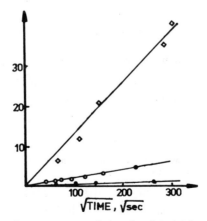

Figure 6.10 Plots of Q vs. the square root of time for determining the diffusivity of calcium fluoride in polymeric films: ($\diamondsuit$) calcium fluoride–ethylcellulose–povidone (1 : 4 : 1); ($\bigcirc$) calcium fluoride–ethylcellulose (1 : 5); ($\bullet$) calcium fluoride–ethylcellulose–glycerin (1 : 5 : 0.22). (From Yonese, M., Iyer, B. V., Fox, J. L., Hefferren, J. J., and Higuchi, W. I., *J. Pharm. Sci.*, 70, 907, 1981. Reproduced with permission of the copyright owner, the American Pharmaceutical Association.)

strated that the last film had a higher effective diffusivity and adhered to the tooth better, it was selected for delivery device A.

To measure the diffusivities of **I** through these three films, calcium fluoride (**I**) was suspended in 0.3 cm^3 of an ethanolic solution of the polymer or polymer mixture. The suspension was poured into a shallow (1-mm depth) cylindrical depression (0.88-cm radius) on a Polytef $^{\circledR}$ disk and dried for 13 min in an electric oven at 50°C. The film on the disk was then immersed into water that was stirred constantly and kept at 30°C. The fluoride released from the film was determined at several sampling times with a fluoride-selective membrane sensor (Orion, Model 96-09).

References

1. C. Graffner and J. Sjögren, *Acta Pharm. Suecica*, **8**, 13 (1971).

2. W. H. Thomas, *J. Pharm. Pharmacol.*, **25**, 27 (1973).

3. B. P. Block and M. B. Thomas, *J. Pharm. Pharmacol.*, *Suppl.*, **30**, 70 (1978).

4. K. Marshall and D. B. Brook, *J. Pharm. Pharmacol.*, **21**, 790 (1969).

5. G. Levy and B. A. Hayes, *New Engl. J. Med.*, **262**, 1053 (1960).

6. W. H. Thomas and R. McCormack, *J. Pharm. Pharmacol.*, **23**, 490 (1971).

7. S. Reigelman and W. J. Crowell, *J. Am. Pharm. Assoc.*, (*Sci. Edn.*), **42**, 96 (1958).

8. I. Setnikar and J. Fantelli, *J. Pharm. Sci.*, **51**, 566 (1962).

9. W. D. Masson, T. E. Needham, and J. C. Price, *J. Pharm. Sci.*, **60**, 1756 (1971).

10. T. E. Needham, Jr., L. A. Luzzi, and W. D. Masson, *J. Pharm. Sci.*, **62**, 1860 (1973).

11. J. E. Hoover, Ed., *Remington's Pharmaceutical Sciences*, 14th ed., Mack, Easton, PA, 1970, p. 284.

12. G. Levy and B. Sahli, *J. Pharm. Sci.*, **51**, 58 (1962).

13. R. E. Sheperd, J. C. Price, and L. A. Luzzi, *J. Pharm. Sci.*, **61**, 1152 (1972).

14. T. E. Needham, L. A. Luzzi, and R. E. Sheperd, *J. Pharm. Sci.*, **62**, 470 (1973).

15. S. T. Chen, R. C. Thompson, and R. I. Poust, *J. Pharm. Sci.*, **70**, 1288 (1981).

16. *The United States Pharmacopeia*, 20th rev., U.S. Pharmacopeial Convention, Inc., Rockville, MD, 1980, p. 959.

17. M. Yonese, J. L. Fox, and W. I. Higuchi, AADR Abstract No. 50 (1978).

18. M. Yonese, B. V. Iyer, J. L. Fox, J. J. Hefferren, and W. I. Higuchi, *J. Pharm. Sci.*, **70**, 907 (1981).

19. J. L. Fox, M. Yonese, B. V. Iyer, L. J. Abrahams, and W. I. Higuchi, *J. Pharm. Sci.*, **70**, 910 (1981).

20. T. Higuchi, *J. Pharm. Sci.*, **52**, 1145 (1963).

Chapter 7

IN VIVO MONITORING: DRUGS IN BIOLOGICAL FLUIDS

7.1 Plastic Membrane Sensor Selective to Hydrophobic Amine Antimalarials

Plastic ion-selective membrane sensor analysis of the hydrophobic amine antimalarial mefloquine and related drugs in blood samples was investigated by Mendenhall et al.[1] These drugs (mefloquine **I**, **II**, and **III**) are used for the treatment of persistent disease caused by *Plasmodia* strains (see also Section 5.66 in Part II).

$$\textbf{I} \qquad \textbf{II}$$

$$\textbf{III}$$

Problems associated with the blood concentration determination of such hydrophobic amines include glass adsorption, protein binding, and poor sensitivity by conventional detection methods (e.g., high-performance liquid chromatography using UV detection).[1]

Solutions containing **I**, **II**, or **III** prepared in dilute aqueous acid (10^{-4} M H_2SO_4) could be quantified by direct measurement with the plastic ion-selective sensor (coated-wire type, prepared as described in Srianujata et al.[2]; see also Coşofreţ[3]). The potential response was linearly related to analyte concentrations over three orders of magnitude. In this simple aqueous matrix, measurements could be made with $\pm 4\%$ accuracy and $\pm 2\%$ precision over the linear concentration range.[1]

The sensor response increased with the analyte hydrophobicity. For the more hydrophobic drugs (**II** and **III**), sensitivity approaching 10^{-9} M could be achieved; mefloquine could be quantified to 10^{-7} M (in all cases, near-Nernstian slopes).

Compared with aqueous solutions, a loss of two orders of magnitude in sensitivity was observed for plastic membrane sensor response to mefloquine and **II** in buffered plasma. In both cases, linear responses were observed, although slopes were sub-Nernstian (**I**, 49 mV decade^{-1}; **II**, 53 mV decade^{-1}). This loss in sensitivity is accounted for by extensive protein binding observed with hydrophobic molecules (mefloquine is 98% protein bound in plasma[5]). Attempts to dislodge mefloquine from protein-binding sites and thereby to allow detection at lower levels by pH manipulation or heat denaturation were unsuccessful. A selective ion-pair extraction was used for mefloquine isolation (Scheme 7.1) prior to plastic selective membrane sensor monitoring.[1] The extraction scheme utilized ether, an electron-donating species, as the extractant instead of the proton-donating solvent (e.g., chloroform) normally indicated for solvation (extraction) of a large-cation–small-anion ion pair. This procedure minimized extraction of potentially interfering, less-hydrophobic amines while maintaining efficient analyte recovery.

The extracted drug was reconstituted in aqueous buffer (pH 6.0), and the concentration was determined using the plastic membrane sensor (linear response up to 10^{-6} M).

Measurements could be made with $\pm 8\%$ accuracy and $\pm 6\%$ precision. These statistics also reflect the sample-to-sample variation ($\pm 5\%$) in mefloquine extraction efficiency from a biological sample.[1]

The use of this method for bioavailability studies evaluating various mefloquine formulations in dogs was demonstrated by analyzing blood samples drawn from two beagles after administration of two experimental formulations. The results (Figures 7.1 and 7.2) indicate adequate sensitivity for formulations having good availability but limited accuracy beyond peak levels for poorly absorbed formulations.

Mefloquine alkylation with various relatively lipophilic alkyl halides (dodecyl-, octyl-, hexyl-, and benzyl-) produced derivatives detectable by

Scheme 7.1

Ion-Pair Extraction Scheme for Mefloquine from Whole Blood[1]

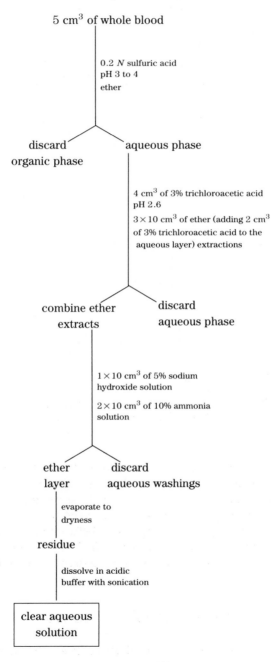

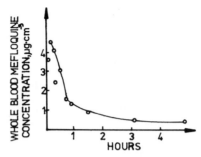

Figure 7.1 Profile of blood mefloquine level vs. time after intravenous administration of 50 mg of the drug (as hydrochloride) to a beagle dog. (From Mendenhall, D. W., Higuchi, T., and Sternson, L. A., *J. Pharm. Sci.*, 68, 746, 1979. Reproduced with permission of the copyright owner, the American Pharmaceutical Association.)

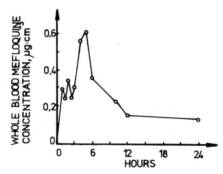

Figure 7.2 Profile of blood mefloquine level vs. time after oral administration of 250-mg mefloquine hydrochloride tablet to a beagle dog. (From Mendenhall, D. W., Higuchi, T., and Sternson, L. A., *J. Pharm. Sci.*, 68, 746, 1979. Reproduced with permission of the copyright owner, the American Pharmaceutical Association.)

the membrane sensor at much lower levels (up to two orders of magnitude) than the parent. A kinetic study of this alkylation reaction revealed that an alkyl amine base was necessary to scavenge the acid produced during reaction to go to completion. At room temperature, with benzyl bromide as the reagent, reaction was 99% complete in 30 min and mefloquine could be detected to about 10^{-8} M, a 100-fold improvement in sensitivity over sensor monitoring of underivatized mefloquine.

7.2   Methadone Membrane Sensor

The use of methadone in opiate-addiction treatment programs has created a need for rapid, simple, yet accurate analytical methods for monitoring drug levels in biological fluids, in order to evaluate patient compliance. Such a method was proposed by Srianujata et al.[2] Methadone, which is a γ-keto-tertiary amine (see Section 5.70 in Part II) was determined in acidified (pH 2 to 3) urine samples by using a miniaturized hydrophobic-cation-selective plastic membrane sensor. Urine was chosen as the test biological fluid because of its ready accessibility and because a significant fraction of the drug is eliminated intact by the kidney.

Higuchi et al.[6] have described an ion-selective membrane sensor that responds preferentially to hydrophobic cations. The membrane was composed of a poly(vinyl chloride)–dioctylphthalate mixture. The selective response of the sensor for cations is made possible by specific anion sites in the membrane, whereas the preference for hydrophobic species is derived from the hydrophobicity of the membrane itself, i.e., response is related to the difference in chemical potential of the analyte in the aqueous vs. membrane phase. The potential of this sensor is directly related to the activity of hydrophobic cations in aqueous solutions,[6] and response is Nernstian. Although these characteristics seem to make the sensor a useful probe for drug analysis, the cumbersome nature of the original assembly prevented application of the sensor to monitor methadone in urine samples. A modified version of this sensor that is smaller, less fragile, and more stable and provides a faster response time was described by Srianujata et al.[2] The selectivity of this membrane sensor permitted the determination to be made at 5×10^{-5} M methadone in urine samples without prior separation. Determinations of about $\pm 15\%$ accuracy were possible with a single potential measurement; better than $\pm 2\%$ accuracy could be obtained by titrating the sample with a solution of sodium tetraphenylborate and monitoring the decrease in methadone activity potentiometrically.

For analysis of urine samples containing methadone, these were diluted with one volume of distilled water and acidified to pH 2 to 3 with 5 M H_2SO_4. The sensor assembly was then introduced into the solution and the mixture titrated with 4.98×10^{-4} M sodium tetraphenylborate solution. Alternatively, the urine samples were assayed by a single direct potentiometric measurement (linear range for calibration curve up to 10^{-6} M with a slope of 59 mV decade^{-1}).

7.3 Penicillin Membrane Sensor

As was shown in Section 5.79 in Part II, many penicillin-selective membrane sensors have been constructed, on the basis of hydrolysis reaction of these compounds, by immobilizing penicillinase on a pH glass electrode.[7-15] Other kinds of penicillin sensors were developed by Campanella and co-workers.[16, 17]

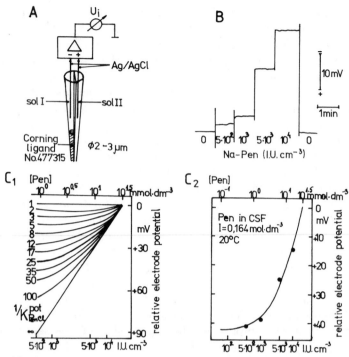

Figure 7.3   Schematic drawing of the penicillin-selective microsensor (A) and the response of the sensor to penicillin (B and C): (A) solution I, 100 mmol dm^{-3} KCl and 1 mmol dm^{-3} Na-penicillin G, solution II, 150 mmol dm^{-3} NaCl or artificial cerebrospinal fluid; (B) changes of the relative electrode potential with increasing penicillin concentration; (C) iso-selectivity curves computed on the basis of the Nicolsky–Eisenman formula and plot of calibration values taken from B; (C_1) various selectivity ratios of penicillin vs. chloride as marked on the left margin; (C_2) selectivity ratio of 25, larger scale. (Reproduced from Speckmann, E. J ., Elger, C. E., and Lehmenkuehler, A., *Electroencephalogr. Clin. Neurophysiol.*, 56, 664, 1983. With permission.)

A double-barreled microelectrode described by Speckmann et al.[18] permitted the continuous and simultaneous measurement of penicillin concentration and of the local bioelectric activity in nervous tissue (when studying epileptic phenomena, the topical or systemic application of penicillin is often used to induce epileptiform activity in animal experiments). The electrode design and the electronic equipment used correspond to those described by Neher and Lux,[19] and the former is shown schematically in Figure 7.3A.

The sensors were calibrated in cerebrospinal fluid (CSF) solutions of constant ionic strength ($I = 0.164\ M$). In these salines, penicillin was substituted for chloride. The solutions used for calibration contained 0, 500, 1000, 5000, and 10,000 international units (IU) of penicillin (Na-Penicillin G) per cubic centimeter of CSF, respectively. More than 50

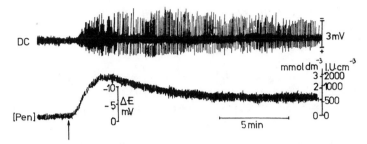

Figure 7.4 Simultaneous measurements of the extracellular penicillin concentration (Pen) and the local DC potential in the rat's motor cortex. Penicillin was applied to the cortical surface (arrow). (Reproduced from Speckman, E. J., Elger, C. E., and Lehmenkuehler, A., *Electroencephalogr. Clin. Neurophysiol.*, 56, 664, 1983. With permission.)

sensors were tested, and they showed an average lifetime of 2 to 3 days. As can be seen in Figure 7.3 C_2 the sensor is able to detect a penicillin concentration even below 500 IU cm^{-3} (response time between 50 and 100 ms).

An example of pilot measurements of the extracellular penicillin concentration using this micro-sensor is shown in Figure 7.4. The micro-sensor was inserted into the rat's motor cortex with the tip finally located about 300 μm below the surface. Beside the penicillin signal, the local bioelectric activity was recorded DC-coupled against a reference electrode placed on the frontal nasal bone.[18] Penicillin was applied to the cortical surface via a second micro-sensor by pressure ejection. The tip of this sensor was placed in the immediate vicinity of the recording sensor just above the cortical surface, which was covered by thermostabilized mineral oil.[18] From the application sensor an amount at most 1 μl penicillin solution (50,000 IU cm^{-3} CSF) was ejected.

After the application of penicillin, a steep increase of penicillin concentration occurred (Figure 7.4). During this increase typical epileptiform potentials developed in the DC record. A few minutes later the penicillin concentration decreased, with the amplitudes of the epileptiform potentials also were reduced.

References

1. D. W. Mendenhall, T. Higuchi, and L. A. Sternson, *J. Pharm. Sci.*, **68**, 746 (1979).

2. S. Srianujata, W. R. White, T. Higuchi, and L. A. Sternson, *Anal. Chem.*, **50**, 232 (1978).

3. V. V. Coşofreţ, *Membrane Electrodes in Drug-Substances Analysis*, Pergamon, London, 1982, pp. 144–146.

4. B. B. Brodie, in *Transport Function of Plasma Proteins*, P. Desgrez and P. M. DeTraverse, Eds., Elsevier, Amsterdam, 1966, pp. 137–145.

5. J. Y. Mu, Z. H. Iraili, and P. G. Dayton, *Drug Metab. Disposition*, **3**, 198 (1976).

6. T. Higuchi, C. R. Illian, and J. L. Tossounian, *Anal. Chem.*, **42**, 1674 (1970).

7. R. Tor and A. Freeman, *Anal. Chem.*, **58**, 1042 (1986).

8. G. J. Papariello, A. K. Mukherji, and C. M. Shearer, *Anal. Chem.*, **45**, 790 (1973).

9. S. O. Enfors and N. Molin, *Process Biochem.*, **13**, 9 (1978).

10. S. O. Enfors and H. Nilsson, *Enzyme Microbiol. Technol.*, **1**, 260 (1979).

11. J. P. Hou and J. W. Poole, *J. Pharm. Sci.*, **61**, 1594 (1972).

12. G. J. Olliff, R. T. Williams, and J. M. Wright, *J. Pharm. Pharmacol.*, **30**, 45 (1978).

13. J. Anzai, M. Shimada, T. Osa, and C. Chen, *Bull. Chem. Soc. Jpn.*, **60**, 4133 (1987).

14. S. Caras and J. Janata, *Anal. Chem.*, **52**, 1935 (1980).

15. S. Caras and J. Janata, *Anal. Chem.*, **57**, 1924 (1985).

16. L. Campanella, M. Tomassetti, and R. Sbrilli, *Ann. Chim. (Rome)*, **77**, 483 (1986).

17. L. Campanella, F. Mazzei, R. Sbrilli, and M. Tomassetti, *J. Pharm. Biomed. Anal.*, **6**, 299 (1988).

18. E. J. Speckmann, C. E. Elger, and A. Lehmenkuehler, *Electroencephalogr. Clin. Neurophysiol.*, **56**, 664 (1983).

19. E. Neher and H. D. Lux, *J. Gen. Physiol.*, **61**, 385 (1973).

Appendix 1

GENERAL CHARACTERISTICS OF SOME COMMERCIALLY AVAILABLE MEMBRANE SENSORS

Table Ia Some Fluoride-Selective Membrane Sensors

	Orion	Philips	Corning	Radiometer
Type (model no.)	94-09	IS-550-F	EE-F	F 1052-F
Measuring range (M)	10^0–10^{-6}	10^{-1}–10^{-6}	10^0–10^{-6}	10^0–3×10^{-7}
Slope (mV decade^{-1})	56	56 ± 3	Theoretical	Almost theoretical
Temperature range (°C)	0–80	0–50	Over 3–35	0–60
Interfering ions	OH$^-$ < 0.1F$^-$	OH$^-$	OH$^-$	OH$^-$ ($k_{F,OH}^{pot} = 0.2$)
Operational life (months)	> 12	12	> 12	—
pH range	5–7 at 10^{-6} M to pH 11 at 10^{-1} M	4–8	—	3–13
Response time	99% response in 1 min or less.	< 60 s	Usually milliseconds, occasionally a few seconds	60 s

Table Ib   More Fluoride-Selective Membrane Sensors

	Radelkis	Metrohm	Beckman	Coleman
Type (model no.)	OP-F-711D OP-F-7111D OP-F-7112D OP-F-7113D	EA-306F	39600	3-803
Measuring range (M)	10^0–10^{-6}	10^0–10^{-6}	10^0–10^{-6}	10^0–10^{-6}
Temperature range (°C)	0–80	0–80	-5–100	-5–100
Interfering ions	OH^- ($k_{F,OH}^{pot} = 0.1$)	$OH^- < 0.1F^-$	OH^- ($k_{F,OH}^{pot} = 0.1$)	OH^-
pH range	—	—	4–8	4–8
Response time	A few seconds in all cases			

Table IIa   Some Chloride-Selective Membrane Sensors

	Orion	Philips	Corning	Radiometer
Type (model no.)	94-17	IS-550-Cl	EE-Cl	F 1012 Cl
Measuring range (M)	10^0–5×10^{-5}	10^{-1}–10^{-5}	10^0–10^{-4}	10^0–10^{-5}
Slope (mV decade^{-1})	57	56 ± 3	Near theoretical	Near theoretical
Temperature range (°C)	0–80	0–50	Over 3–35	-5–60
Selectivity coefficients and interfering ions	$S^{2-} \leq 10^{-7} M$; Br^-, I^- CN^- must be present in trace amounts only	S^{2-} is a contaminant; CN^- 400, I^- 86.5, $S_2O_3^{2-}$ 60; Br^- 1.2	$S^{2-} \leq 10^{-7}$; Br^-, I^-, CN^-	$S^{2-} < 10^{-4} M$; Br^- 2, I^- 2, CN^- 8
Operational lifetime (months)	More than 12 in all cases			
pH range	0–13	1–10	—	0–14
Response time	A few seconds in all cases			

Table IIb More Chloride-Selective Membrane Sensors

	Metrohm	Beckman	Schott	Coleman
Type (model no.)	EA-306-Cl	39604	Cl 690	3-802
Measuring range (M)	10^0–5×10^{-5}	10^0–5×10^{-5}	10^0–10^{-5}	10^0–10^{-6}
Temperature range (°C)	0–50	-5–100	0–50	-5–100
Selectivity coefficients and interfering ions	$S^{2-} \leq 10^{-7}\ M$; Br^-, I^-, CN^-	$S^{2-}, CN^-,$ Br^-, I^-	$S^{2-}, Br^-,$ $I^-, CN^-,$ $S_2O_3^{2-},$ SCN^-, NH_3	S^{2-}, CN^-
pH range	0–14 in all cases			
Response time	A few seconds in all cases			

Table IIIa Some Bromide-Selective Membrane Sensors

	Orion	Philips	Corning	Radiometer
Type (model no.)	94-35	IS-550-Br	EE-Br	F 1022 Br
Measuring range (M)	10^0–5×10^{-6}	10^0–10^{-6}	10^0–5×10^{-6}	10^0–10^{-6}
Slope (mV decade^{-1})	57	56 ± 3	Theoretical	Theoretical
Temperature range (°C)	0–80	0–50	Over 3–35	-5–60
Selectivity coefficients and interfering ions	$I^-\ 2 \times 10^{-4}$	S^{2-} interferes $CN^-\ 25,$ $I^-\ 20,$ $S_2O_3^{2-}\ 1.5$	S^{2-} and I^- must be absent	$S^{2-}\ 10^{-4},$ $I^-\ 2, CN^-\ 1,$ strongly reducing agents
Operational lifetime (months)	More than 12 in all cases			
pH range	0–14	1–11	—	0–14
Response time	99% response in 1 min or less for Br^- conc. $> 10^{-5}\ M$	30 s	Usually milliseconds, occasionally a few seconds	60 s

Table IIIb More Bromide-Selective Membrane Sensors

	Metrohm	Beckman	Schott	Coleman
Type (model no.)	EA 306 Br	39602	Br 690	3-801
Measuring range (M)	10^0–5×10^{-6}	10^0–5×10^{-5}	10^0–10^{-6}	10^0–10^{-7}
Temperature range (°C)	0–50	-5–100	0–50	-5–100
Selectivity coefficients and interfering ions	$S^{-2} \leq 10^{-7}$ $I^- < 2 \times 10^{-4}$	I^-, S^{2-} Br^-, CN^-	$I^-, S^{2-},$ $CN^-, S_2O_3^{2-},$ NH_3	S^{2-}, CN^-
pH range	0–14	0–14	—	0–14
Response time		A few seconds in all cases		

Table IVa Some Iodide-Selective Membrane Sensors

	Orion	Philips	Corning	Radiometer
Type (model no.)	94-53	IS-550-I	EE-I	F 1032 I
Measuring range (M)	10^0–2×10^{-7}	10^0–10^{-7}	10^0–10^{-6}	10^0–10^{-6}
Slope (mV decade^{-1})	57	56 ± 3	Near theoretical	Near theoretical
Temperature range (°C)	0–80	0–50	Over 3–35	-5–60
Selectivity coefficients and interfering ions	$S^{2-} < 10^{-7}$	$S^{2-}, CN^-,$ strongly reducing agents	S^{2-} must be absent; I^- 1.0	S^{2-} 30, CN^- 10^{-4}, reducing agents
Operational lifetime (months)		More than 12 in all cases		
Response time		A few seconds in all cases		

Table IVb More Iodide-Selective Membrane Sensors

	Radelkis	Metrohm	Beckman	Schott
Type (model no.)	OP-I-711D OP-I-7111D OP-I-7112D OP-I-7113D	EA-306-I	39606	I/CN 690
Measuring range (M)	10^0–5×10^{-8}	10^0–10^{-8}	10^0–10^{-6}	10^0–10^{-7}
Temperature range (°C)	0–80	0–50	-5–100	0–50
Selectivity coefficients and interfering ions	S^{2-} even trace amounts interfere	$S^{2-} < 10^{-7}\ M$	S^{2-}	S^{2-}, CN^-, $S_2O_3^{2-}$
pH range	0–14 in all cases			
Response time	A few seconds in all cases			

Table V Sulfide-Selective Membrane Sensors[a]

	Orion	Philips	Corning	Radiometer
Type (model no.)	94-16	IS-550-S	EE-S	F 1212 S
Measuring range (M)	10^0–10^{-7}	10^0–10^{-8}	10^0–10^{-7}	10^0–10^{-6}
Slope (mV decade^{-1})	Almost theoretical in all cases			
Temperature range (°C)	0–80	0–50	Over 3–35	-5–60
Selectivity coefficients and interfering ions	$Hg^{2+} < 10^{-7}\ M$	Hg^{2+}	Hg^{2+}	$CN^- < 10^{-3}\ M$
Operational lifetime (months)	More than 12 in all cases			
pH range	0–14	0–14	Measurements must be made at pH 12	8–14
Response time	Varies from several seconds in concentrated solutions to several minutes near the limit of detection	20 s	Usually milliseconds, occasionally a few seconds	Typically 1–3 min

[a] Other sulfide-selective membrane sensors are produced by Beckman (39610), Coleman (3-805), EDT Research (ISE 305), Schott (S/Ag 690), Tacussell (PS-3), Radelkis (OP-S-711D, OP-S-7111D, OP-S-7112D, and OP-S-7113D), Metrohm (EA-306-S/Ag), Leeds & Northrup (117508), etc.

Table VI Lead-Selective Membrane Sensors[a]

	Orion	Metrohm	Ingold
Type (model no.)	94-82	EA-306 Pb	157200
Measuring range (M)	10^{-1}–10^{-6}	10^{0}–10^{-7}	10^{0}–10^{-7}
Temperature range (°C)	0–80	0–80	—
Selectivity coefficients and interfering ions	Ag^+, Hg^{2+}, Cu^{2+} $\ll 10^{-7}\ M$; high levels of Cd^{2+} and Fe^{3+} interfere	Ag^+, Cu^{2+}, Hg^{2+} must not be high levels of present in detectable amounts	Ag^+, Cu^{2+}, Hg^{2+} $\leq 10^{-7}\ M$; Cd^{2+} and Fe^{3+} interfere
pH range	2–14	2–14	4–7

[a]Other lead-selective membrane sensors are produced by Radiometer (F 3004), Leeds & Northrup (117407), Tacussel (PPB 1), etc.

Table VII Copper-Selective Membrane Sensors[a]

	Orion	Philips	Corning	Radiometer	Metrohm
Type (model no.)	94-29	IS-550-Cu	EE-Cu	F 1112 Cu	EA-306-Cu
Measuring range (M)	Saturated to 10^{-8}	10^{0}–10^{-7}	10^{0}–10^{-6}	10^{0}–10^{-6}	10^{0}–10^{-6}
Slope (mV decade^{-1})	Almost theoretical in all cases				
Temperature range (°C)	0–80	0–50	Over 3–35	-5–60	0–80
Selectivity coefficients and interfering ions	S^{2-}, Ag^+, Hg^{2+} $\leq 10^{-7}\ M$	Ag^+ and Hg^{2+}	S^{2-} must be absent	$Cu^+, Ag^+,$ Hg^{2+}	S^{2-}, Ag^+, Hg^{2+} $\leq 10^{-7}\ M$
Operational lifetime	12 in all cases				
pH range	0–14	1–14	0–14	0–14	0–14
Response time	A few seconds in all cases				

[a]Other copper-selective membrane sensors are produced by Radiometer (F 3002), Beckman (39612), Leeds & Northrup (117403), Coleman (3-804), Tacussel (PCU 2), etc.

Table VIII   Silver-Selective Membrane Sensors[a]

	Orion	Philips	Corning	Radiometer
Type (model no.)	94-16	IS-550-Ag	EE-Ag	F 1212 S
Measuring range (M)	10^0–10^{-7}	10^0–10^{-7}	10^0–10^{-7}	10^0–10^{-6}
Slope (mV decade^{-1})	Almost theoretical in all cases			
Temperature range (°C)	0–80	0–50	Over 3–35	-5–60
Selectivity coefficients and interfering ions	Hg^{2+} $< 10^{-7}$ M	Hg^{2+}	Hg^{2+}, S^{2-}	Hg^{2+} ($k^{pot}_{Ag^+, Hg^{2+}}$ = 0.1)

[a]Other silver-selective membrane sensors are produced by Beckman (39610), Coleman (3-805), Schott (S/Ag 690), Radelkis (OP-Ag-711D, OP-Ag-7111D, OP-Ag-7112D, and OP-Ag-7113D), Metrohm (EA-306-S/Ag), Radiometer (F 3001), Leeds & Northrup (117408), etc.

Table IXa   Some Calcium-Selective Membrane Sensors

	Orion	Corning	Radiometer
Type (model no.)	93-20	EE-Ca	F-2110 Ca; F 2112 Ca
Measuring range (M)	10^0–10^{-5}	10^0–10^{-5}	10^0–5×10^{-6}
Slope (mV decade^{-1})	24	Almost theoretical	Almost theoretical
Temperature range (°C)	0–50	Over 3–35	0–60
Selectivity coefficients and interfering ions	Max. level (M) at 10^{-3} M Ca^{2+}: Na^+ 0.3; Zn^{2+} 2×10^{-6}; Pb^{2+} 5×10^{-6}; Fe^{2+}, Cu^{2+} 7×10^{-5}; Sr^{2+}, Mg^{2+} 8×10^{-3}; Ba^{2+} 3×10^{-2}; Ni^{2+} 5×10^{-2}	Mg^{2+} 1.5×10^{-2}; Ba^{2+} 10^{-2}; Pb^{2+} 0.60; Zn^{2+} 1.20; Na^+ 3×10^{-3}; K^+ 2.2×10^{-5}	Mg^{2+} 80; Sr^{2+} 3; Ba^{2+} 500; Cu^{2+} 1; Zn^{2+} 10^{-2}; Cd^{2+} 6; $Na^+, K^+, Li^+ < 10^{-5}$
Response time	Exhibits good response time for conc. $> 5 \times 10^{-5}$ M	—	Max. 60 s at low calcium conc.

Table IXb Philips Calcium-Selective Membrane Sensors

	Philips	
Type	IS-561-Ca (plastic)	IS-560-Ca (liquid)
Measuring range (M)	10^1–10^{-6}	10^0–10^{-5}
Slope (mV decade^{-1})	29 ± 2	30
Temperature range (°C)	0–50	0–50
Selectivity coefficients	NH_4^+ 10^{-5}; Li^+ 10^{-3}; Na^+ 3×10^{-4}; Ba^{2+} 4×10^{-4}; K^+ 2×10^{-4}; Rb^+ 10^{-4}; Mg^{2+} 4×10^{-5}	Al^{3+} 0.90; Ba^{2+} 0.02; Ca^{2+} 0.042; Cu^{2+} 0.070; Fe^{2+} 0.045; Mg^{2+} 0.032; Mn^{2+} 0.38; Zn^{2+} 0.9
Operational lifetime (months)	4–6	6

Table Xa Potassium-Selective Membrane Sensors

	Orion	Corning	Radiometer
Type (model no.)	93-19	EE-K-47613200	F-2312 K
Measuring range (M)	10^0–10^{-5}	10^0–10^{-5}	10^0–10^{-6}
Slope (mV decade^{-1})	54	Near theoretical	Near theoretical
Temperature range (°C)	0–50	Over 3–35	0–60
Selectivity coefficients	Max. level (M) at 10^{-3} M K^+: Cs^+ 10^{-4}; Na^+ 0.5; H^+, Tl^+ 10^{-2}; NH_4^+ 3×10^{-3}; Ag^+ 0.1	Na^+ 2.6×10^{-3}; Li^+ 2.1×10^{-3}; Rb^+ 1.9; Cs^+ 0.38; NH_4^+ 0.30	Na^+ 7×10^{-5}; NH_4^+ 1.9×10^{-2}; Li^+ 4×10^{-5}; Rb^+ 2.8; Cs^+ 0.5
Response time		Max. 60 s at low potassium conc.	

Table Xb More Potassium-Selective Membrane Sensors

	Radelkis	Philips	
Type (model no.)	OP-K-711D OP-K-7111D OP-K-7112D OP-K-7113D	IS-561-K (plastic)	IS-560-K (liquid)
Measuring range (M)	10^0–10^{-6}	10^1–10^{-6}	10^0–10^{-6}
Slope (mV decade^{-1})	Theoretical	58 ± 3	59.2
Temperature range (°C)	5–60	0–50	0–50
Operational lifetime (months)	12	4–6	—
Response time	Max. 60 s at low potassium conc.		

Table XI Ammonia-Gas-Sensing Membrane Sensors

	Orion	Beckman	EIL
Type (model no.)	95-10	39565	8002-8
Measuring range (M)	10^0–10^{-6}	10^0–10^{-6}	5×10^{-2}–5×10^{-6}
Temperature range (°C)	0–50	0–50	5–40
Interferences	Volatile amines in all cases		
pH range	Sample and standard must be adjusted to fixed pH or to above 11	11	7–14
Response time	60 s	—	Less than 1 min for a decade increase in conc.

Appendix 2

KNOWN ADDITION TABLES

Table XII Known Addition Table, Values for Q vs. ΔE at 25°C for 10 cm^3 Added to 100 cm^3 (from Orion Research, Inc.) (Slope 59 mV decade^{-1})

ΔE	Q	ΔE	Q	ΔE	Q
−5.0	0.297	−7.1	0.222	−9.2	0.174
−5.1	0.293	−7.2	0.219	−9.3	0.173
−5.2	0.288	−7.3	0.217	−9.4	0.171
−5.3	0.284	−7.4	0.214	−9.5	0.169
−5.4	0.280	−7.5	0.212	−9.6	0.167
−5.5	0.276	−7.6	0.209	−9.7	0.165
−5.6	0.272	−7.7	0.207	−9.8	0.164
−5.7	0.268	−7.8	0.204	−9.9	0.162
−5.8	0.264	−7.9	0.202	−10.0	0.160
−5.9	0.260	−8.0	0.199	−10.2	0.157
−6.0	0.257	−8.1	0.197	−10.4	0.154
−6.1	0.253	−8.2	0.195	−10.6	0.151
−6.2	0.250	−8.3	0.193	−10.8	0.148
−6.3	0.247	−8.4	0.190	−11.0	0.145
−6.4	0.243	−8.5	0.188	−11.2	0.143
−6.5	0.240	−8.6	0.186	−11.4	0.140
−6.6	0.237	−8.7	0.184	−11.6	0.137
−6.7	0.234	−8.8	0.182	−11.8	0.135
−6.8	0.231	−8.9	0.180	−12.0	0.133
−6.9	0.228	−9.0	0.178	−12.2	0.130
−7.0	0.225	−9.1	0.176	−12.4	0.128

Table XII *Continued*

ΔE	Q	ΔE	Q	ΔE	Q
-12.6	0.126	-21.8	0.0637	-31.0	0.0374
-12.8	0.123	-22.0	0.0629	-31.2	0.0370
-13.0	0.121	-22.2	0.0621	-31.4	0.0366
-13.2	0.119	-22.4	0.0613	-31.6	0.0362
-13.4	0.117	-22.6	0.0606	-31.8	0.0358
-13.6	0.115	-22.8	0.0598	-32.0	0.0354
-13.8	0.113	-23.0	0.0591	-32.2	0.0351
-14.0	0.112	-23.2	0.0584	-32.4	0.0347
-14.2	0.110	-23.4	0.0576	-32.6	0.0343
-14.4	0.108	-23.6	0.0569	-32.8	0.0340
-14.6	0.106	-23.8	0.0563	-33.0	0.0336
-14.8	0.105	-24.0	0.0556	-33.2	0.0333
-15.0	0.103	-24.2	0.0549	-33.4	0.0329
-15.2	0.1013	-24.4	0.0543	-33.6	0.0326
-15.4	0.0997	-24.6	0.0536	-33.8	0.0323
-15.6	0.0982	-24.8	0.0530	-34.0	0.0319
-15.8	0.0967	-25.0	0.0523	-34.2	0.0316
-16.0	0.0952	-25.2	0.0517	-34.4	0.0313
-16.2	0.0938	-25.4	0.0511	-34.6	0.0310
-16.4	0.0924	-25.6	0.0505	-34.8	0.0307
-16.6	0.0910	-25.8	0.0499	-35.0	0.0304
-16.8	0.0897	-26.0	0.0494	-36.0	0.0289
-17.0	0.0884	-26.2	0.0488	-37.0	0.0275
-17.2	0.0871	-26.4	0.0482	-38.0	0.0261
-17.4	0.0858	-26.6	0.0477	-39.0	0.0249
-17.6	0.0846	-26.8	0.0471	-40.0	0.0237
-17.8	0.0834	-27.0	0.0466	-41.0	0.0226
-18.0	0.0822	-27.2	0.0461	-42.0	0.0216
-18.2	0.0811	-27.4	0.0456	-43.0	0.0206
-18.4	0.0799	-27.6	0.0450	-44.0	0.0196
-18.6	0.0788	-27.8	0.0445	-45.0	0.0187
-18.8	0.0777	-28.0	0.0440	-46.0	0.0179
-19.0	0.0767	-28.2	0.0435	-47.0	0.0171
-19.2	0.0756	-28.4	0.0431	-48.0	0.0163
-19.4	0.0746	-28.6	0.0426	-49.0	0.0156
-19.6	0.0736	-28.8	0.0421	-50.0	0.0149
-19.8	0.0726	-29.0	0.0417	-51.0	0.0143
-20.0	0.0716	-29.2	0.0412	-52.0	0.0137
-20.2	0.0707	-29.4	0.0408	-53.0	0.0131
-20.4	0.0698	-29.6	0.0403	-54.0	0.0125
-20.6	0.0689	-29.8	0.0399	-55.0	0.0120
-20.8	0.0680	-30.0	0.0394	-56.0	0.0115
-21.0	0.0671	-30.2	0.0390	-57.0	0.0110
-21.2	0.0662	-30.4	0.0386	-58.0	0.0105
-21.4	0.0654	-30.6	0.0382	-59.0	0.0101
-21.6	0.0645	-30.8	0.0378		

Table XIII Known Addition Table, Values for Q vs. ΔE at 25°C for 10% Volume Change (from Orion Research, Inc.) (Slope 29.6 mV decade^{-1})

ΔE	Q	ΔE	Q	ΔE	Q
2.5	0.297	6.6	0.119	10.7	0.0654
2.6	0.288	6.7	0.117	10.8	0.0645
2.7	0.280	6.8	0.115	10.9	0.0637
2.8	0.272	6.9	0.113	11.0	0.0629
2.9	0.264	7.0	0.112	11.1	0.0621
3.0	0.257	7.1	0.110	11.2	0.0613
3.1	0.250	7.2	0.108	11.3	0.0606
3.2	0.243	7.3	0.106	11.4	0.0598
3.3	0.237	7.4	0.105	11.5	0.0591
3.4	0.231	7.5	0.103	11.6	0.0584
3.5	0.225	7.6	0.1013	11.7	0.0576
3.6	0.219	7.7	0.0997	11.8	0.0569
3.7	0.214	7.8	0.0982	11.9	0.0563
3.8	0.209	7.9	0.0967	12.0	0.0556
3.9	0.204	8.0	0.0952	12.1	0.0549
4.0	0.199	8.1	0.0938	12.2	0.0543
4.1	0.195	8.2	0.0924	12.3	0.0536
4.2	0.190	8.3	0.0910	12.4	0.0530
4.3	0.186	8.4	0.0897	12.5	0.0523
4.4	0.182	8.5	0.0884	12.6	0.0517
4.5	0.178	8.6	0.0871	12.7	0.0511
4.6	0.174	8.7	0.0858	12.8	0.0505
4.7	0.171	8.8	0.0846	12.9	0.0499
4.8	0.167	8.9	0.0834	13.0	0.0494
4.9	0.164	9.0	0.0822	13.1	0.0488
5.0	0.160	9.1	0.0811	13.2	0.0482
5.1	0.157	9.2	0.0799	13.3	0.0477
5.2	0.154	9.3	0.0788	13.4	0.0471
5.3	0.151	9.4	0.0777	13.5	0.0466
5.4	0.148	9.5	0.0767	13.6	0.0461
5.5	0.145	9.6	0.0756	13.7	0.0456
5.6	0.143	9.7	0.0746	13.8	0.0450
5.7	0.140	9.8	0.0736	13.9	0.0445
5.8	0.137	9.9	0.0726	14.0	0.0440
5.9	0.135	10.0	0.0716	14.1	0.0435
6.0	0.133	10.1	0.0707	14.2	0.0431
6.1	0.130	10.2	0.0698	14.3	0.0426
6.2	0.128	10.3	0.0689	14.4	0.0421
6.3	0.126	10.4	0.0680	14.5	0.0417
6.4	0.123	10.5	0.0671	14.6	0.0412
6.5	0.121	10.6	0.0662	14.7	0.0408

Table XIII *Continued*

ΔE	Q	ΔE	Q	ΔE	Q
14.8	0.0403	16.6	0.0333	21.5	0.0206
14.9	0.0399	16.7	0.0329	22.0	0.0196
15.0	0.0394	16.8	0.0326	22.5	0.0187
15.1	0.0390	16.9	0.0323	23.0	0.0179
15.2	0.0386	17.0	0.0319	23.5	0.0171
15.3	0.0382	17.1	0.0316	24.0	0.0163
15.4	0.0378	17.2	0.0313	24.5	0.0156
15.5	0.0374	17.3	0.0310	25.0	0.0149
15.6	0.0370	17.4	0.0307	25.5	0.0143
15.7	0.0366	17.5	0.0304	26.0	0.0137
15.8	0.0362	18.0	0.0289	26.5	0.0131
15.9	0.0358	18.5	0.0275	27.0	0.0125
16.0	0.0354	19.0	0.0261	27.5	0.0120
16.1	0.0351	19.5	0.0249	28.0	0.0115
16.2	0.0347	20.0	0.0237	28.5	0.0110
16.3	0.0343	20.5	0.0226	29.0	0.0105
16.4	0.0340	21.0	0.0216	29.5	0.0101
16.5	0.0336				

Appendix 3

GENERAL METHODS FOR INORGANIC ANIONS ANALYSIS

i. *Ionizable Fluoride:*

    a. *Direct measurements*—Three standards (10^{-2}, 10^{-3}, and 10^{-4} M) are prepared by serial dilution of 0.1 M sodium fluoride stock solution. Ionic strength and pH are kept constant with TISAB (50 cm^3 TISAB to each 50 cm^3 standard). The electrodes (fluoride-selective and SCE) are immersed successively in the standards, and the EMF readings (linear axis) are plotted against concentration (logarithmic axis). The EMF measurements are made under stirring and the unknown concentration is determined from the calibration curve.

    b. *Known addition*—For measurements on an unknown sample of fluoride, the electrodes (fluoride-selective and SCE) are placed in 100 cm^3 of solution (50-cm^3 sample containing fluoride, and 50 cm^3 TISAB) to give a stable reading of E_1 (in millivolts). A standard fluoride solution that is about 10 times as concentrated as the sample concentration is prepared by diluting 0.1 M fluoride standard (50 cm^3 TISAB is added to each 50 cm^3 standard) and 10 cm^3 of this standard are pipetted into the cell solution. This solution is thoroughly stirred and the new reading E_2 (in millivolts) is recorded. The value Q that corresponds to the change in potential ΔE ($\Delta E = E_1 - E_2$) is given in Table XII, Appendix 2. To determine the original sample concentration, Q is multiplied by the concentration of the added standard (see Equation 4.7 in Part II).

    c. *Potentiometric titration*—The accurately weighed sample (3 to 8 mg) is dissolved in 100 cm^3 of 1 : 1 (v/v) aqueous dioxan contained in a 200 cm^3 polyethylene beaker. The pH of the solution is adjusted to 5 to 6 with 0.02 M sodium hydroxide solution. The electrode pair (see the preceding text) is immersed in the sample solution and this solution is potentiometrically titrated with 0.02 M thorium nitrate solution. The electrode potential is recorded as a function of added titrant volume, and the electrode potential is plotted vs. the titrant volume on linear graph paper. The end point corresponds to the maximum slope $(\Delta E/\Delta V)$ on the titration curve.

(1 cm^3 0.02 M thorium nitrate solution corresponds to 1.52 mg F$^-$.)

ii. *Ionizable Chloride, Bromide, and Iodide:*

a. *Direct measurement*—For measurements in units of moles per cubic decimeter, 10^{-2}, 10^{-3}, and 10^{-4} M standards are prepared by serial dilution of appropriate 0.1 M halide solutions (sodium chloride, sodium bromide and sodium iodide, respectively). The ionic strength is kept constant (e.g., to 0.1 M) with sodium nitrate solution. The halide-selective and reference electrodes (single-junction reference electrode for bromide and iodide and double-junction reference electrode for chloride) are placed in the standard solutions in the order 10^{-3}, 10^{-4}, and 10^{-2} M. The EMF readings (linear axis) are plotted against concentration (logarithmic axis). The EMF measurements are made under stirring, and the unknown concentration is determined from the calibration curve.

b. *Known addition*—To measure an unknown sample of halide, the appropriate halide-ion-selective membrane sensor and reference electrode are placed in 100 cm^3 of sample and 2 cm^3 5 M NaNO$_3$ solution (ionic strength adjuster, after Orion Research, Inc.) are added. The reading of E_1 (in millivolts) is recorded. A standard solution of about 10 times the concentration of the sample is prepared by diluting 0.1 M of the appropriate halide standard (2 cm^3 5 M NaNO$_3$ solution is added to each 100 cm^3 standard) and 10 cm^3 of this standard is pipetted into the cell solution. The solution is thoroughly stirred and E_2 (in millivolts) is recorded. The value Q that corresponds to the change in potential ΔE ($\Delta E = E_1 - E_2$) is given in Table XII, Appendix 2. To determine the original sample concentration, Q is multiplied by the concentration of added standard (Equation 4.7 in Part II).

c. *Potentiometric titration*—The sample (total volume approximately 50 cm^3, concentration approximately 10^{-3} M) is potentiometrically titrated with 10^{-2} M silver nitrate solution. The titrant is added from a 10-cm^3 burette in 0.5-cm^3 increments. When the potential change per increment begins to increase, 0.1- to 0.2-cm^3 increments are added. About 1 cm^3 titrant is added beyond the end point. The electrode potential is recorded as a function of added titrant volume and the electrode potential is plotted vs. the titrant volume on linear graph paper. The end points correspond to the maximum slopes on the titration curves.

(1 cm^3 10^{-2} M AgNO$_3$ solution corresponds to 0.355 mg Cl$^-$, 0.799 mg Br$^-$, and 1.269 mg I$^-$, respectively.)

iii. *Ionizable Sulfate:*

a. *Potentiometric titration with lead(II) solution*—The sample solution (total volume approximately 50 cm^3, concentration approximately 10^{-3} M in 50% dioxan, pH 4 to 6.5) is potentiometrically titrated with 10^{-2} M lead(II) perchlorate (pH 4.4). The end point corresponds to the maximum slope on the titration curve.

(1 cm^3 10^{-2} M lead(II) perchlorate solution corresponds to 0.96 mg sulfate.)

A lead(II)-ion-selective membrane sensor is used as indicator.

b. *Potentiometric titration with barium(II) solution*—The sample solution (total volume approximately 60 cm^3, concentration approximately 10^{-3} M in 20% ethanol, pH 4 to 5) is potentiometrically titrated with 2 × 10^{-2} M barium perchlorate in 80% ethanol. The end point corresponds to the maximum slope on the titration curve.

(1 cm^3 0.02 M barium perchlorate solution corresponds to 1.92 mg sulfate.)

A barium(II)-ion-selective membrane sensor is used as indicator.

iv. *Ionizable Phosphate:*

To the sample solution containing 0.5 to 1.5 mg phosphorus as orthophosphate, 4.0 cm^3 of buffer solution (0.5 M ammonium acetate, adjusted to pH 8.9 with ammonia) is added. The sample is diluted to about 50 cm^3 with distilled water. The pair of electrodes (lead(II)-ion-selective indicator and SCE double-junction as reference, with outer chamber filled with 1 M sodium nitrate solution) is immersed in the sample solution, which is potentiometrically titrated with 0.01 M lead perchlorate solution (adjusted to pH 4.8 to 5.0 with dilute perchloric acid). The end point corresponds to the maximum slope on the titration curve.

(1 cm^3 10^{-2} lead(II) perchlorate corresponds to 0.206 mg phosphorus.)

BIBLIOGRAPHIES

Selected Books on Ion-Selective Membrane Electrodes

Ammann, D., *Ion-Selective Microelectrodes—Principles, Design and Application*, Springer-Verlag, Berlin, 1986.

Arnold, M. A. and Rechnitz, G. A., *Biosensors: Fundamentals and Applications*, Oxford Science Publications, Oxford, 1987.

Bailey, E. L., *Analysis with Ion-Selective Electrodes*, Heiden, London, 1976.

Baiulescu, G. E. and Coşofreţ, V. V., *Applications of Ion-Selective Membrane Electrodes in Organic Analysis*, John Wiley & Sons, New York, 1977 (translated into Russian by MIR, 1981).

Berman, H. J. and Hebert, N. C., Eds., *Ion-Selective Microelectrodes*, Plenum Press, New York, 1974.

Cammann, K., *Das Arbeiten mit Ionnenselektiven Elektroden*, Springer-Verlag, Berlin, 1973 and 1977.

Chen, Z. Z. and Qin, Z., *Applications of Ion-Selective Electrodes in Pharmaceutical Analysis*, Renmin Weisheng Publ. House, Beijing, 1985 (in Chinese).

Coşofreţ, V. V., *Membrane Electrodes in Drug-Substances Analysis*, Pergamon Press, Oxford, 1982.

Covington, A. K., Ed., *Ion-Selective Electrode Methodology*, Vols. 1 and 2, CRC Press, Boca Raton, 1979.

Freiser, H., Ed., *Ion-Selective Electrodes in Analytical Chemistry*, Vols. 1 and 2, Plenum Press, New York, 1978 and 1979.

423

Janata, J., *Principles of Chemical Sensors*, Plenum Press, New York, 1989.

Kessler, M., Clark, L. C., Jr., Lübbers, D. W., Silver, I. A., and Simon, W., Eds., *Ion and Enzyme Electrodes in Biology and Medicine*, Urban and Schwarzenberg, Munich, 1976.

Kessler, M., Harrison, D. K., and Höper, J., *Ion Measurements in Physiology and Medicine*, Springer-Verlag, Berlin, 1985.

Koryta, J., *Ion-Selective Electrodes*, Cambridge University Press, London, 1975.

Koryta, J., *Ions, Electrodes and Membranes*, John Wiley & Sons, New York, 1982.

Koryta, J., and Stulik, K., *Ion-Selective Electrodes*, 2nd ed., Cambridge University Press, Cambridge, 1983.

Koryta, J., Ed., *Medical and Biological Applications of Electrochemical Devices*, John Wiley & Sons, New York, 1980.

Lakshminarayanaiah, N., *Membrane Electrodes*, Academic Press, New York, 1976.

Lindner, E., Tóth, K., and Pungor, E., *Dynamic Characteristics of Ion-Selective Electrodes*, CRC Press, Boca Raton, 1988.

Lübbers, D. W., Acker, H., Buck, R. P., Eisenman, G., Kessler, M., and Simon, W., Eds., *Progress in Enzyme and Ion-Selective Electrodes*, Springer-Verlag, Berlin, 1981.

Ma, T. S. and Hassan, S. S. M., *Organic Analysis Using Ion-Selective Electrodes*, Vols. 1 and 2, Academic Press, London, 1982.

Moody, G. J. and Thomas, J. D. R., *Selective Ion Sensitive Electrodes*, Merrow, Watford, 1971.

Morf, W. E., *The Principles of Ion-Selective Electrodes and of Membrane Transport*, Elsevier Studies in Analytical Chemisrty, Vol. 2, Elsevier, Amsterdam, 1981.

Ngo, T. T., Ed., *Electrochemical Sensors in Immunological Analysis*, Plenum Press, New York, 1987.

Schindler, J. G. and Schindler, M. M., *Bioelectrochemical Membrane Electrodes*, de Gruyter, Berlin, 1983.

Schuetzle, D. and Hammerle, R., Eds., *Fundamentals and Applications of Chemical Sensors*, American Chemical Society, Washington, DC, 1986.

Vesely, J., Weiss, D., and Stulik, K., *Analysis with Ion-Selective Electrodes*, Ellis Horwood, Chichester, 1978.

Selected Books on Pharmaceutical Analysis

Ahuja, S., Ed., *Chemical Analysis*, Vol. 85: *Ultratrace Analysis of Pharmaceuticals and Other Compounds of Interest*, John Wiley & Sons, New York, 1986.

Aszalos, A., Ed., *Modern Analysis of Antibiotics*, Marcel Dekker, New York, 1986.

Berman, E., *Analysis of Drugs of Abuse*, Heyden and Son, London, 1977.

Connors, K. A., *A Textbook of Pharmaceutical Analysis*, 3rd ed., Wiley Interscience, New York, 1982.

Deasy, P. B. and Timoney, R. F., Eds., *The Quality Control of Medicines*, Elsevier Science Publishers, Amsterdam, 1976.

Deasy, P. B. and Timoney, R. F., Eds., *Progress in the Quality Control of Medicines*, Elsevier Biomedical Press, Amsterdam, 1981.

Ebel, S., *Handbook of Drug Analysis*, Verlag Chemie, Weinheim, 1977.

Florey, K., Ed., *Analytical Profiles of Drug Substances*, Vols. 1 through 19, 1990 (Vol. 19). Academic Press, New York.

Fong, G. W. and Lane, S. K., Eds., *HPLC in the Pharmaceutical Industry. Drugs and the Pharmaceutical Sciences*, Vol. 47, Marcel Dekker, New York, 1991.

Garratt, D. C., *The Quantitative Analysis of Drugs*, 3rd ed., Chapman and Hall, London, 1976.

Görög, S. and Heftmann, E., Eds., *Advances in Steroid Analysis 1987*, Akad. Kiadó, es Nyonda Valalat, Budapest, 1988.

Görög, S. and Heftmann, E., Eds., *Advances in Steroid Analysis '90*, Akad. Kiadó, Budapest, 1991.

Jack, D. B., *Drug Analysis by Gas Chromatography*, Academic Press, Orlando, 1984.

Jenkins, G. L., Knevel, A. M., and DiGangi, F. E., *Jenkins' Quantitative Pharmaceutical Chemistry*, McGraw-Hill, New York, 1977.

Mills, T., Price, W. N., Price, P. T. and Roberson, J. C., *Instrumental Data for Drug Analysis*, Elsevier, New York, 1982.

Munson, J. N., Ed., *Modern Methods of Pharmaceutical Analysis*, Marcel Dekker, New York, 1981.

Roth, H. J. and Blaschke, G., *Pharmaceutical Analysis*, Thieme, Stuttgart, 1978.

Schrimer, R., *Modern Methods of Pharmaceutical Analysis*, Vols. 1 and 2, 2nd ed., CRC Press, Boca Raton, 1990.

Wagner, H., Bladt, S., and Zgainski, E. M., *Plant-Drug Analysis*, Springer-Verlag, Berlin, 1984.

Weiner, I. W., Ed., *Liquid Chromatography in Pharmaceutical Development: An Introduction*, Aster, Springfield, OR, 1985.

Selected Chapters and Reviews on Membrane Sensors in Pharmaceutical Analysis

Buck, R. P. and Coşofreţ, V. V., Design of sensitive drug sensors: principles and practice, in *Fundamentals and Applications of Chemical Sensors*, D. Schuetzle and R. Hammerle, Eds., American Chemical Society, Washington, DC, 1986, chap. 22.

Campanella, L. and Tommassetti, M., Sensors in pharmaceutical analysis, *Selective Electrode Rev.*, **11**, 69–109 (1989).

Coşofreţ, V. V., Analytical control of drug-type substances with membrane electrodes, *Ion-Selective Electrode Rev.*, **2**, 159–218 (1981).

Coşofreţ, V. V., The use of membrane electrodes for the determination of inorganic species in pharmaceutical analysis, *Trends in Anal. Chem.*, **10**, 261–265 (1991).

Coşofreţ, V. V., Drug membrane sensors and their pharmaceutical applications, *Trends in Anal. Chem.*, **10**, 298–301 (1991).

Coşofreţ, V. V. and Buck, R. P., Drug-type substances analysis with membrane electrodes, *Ion-Selective Electrode Rev.*, **6**, 59–121 (1984).

Pilkington, T. C. and Lawson, B. L., Analytical problems facing the development of electrochemical transducers for in vivo drug monitoring, *Clin. Chem.*, **28**, 1946–1955 (1982).

Pungor, E., Feher, Z., Nagy, G., Lindner, E., and Tóth, K., Trends in the application of electroanalytical techniques to the analytical control of pharmaceuticals, *Anal. Proc.* (*London*), **19**, 79–82 (1982).

Solsky, R. L., Ion-selective electrodes in biomedical analysis, *CRC Crit. Rev. Anal. Chem.*, **14**, 1–52 (1982).

Zhang, Z. R. and Coşofreţ, V. V., New developments in pharmaceutical analysis with membrane sensors, *Selective Electrode Rev.*, **12**, 35–135 (1990).

INDEX
Drug-Type Substances Assayed by Membrane Sensors